Fundamentals of
Optical Fiber
Communications

Prentice Hall International Series in Optoelectronics

Consultant editors: John Midwinter, University College London, UK
Alan Snyder, Australian National University
Bernard Weiss, University of Surrey, UK

Fundamentals of Optical Fiber Communications
W. van Etten and J. van der Plaats

Light Emitting Diodes
K. Gillessen and W. Schairer

Optical Communication Systems
J. Gowar

Optical Sensing Techniques and Signal Processing
T. E. Jenkins

Optical Fiber Communications
J. M. Senior

Lasers: Principles and Applications
J. Wilson and J. F. B. Hawkes

Optoelectronics: An Introduction (Second Edition)
J. Wilson and J. F. B. Hawkes

Fundamentals of Optical Fiber Communications

Wim van Etten and Jan van der Plaats

Eindhoven University of Technology, The Netherlands

Prentice Hall

New York   London   Toronto   Sydney   Tokyo   Singapore

First published 1991 by
Prentice Hall International (UK) Ltd
66 Wood Lane End, Hemel Hempstead
Hertfordshire HP2 4RG
A division of
Simon & Schuster International Group

Typeset in 10/12pt Times
by Keytec Typesetting Ltd, Bridport, Dorset

Printed and bound in Great Britain at the
University Press, Cambridge

Library of Congress Cataloging-in-Publication Data

Etten, Wim van.
  Principles of optical fiber communications / Wim van Etten, Jan van der Plaats.
    p.   cm. — (Prentice-Hall international series in optoelectronics)
  Includes bibliographical references and index.
  ISBN 0–13–717513–2 : $80.00
    1. Optical communications.   2. Optical fibers.   I. Plaats, Jan van der.   II.
Title.   III. Series.
  TK5103.59.E77   1990
  621.382′75 — dc20   90-36827
                                                                                          CIP

British Library Cataloguing in Publication Data

Van Etten, Wim
  Principles of optical fiber communications.
  1. Optical communications systems
  I. Title   II. Van der Plaats, Jan
  621. 38275

  ISBN 0–13–717513–2

---

1 2 3 4 5   95 94 93 92 91

Contents

Preface   xi

1   Introduction   1
    1.1   The configuration of a glassfiber   4
    1.2   The manufacture of fibers   6
    1.3   The attenuation of light in glassfibers   9
    1.4   Light sources   13
    1.5   Detectors   17
    1.6   System model   17
          References   18
          Problem   19

2   Analysis of the slab waveguide   20
    2.1   Theoretical models   20
    2.2   The slab waveguide analyzed with wave optics   21
    2.3   Transverse electric waves   22
    2.4   Transverse magnetic waves   28
    2.5   The propagation constant of a mode   28
    2.6   Geometric optics interpretation   29
          References   32
          Problems   32

3   Analysis of the step index fiber   33
    3.1   The general solution of the wave equation   34
    3.2   Unattenuated waves   36
    3.3   Transverse and hybrid waves and modes   40
    3.4   Transverse waves and modes   40
    3.5   Hybrid modes   45
    3.6   Leaky modes   52
    3.7   Linearly polarized (LP) modes   52
          References   58
          Problems   58

4   Dispersion in the step index fiber   60
    4.1   Phase characteristic and modulation bandwidth   62

4.2 The propagation constant of a mode in a step index fiber   68
4.3 Waveguide dispersion   69
4.4 Material dispersion   72
4.5 Waveguide and material dispersion in a step index fiber   73
4.6 Multimode dispersion   75
4.7 Dispersion caused by a non-monochromatic source   76
     References   76
     Problems   76

5 The monomode fiber   78
5.1 The electromagnetic field in a monomode step index fiber   79
5.2 Power flow in the z-direction   84
5.3 The mode field diameter   86
5.4 Waveguide dispersion in monomode fibers   88
5.5 Waveguide and material dispersion in monomode fibers   90
5.6 Dispersion-shifted and dispersion-flattened fibers   92
     References   94
     Problems   95

6 Propagation of light rays in multimode graded index fibers   96
6.1 The eikonal equation   97
6.2 Solutions of the eikonal equation for a cylindrical symmetrical
     fiber and the resulting ray equations   99
6.3 Some analytical solutions   101
6.4 Numerical solutions   109
6.5 Ray congruencies, the h, g-coordinate system   110
6.6 The local numerical aperture   114
6.7 The relationship between ray congruencies and modes   115
6.8 The WKB method   117
     References   119
     Problems   119

7 Dispersion in graded index fibers   121
7.1 Mode model   122
7.2 Ray model   129
     References   135
     Problems   135

8 Light sources and detectors   137
8.1 Choosing the wavelength region   137

8.2   The light-emitting diode (LED)   139
8.3   The semiconductor laser diode (LD)   147
8.4   Semiconductor laser versus LED   154
8.5   Photodiodes   156
      References   166
      Problems   168

9 **Modulation of semiconductor light sources**   **169**
9.1   The rate equations   170
9.2   The laser condition   173
9.3   The efficiency of lasers   175
9.4   The turn-on delay of a laser and the behaviour of an LED   177
9.5   Transient behaviour of a laser   178
9.6   Modulation of a laser by small signals   179
9.7   Amplitude noise of lasers   183
      References   183
      Problems   184

10 **Transfer characteristic and impulse response of fiber communication systems**   **185**
10.1   Transmission via a single-mode fiber   185
10.2   Transmission via multimode fibers   195
       References   211
       Problems   212

11 **Power launching and coupling efficiency**   **213**
11.1   The ray density of a Lambertian source in the phase space   214
11.2   Power launching from the source into a multimode fiber   216
11.3   Multimode fiber–fiber coupling   221
11.4   Coupling model for single-mode fibers   243
11.5   Power coupling from the source into single-mode fibers   247
11.6   Single-mode fiber–fiber coupling   251
11.7   Fiber–detector coupling   254
       References   255
       Problems   256

12 **Receiver principles and signal-to-noise ratio in analog receivers**   **257**
12.1   Connection diagram and equivalent scheme of photodetectors   258
12.2   The impulse response of a PIN photodiode   260
12.3   Signal-to-noise ratio in analog receivers   264

12.4   The thermal noise in front-end amplifiers   273
       References   280
       Problems   280

13   Receivers for digital optical fiber communication systems   283
       13.1   Introduction   283
       13.2   Analysis of the simplified receiver model   284
       13.3   The quantum limit   289
       13.4   The general receiver model   290
       References   302
       Problems   302

14   System noise   304
       14.1   Intensity noise of the light source   305
       14.2   Competition noise   306
       14.3   Partition noise   306
       14.4   Modal noise   312
       14.5   The signal-to-noise ratio due to system noise and receiver noise   320
       References   322
       Problems   322

15   System components and aspects of system design   324
       15.1   Introduction   324
       15.2   Comparison of optical fibers and copper cables   325
       15.3   Optical fiber cables   326
       15.4   Splices and connectors   328
       15.5   Optical isolators   333
       15.6   Polarization-maintaining fiber   334
       15.7   Wavelength multiplexing   336
       15.8   Repeater distance and link budget   341
       15.9   Line coding   344
       15.10 Selection of the system components   348
       References   350

16   Coherent optical fiber communication   351
       16.1   Introduction   351
       16.2   Basic principles of coherent optical systems   352
       16.3   Signal-to-noise ratio of coherent optical receivers   356
       16.4   Balanced mixing and phase diversity reception   358
       16.5   Polarization aspects of coherent systems   365

16.6   Concluding remarks   367
        References   367

Appendix 1   Bessel functions   369

Appendix 2   Transmission of modulated signals via bandpass systems   374

Appendix 3   The propagation of Gaussian beams in free space and
                 optical systems   378

Appendix 4   Poisson processes   386

Appendix 5   Some physical constants   400

Index   **401**

Preface

Since the early 1970s, the authors of this book have given courses in optical glassfiber communications at the Department of Electrical Engineering of the Eindhoven University of Technology, as a part of their general teaching activities in telecommunications. Over the years they have selected topics that, on the one hand, give a good overview of the main principles of optical fiber communications and, on the other hand, are well suited to being taught in the classroom. This book is a compilation of the class notes originally established by the authors and updated by continuing use. It assumes some basic knowledge of the principles of electromagnetic fields, optics, semiconductor physics, Fourier analysis and noise calculations. Chapters 1, 2, 6, 7, 8, Sections 12.1–12.3 and 13.1–13.3 and Chapter 15 contain material for an undergraduate course, while the rest of the book can be taught in a graduate course. In addition, the text is well suited as a reference for scientists and engineers in research and development laboratories.

Chapter 1 starts with a comparison of optical communication systems with other communication systems of different kinds. Attention is paid to the fabrication processes that are used to produce glassfibers, to the different causes of attenuation and to a method to measure this attenuation. The most important properties of the light sources and the detectors, the elements necessary to form a simple communication link with an optical waveguide as the transmission medium, are reviewed and a general model of a link is established. The first part of the book, Chapters 2–7, deals with the propagation of light through optical waveguides. Chapter 2 starts with the treatment of slab waveguides, on the one hand as an introduction to the solution techniques used in subsequent chapters and on the other hand because these waveguides play an important role in integrated optics. The waveguides are analyzed with the help of Maxwell's equations, the characteristic equation with its discrete solutions is derived and the mode concept is introduced. In Chapter 3 this is repeated for the round step index waveguide. The E, H and hybrid modes are derived and attention is paid to LP modes. The expressions for the phase characteristics that follow from the solutions are further developed in Chapter 4, which treats the dispersion of the waveguides and the influence of the fiber on the bandwidth. A separate chapter, Chapter 5, is devoted to single-mode fibers. The principles derived in the earlier chapters are applied for this special case. Chapter 6 treats graded index fibers. In the previous chapters the wave optics model was used. In this chapter the waveguides are analyzed with the geometric optics model, using the eikonal equation as the starting point. At the end of the chapter the results are compared with the results obtained by the WKB method. In Chapter 7 two different models are used to derive the dispersion in graded index fibers. Chapter 8 deals with the semiconductor light sources and detectors that are used in optical fiber communication systems. Little attention is paid to the physical background, the emphasis being on the external characteristics. This approach is further elaborated in Chapter 9 for the modulation aspects of the sources. Once the optical fiber waveguiding and the light sources and detectors are introduced, a description of an entire link can be given; this is done in Chapter 10. A great deal of the total link loss is due to coupling losses; Chapter 11 gives an extensive treatment of these coupling losses, both in multimode and single-mode fiber links. The receivers in an optical transmission system require special attention. Noise phenomena are quite different compared to the classical communication model, where the noise is assumed to be additive, stationary and Gaussian. The shot noise, or Poisson noise, does not show these elegant properties. That is why analog receivers (Chapter 12) require a different approach compared to digital receivers (Chapter 13). Apart from the aforementioned shot noise, multimode optical fiber systems can suffer from system noise (Chapter 14), which arises from non-ideal matching of system components. In general, realization of an optical fiber link or network requires more components than standard optical fiber, light source(s) and detector(s). Such components may be: wavelength division multiplexers, optical isolators, polarization-maintaining fibers, etc. These components and other system aspects are treated in Chapter 15. Finally, Chapter 16 has been devoted to coherent optical fiber communication, a subject to which much attention is now being paid in laboratories and which may lead to very promising applications in the future.

A number of subjects closely related to optical fiber communications, but not specific to it, are dealt with in the appendices. These include: Bessel functions, transmission via bandpass systems, Gaussian beams and Poisson processes.

At the end of most of the chapters some exercises are provided, giving the reader the opportunity to check his or her knowledge by means of practical problems.

A task as extensive as writing a book always requires support. We thank Dr Peter Attwood for correcting the English text, and Gerard Baten for producing a number of the figures. One of the authors (W. v. E.) thanks his wife Kitty for typing his part of the manuscript.

1

Introduction

1.1   The configuration of a glass fiber
1.2   The manufacture of fibers
1.3   The attenuation of light in glassfibers
1.4   Light sources
1.5   Detectors
1.6   System model
      References
      Problems

The transmission of information by electromagnetic waves is one of the main characteristics of telecommunication systems. Electromagnetic waves include visible and infra-red light, which is used for optical communication. There is fundamentally no difference between light, radio waves, X-rays and gamma rays: the different properties of these various waves depend on their different frequencies/wavelengths. Figure 1.1. gives a general view of the spectrum of electromagnetic waves.

   The simplest electromagnetic wave is a plane, linearly polarized wave in a homogeneous, isotropic medium. An orthogonal system of coordinates x, y, z is depicted in Figure 1.2. and a plane electromagnetic wave propagates in the positive z direction. The vectors $\mathbf{E}$ and $\mathbf{H}$, perpendicular to each other and to the direction of propagation, are independent of x and y. For points along the z

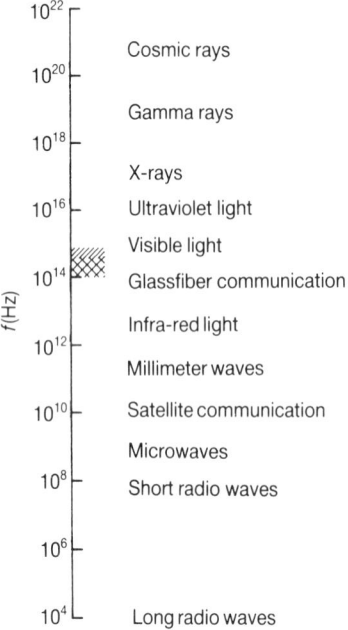

Figure 1.1 A general view of the spectrum of electromagnetic waves.

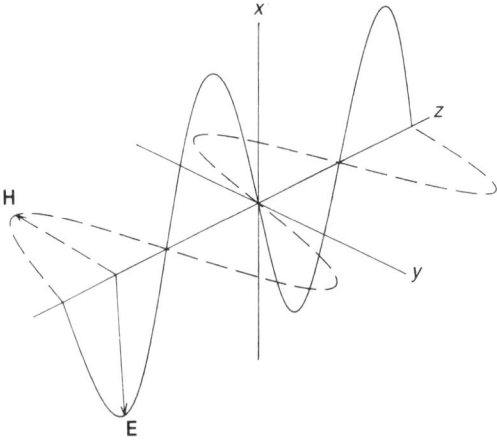

Figure 1.2 The electric and magnetic field vectors of a plane harmonic wave in homogeneous, isotropic medium.

axis, it can be seen how the vectors **E** and **H** depend on z in the case of a harmonic wave.

The frequency used is only one way of distinguishing between the various communication systems. Another way that determines the nature of a system concerns the use of guided or unguided (free space) transmission of information. The earliest communication systems used visible light as a carrier of information, e.g. smoke and fire signals, flags and semaphore. A common characteristic of all these systems is the absence of wires, i.e. no man-made waveguides between the transmitter and the receiver. Such systems use unguided transmission of information like modern radio communication systems. The use of wires to guide the electromagnetic waves from their source to the destination was first introduced on a large scale after 1840, with the rise of telegraphy and the introduction of Morse code. The waveguides consisted of two insulated metal wires, or a single bare wire and an earth. Relatively very low fequencies were used. After the invention of telephony, guides were used in the frequency range of 300–3400 Hz. In the course of time there has been an exponential increase in the range of frequencies used for the guided transmission of information. We can mention in this context carrier wave telephony, coaxial cable television and high-speed digital transmission. This development is sketched in Figure 1.3.

After the discovery of electromagnetic waves by Hertz, there was a similar development in the use of unguided waves, from the low frequencies of the first broadcasts to the high frequencies of satellite communications. The guidance of the high-frequency electromagnetic waves used in optical telecommunications nowadays is provided by glassfibers. Just as in the case of communication with metallic conductors, this is a form of guided transmission of information. Communication with metallic conductors is adequate for the frequency range from 0 Hz to a few gigahertz, but when light is used to carry information in glassfiber communication systems, it has a much higher frequency. The frequency range used for glassfiber communication extends from about 2×10^{14} Hz

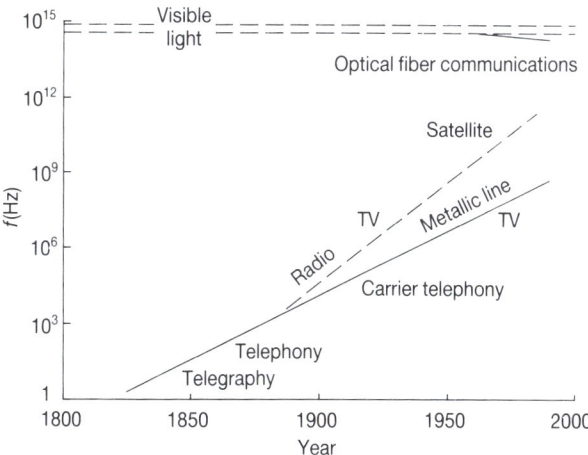

Figure 1.3 The development of unguided and guided transmission of information as a function of the frequency.

to about 4×10^{14} Hz, this corresponds to a wavelength range from 750 to 1500 nm, or a frequency band in a single glassfiber of 2×10^{14} Hz. In such a wide frequency band, an unbelievably large amount of information can be transmitted – at least in theory. For instance, such a frequency range can contain two million television signals of 100 Mbit s^{-1} each.

In a communication link, a power source must be available at the transmitter side of the channel. In glassfiber systems, light-emitting diodes (LEDs) and semiconductor lasers are used as the power sources. The light from these sources is coupled directly or indirectly into the fiber by a lens system. Information is added by modulating the power source directly by varying the current through the LED or laser, or indirectly by modulating the light externally.

1.1 The configuration of a glassfiber

Most glassfibers consist of a cylindrical core surrounded by a cladding and the whole is protected mechanically by a coating (see Figure 1.4). In relation to the production process, the glassfiber is a uniform transmission line, i.e. its cross-section is the same along the whole length of the line. Current fibers have a circular cross-section. Optical density is a function of the distance from the fiber axis and this symmetrical function is usually called the refractive index profile. These variations in density allow the fiber to guide the light. In Figure 1.5 the cross-sections and the refractive index profiles of a number of fibers have been drawn, together with their standard dimensions.

The simplest optical waveguide in Figure 1.5(a) consists of a single thread of homogeneous transparent material, such as glass, with a refractive index of n_1. This thread is surrounded by air with a refractive index of n_0. The electromagnetic field of a guided wave is not only present in the thread itself, but is also found partly in the surrounding air, as will be described in Chapter 3. Therefore, this simple waveguide has no practical use in telecommunications,

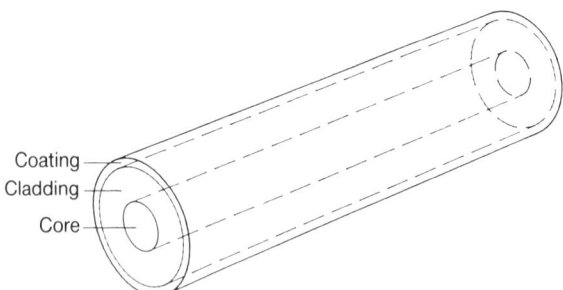

Coating
Cladding
Core

Figure 1.4 The configuration of a glassfiber of the core–cladding type.

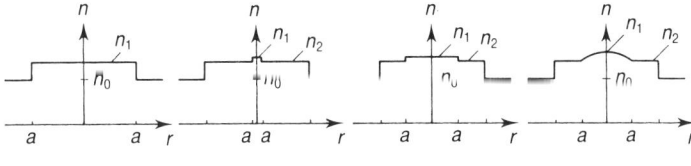

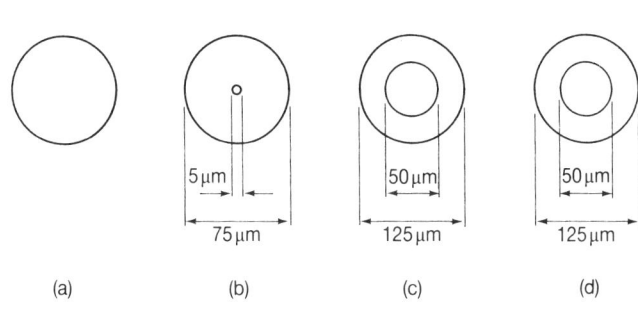

(a)   (b)   (c)   (d)

Figure 1.5 Cross-sections and refractive index profiles of a number of glass-fibers.

because every support of the thread disturbs the field. Nearly all practical fibers are of the core–cladding type, as shown in Figures 1.5(b)–(d). The core and the cladding are made of glass. Figure 1.5(b) and (c) show step index fibers. The refractive index of the core is constant, regardless of the distance from the axis. If the following condition holds for a step index fiber, then light transmission can only take place in a single mode, as will be derived in Chapter 3:

$$\lambda_0 > \frac{2\pi a}{2.405}\sqrt{n_1^2 - n_2^2} \tag{1.1}$$

where

λ_0 = wavelength of light in vacuum
a = radius of the core
n_1 = refractive index of the core
n_2 = refractive index of the cladding

If the condition (1.1) holds, then the fiber is called a single-mode fiber or a monomode fiber. A monomode fiber has a relatively small core, see Figure 1.5(b). Every mode has its own characteristic field distribution and its own propagation constant. On the other hand, if the diameter of the core is relatively large, such as that shown in Figure 1.5(c), then light propagation can take place in a multitude of modes and the fiber is called a multimode fiber. As stated earlier, part of the electromagnetic fields of the propagation modes are found in the cladding. The intensity of the fields in the cladding decreases more or less exponentially with its distance from the axis of the fiber; therefore, if the cladding is sufficiently thick, any fields outside the cladding can be neglected and they have no effect on light propagation in the fiber. In Figure 1.5(d) a fiber is shown with a refractive index that varies with the distance from the axis; such a fiber is called a graded index fiber. The refractive index can be given by the function

$$n(r) = n(0)\left[1 - 2\Delta\left(\frac{r}{a}\right)^x\right]^{1/2}, \qquad 0 \leqslant r \leqslant a \qquad (1.2a)$$

$$n(r) = n(0)(1 - 2\Delta)^{1/2}, \qquad r > a \qquad (1.2b)$$

where

r = distance from the axis
Δ = dimensionless constant
x = a parameter that defines the shape of the profile

A profile described by this function is called a power-law index profile. If $x = 2$, then it becomes a so-called parabolic refractive index profile.

1.2 The manufacture of fibers [1]

Glassfibers can be made from compound glass or from silica [2]. Fibers made from compound glass give a relatively high attenuation compared with fibers made from silica. Fibers made from these materials are produced in the following ways.

1.2.1 Fibers from compound glass

Rod in tube method
A glass rod with a high refractive index is placed inside a glass tube of lower refractive index. These are then heated so that the rod and tube fuse together before they are drawn to form a glassfiber.

Double crucible
Two concentric crucibles, as shown in Figure 1.6, are placed in a resistance oven. The inner crucible contains glass of a higher refractive index than that in the outer crucible. When the liquefied glass flows out of the orifices, it cools down quickly, thus forming the glassfiber.

1.2.2 Fibers from silica

The chemical vapour deposition (CVD) method [3]
A stream of $SiCl_4$ and $GeCl_4$ is passed through a silica tube together with oxygen, as shown in Figure 1.7. In a high-temperature region of about 1000–1400 K, created by a torch, the following reactions take place:

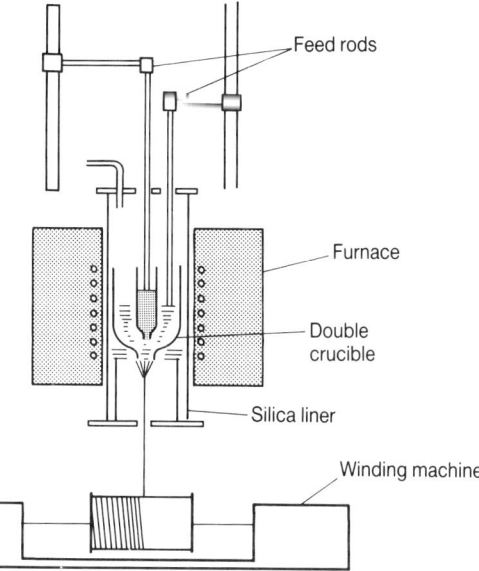

Figure 1.6 The double-crucible method for making glassfibers. (Source: reproduced with permission from *Proceedings of the IEE*, vol. 123, no. 6, June 1976, pp. 591–6, 'Preparation of sodium borosilicate glass fibre for optical communication', by K. J. Beales *et al.*)

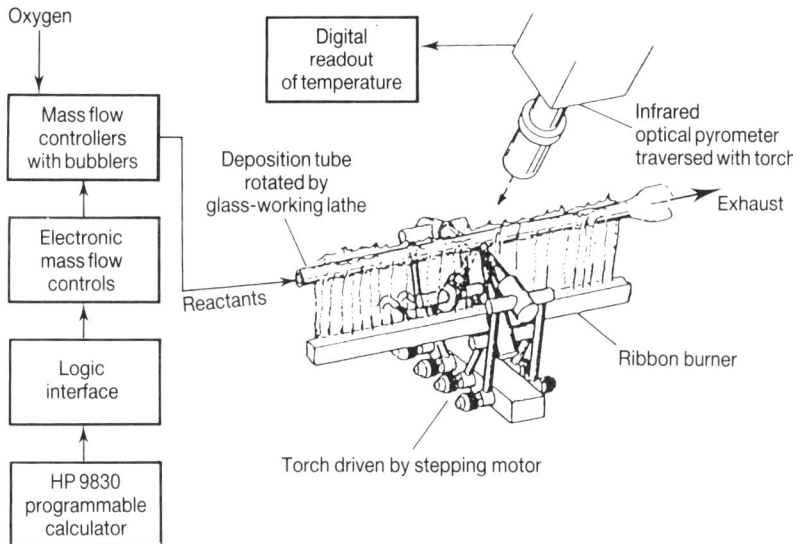

Figure 1.7 The CVD method of fiber production. (Source: reproduced with permission from *Optical Fiber Communications*, by John M. Senior, Prentice Hall, 1985.)

$$SiCl_4 + O_2 \ \rightarrow SiO_2 + 2Cl_2$$

$$GeCl_4 + O_2 \rightarrow GeO_2 + 2Cl_2$$

The higher the concentration of $GeCl_4$, the higher the refractive index. In the hot region, a kind of soot is formed, consisting of fine particles of SiO_2 and GeO_2, and this porous soot is deposited on the wall of the tube. The hot region is progressively moved along the tube; in this way, up to a thousand soot layers can be deposited on the inside of the tube, with every layer having its own refractive index. About 100 g of material is deposited in one hour. Next, the tube is heated in order to bring about the transition of the porous deposits to solid glass and the tube collapses, producing a so-called preform. As the last step in the process, this preform is drawn into a fiber.

The plasma CVD method [4]
The only difference between the PCVD method and the previous one is the way in which the tube is heated; in this case, microwaves are used to form a plasma.

The vapour-phase axial deposition (VAD) method [5]
This method starts with a silica rod. Fine glass particles of $SiCl_4$, $GeCl_4$, BCl_3

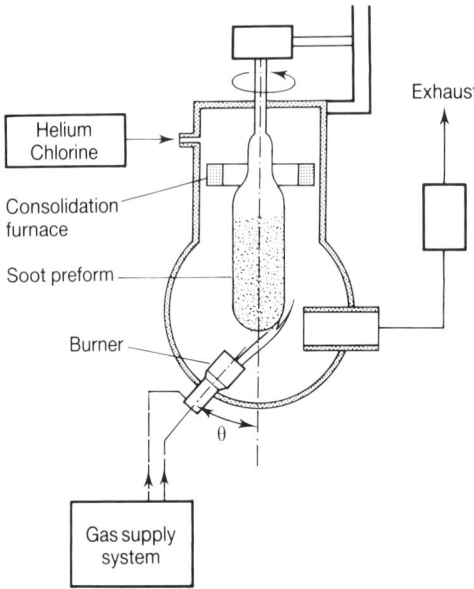

Figure 1.8 The VAD method of fiber production. (Source: reproduced with permission from *Proceedings of the Eighth European Conference on Optical Communication*, Cannes 1982, pp. 1–8, 'Technical and economic aspects of the different fibre fabrication processes', by G. J. Koel.)

and PCl$_3$ are synthesized by hydrolysis with a oxy-hydrogen torch; these particles are deposited on the silica rod, as shown in Figure 1.8.

To produce the refractive index profile (*n*-profile), streams of different raw materials are blown through two or more orifices in order to make a porous preform that grows axially as the whole assembly is drawn upwards. The porous preform is heated in an electric resistance oven until it becomes a transparent glass. The advantages of this method over the (P)CVD method is that there is a steady flow of deposits during the formation part of the process.

1.3 The attenuation of light in glassfibers

As stated before, it is well known that light can be guided to follow the curved path of a transparent guide. This property is often explained as a kind of repetitive total internal reflection. The first well-described scientific demonstration was given by Tyndall in 1870 to the Royal Society [6]. He showed that light followed the curved path of a jet of water, and this principle was used until about 1930 to transmit pictures over relatively short distances in bundles of glassfibers. Interest in glassfibers for telecommunications occurred as soon as glassfibers could be made with sufficient transparency. A milestone in the history of telecommunications was reached at the end of the 1960s with the introduction of optical fibers. The availability of semiconductor lasers at that time stimulated the search for glassfibers with a sufficiently low attenuation (i.e. less than 10 dB km^{-1}). In Figure 1.9 the attenuation curves as a function of

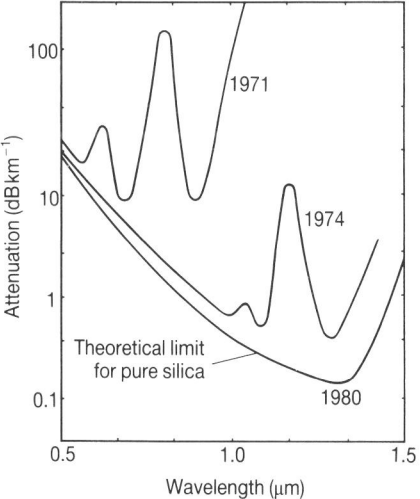

Figure 1.9 The attenuation of glassfibers as a function of the wavelength in the course of time.

wavelength and time illustrate the rapid progress that was made in the 1970s.

Attenuation as a function of the wavelength has decreased almost to its theoretical limit. Originally, wavelengths between 800 and 900 nm were used for optical communication with glassfibers. This wavelength range is often called the first window; thereafter, the second window of about 1300 nm begins. The second wavelength region is especially important for monomode fibers with a very low dispersion. The modern glassfiber does not have regions of low attenuation, but attenuation decreases steadily with increasing wavelength up to a wavelength of about 1500 nm, above which the attenuation increases rapidly.

1.3.1 Attenuation measurements

The attenuation of a glassfiber as a function of wavelength can be measured by the widely used cut-back technique, the principle of which is illustrated in Figure 1.10.

Procedure
A small wavelength band around a wavelength of λ_0 is selected from white light by means of a monochromator, with the help of a diffraction grating. A favourable signal-to-noise ratio at the receiver is obtained when the bundle of light that leaves the monochromator is amplitude modulated by a chopper, e.g.

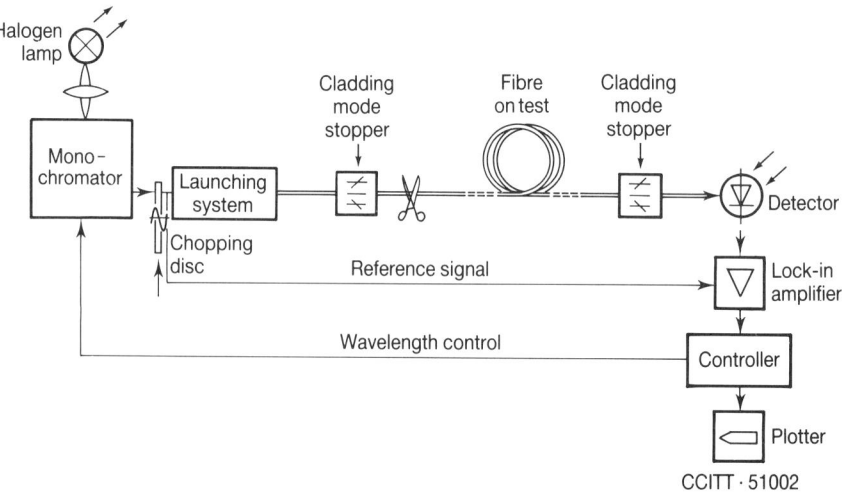

CCITT · 51002

Figure 1.10 The cut-back technique for measuring fiber attenuation. (Source: reproduced from the *CCITT Blue Book*, Fascicle III.3, Recommendation G.651, Figure B-8; the *Blue Book* can be obtained from the ITU General Secretariat, Sales Section, Place des Nations, CH-1211 Geneva 20, Switzerland; reproduced with the prior authorization of the ITU.)

a rotating disc with apertures that interrupts the bundle of light periodically. A reference signal from the chopper gives information about the modulating frequency to the lock in amplifier that is used to measure the signal received. The modulated light is coupled into the fiber by a launching device, usually consisting of a microscope objective and a set of micro-manipulators. The light that is coupled into the cladding is removed by a cladding mode stripper in the following way. A certain length of fiber with its coating removed is laid in a liquid with a higher refractive index than the refractive index of the cladding. The light that leaves the fiber at the far end is shone onto a photodiode in order to produce a current i_1 that is proportional to the power of the light picked up by the photodetector. The photocurrent i_1 is measured as a function of the wavelength. Next, the fiber is cut about 1–3 m from the cladding mode stripper, without altering anything in the launching conditions. The light that now leaves the short length of fiber is shone onto the same photodiode in order to produce a current i_2, proportional to the power of the light received. This photocurrent i_2, is also measured as a function of the wavelength. The attenuation α of the original fiber length minus the short length that was cut off at the beginning is given by

$$\alpha = 10 \log \frac{i_2}{i_1} \, \text{dB} \tag{1.3}$$

The attenuation measured in this way depends partly on the specific mechanism of the wave propagation in the fiber and partly on the way the light is coupled into the fiber. Some of the attenuation is caused by scattering and absorption, but this has nothing to do with the configuration of the fiber and can be measured on the bulk material.

1.3.2 Scattering of the light, Rayleigh scattering

Local, microscopic variations in density and small crystals in the fibers can cause scattering of light in the optical waveguide. The quicker the glass is cooled down during the production process, the smaller the crystals formed. Rayleigh scattering is scattering of light by objects that are smaller than the wavelength of the light. An absorbed photon is reradiated immediately, in a random direction and with the same energy as the absorbed quantum. This is called elastic or coherent scattering, as it is coherent with regard to the phase relationship between the received and transmitted photon. The attenuation caused by Rayleigh scattering is proportional to λ_0^{-4}.

1.3.3 Absorption of light

Light is absorbed mainly by hydroxyl groups and metal ions. Attenuation due to hydroxyl groups, as a function of wavelength, is presented in Figure 1.11, and attenuation due to metals in Figure 1.12.

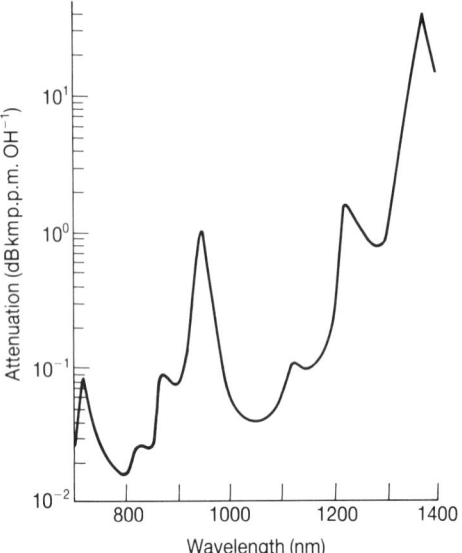

Figure 1.11 Attenuation caused by hydroxyl groups. (Source: reproduced with permission from *Proceedings of the IEEE*, vol. 66, no. 4, April 1978, pp. 429–41, 'Measurements in fiber optics', by M. Barnoski and S. Personick.)

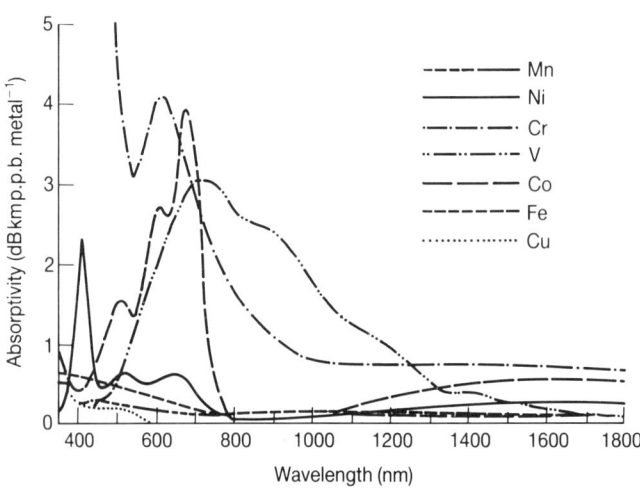

Figure 1.12 Attenuation caused by metals. (Source: reproduced with permission from *Journal of the American Ceramic Society*, vol. 57, no. 7, July 1974, pp. 309–13, 'Optical absorption of the transition elements in vitreous silica', by P. C. Schultz.)

1.3.4 Miscellaneous loss mechanisms

Apart from the losses mentioned already, losses are caused by irregularities in the fiber itself, such as varying core diameter, an elliptic core, curves, etc. In a transmission line comprising a concatenation of fibers, every joint contributes to the total attenuation.

1.4 Light sources

The light source of an optical fiber communication system converts an electrical input signal into an optical signal. Semiconductor LEDs and lasers are mainly used as the light sources in these systems. The physical aspects of these light sources and their external characteristics will be dealt with in Chapter 8, whereas the modulation of the sources will be analyzed in Chapter 9. The coupling efficiency to optical fibers is calculated in Chapter 11.

As far as optical communication is concerned, the following aspects are of interest:

- The static and dynamic relationship between the electrical input and the optical output.
- The dimensions of the light-emitting area and the radiation pattern of the optical bundle.
- The question of whether the light consists of coherent or incoherent light and what the optical spectrum looks like.
- The effect of temperature on its characteristics.
- The efficiency.
- The lifetime.

LEDs

The mode of operation for an LED is based on the emission of photons due to the spontaneous recombination of holes and electrons in a semiconductor device. The number of carriers recombining per unit of time is proportional to the number of carriers present in the active region of the device, and thus is proportional to the forward current through the LED, so that a linear relationship between the input current and the optical output power results.

Although the LED behaves like an incoherent source, its coherence time is nevertheless quite large, namely of the order of 10^{-13} s; the optical power spectrum has a Gaussian shape and its full width at half of the maximum value amounts to 25–100 nm, depending on the maximum emission wavelength.

The dimensions of the emitting area of an LED are similar in magnitude to the core diameter of a multimode fiber. A typical radiation pattern is shown in Figure 1.13. From this figure it can be seen that the shape of the radiation pattern is almost spherical with a $\cos \theta$ dependence, and thus the radiation of an

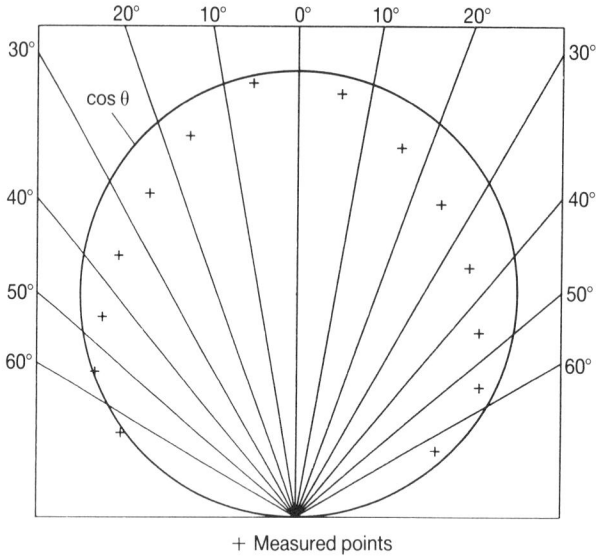

Figure 1.13 Radiation pattern of the Plessy HR1301 LED.

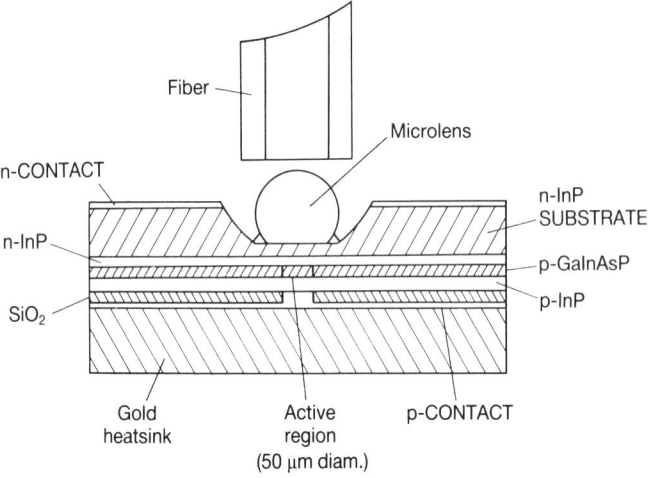

Figure 1.14 An LED with a ball lens coupled to a fiber.

LED can be approximated by that of a Lambertian source. Figure 1.14 shows an LED with a ball lens coupled to a fiber.

Lasers
Figure 1.15 gives an example of a typical semiconductor laser diode; in this

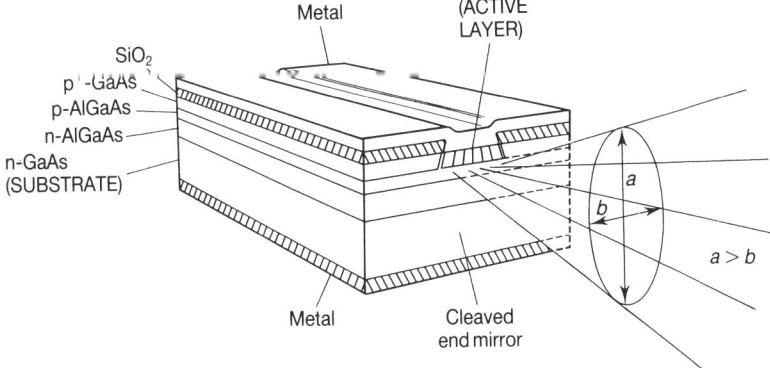

Figure 1.15 Construction of a semiconductor laser with emission bundle.

figure, the shape of the emitted light bundle is shown diagrammatically. Since a laser emits very coherent light, its coupling efficiency to a fiber can be much larger than for an LED (see Chapter 11). Figure 1.16 shows the relationship between the input current and the optical output power of a typical laser diode at various temperatures. The slopes of these characteristics are quite shallow near the origin and, at a certain current, increase suddenly. This current is

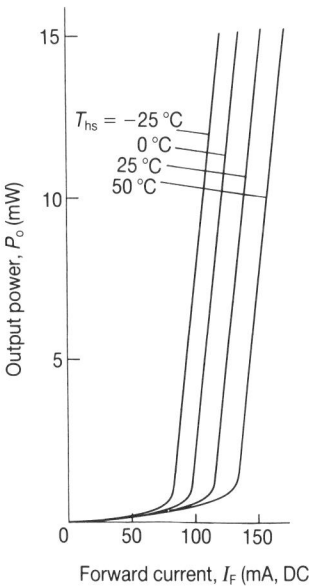

Figure 1.16 The optical output power versus current characteristic of a semiconductor laser with temperature as a parameter.

called the threshold current and below it the device acts as an LED; laser action only occurs above this threshold, in the region where the slope is large. A laser can be modelled as an electromagnetic resonant cavity which, in general, can oscillate in different modes. If that is the case, the spectrum of the laser will contain a number of spectral lines. Typical power spectra of a laser diode are shown in Figure 1.17. The width of a single line can be as small as 1 pm, which corresponds to a coherence length of a few meters.

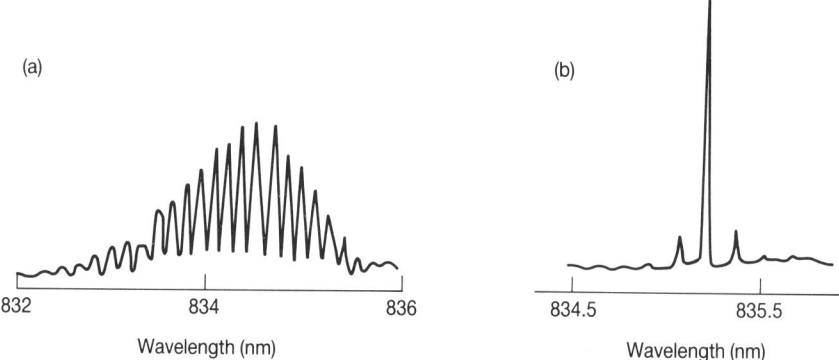

Figure 1.17 The power spectrum of a semiconductor laser: (a) just below the threshold; (b) just above the threshold. (Source: reproduced with permission from *Proceedings of the IEE*, vol. 123, no. 6, June 1976, pp. 633–41, 'Subsystems for optical-fibre-communication field demonstrations', by M. Ramsay *et al.*)

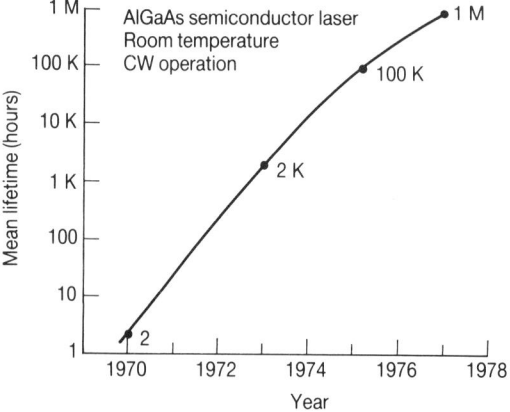

Figure 1.18 The evolution of the lifetime of semiconductor lasers in the 1970s.

The lifetime of semiconductor lasers increased during the 1970s, as illustrated in Figure 1.18. At the present time, the lifetime is large enough to use such lasers in professional communication systems.

1.5 Detectors

In an optical fiber communication system, the detector acts in the opposite way to the light source, i.e. it converts the received optical signal into an electrical signal. Like the light source, the optical detector is usually some form of semiconductor diode. For the short-wavelength region, 800–900 nm, Si photodiodes are suitable, whereas for the long-wavelength region, 1300–1600 nm, Ge or InGaAsP photodiodes have to be used (see Chapter 8). Both diodes that lack an internal gain (the so-called PIN diodes) and diodes that have an internal gain (the so-called avalanche photodiodes or APDs) are available.

Two parameters are important for describing photodiodes: responsivity and quantum efficiency. Responsivity is defined as the amount of photocurrent passing through the device per unit of optical power received; while quantum efficiency is the mean number of carrier pairs generated in a semiconductor per received photon. These parameters will be analyzed in Chapter 8, when it will be shown that the photocurrent is proportional to the optical power received. The advantage of using an APD is dealt with in Chapters 12 and 13. The noise that is generated in the photodetector is analyzed in Chapter 12 and Appendix 4.

1.6 System model

As assumed in the previous section and derived in the next chapter, the propagation of a wave in a multimode fiber occurs in many modes. Each mode has its own characteristic field distribution and its own propagation constant. The propagation constant of the nth mode can be denoted by

$$\gamma_n(\omega) = \alpha_n(\omega) + j\beta_n(\omega) \tag{1.4}$$

where $\alpha(\omega)$ and $\beta(\omega)$ are the attenuation and phase constants, respectively. If different modes have the same propagation constant, then the modes are called degenerate. In the model of the system, we can associate each mode with a transmission channel. The fiber can be depicted as a multitude of parallel transmission channels, as shown in Figure 1.19. At the transmitter side, an optical power p_{in} is fed into the nth transmission channel.

The ratios of the various power inputs depend on the way in which the light from the source is coupled into the fiber. The transmission channels are mutually coupled by irregularities in the fiber, by curves, by the couplers, etc.

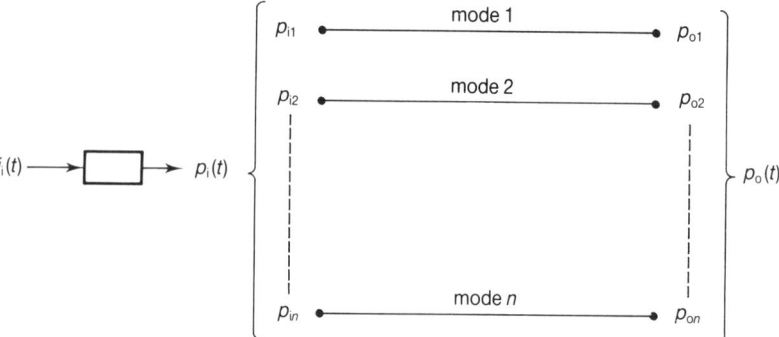

Figure 1.19 The system model.

Therefore there is a constant exchange of power between the different modes. The light that leaves the fiber from the far end is caught by the photodetector. In the model, this means that the individual power outputs of the different channels have to be summated since the modes are orthogonal. In many cases, the carrier cannot be regarded as monochromatic and this means an extra complication in the model. The power spectrum of the unmodulated source has a certain width and therefore the power is transmitted by a multitude of carrier frequencies. It will be clear that the total transmission depends on many factors.

References

[1] C. K. Kao, *Optical Fibre* (London: Peter Peregrinus, 1988)

[2] D. Küppers, H. Lydtin and F. Meijer, 'Preparation methods for optical fibres applied in Philips Research', *International Conference on Integrated Optics and Optical Fiber Communication*, Tokyo, July 1977, pp. 319–22.

[3] F. V. DiMarcello and J. C. Williams, 'Reproducibility of optical fibers prepared by a chemical vapor deposition process', *Proceedings of the First European Conference on Optical Fibre Communication*, London, September 1975, pp. 36–8 (IEE Conference Publication no. 132).

[4] D. Küppers and J. Koenings, 'Preform fabrication by deposition of thousands of layers with the aid of plasma activated CVD', *Proceedings of the Second European Conference on Optical Fibre Communication*, Paris, September 1976, pp 49–54.

[5] T. Izawa, S. Kobayashi, S. Sudo and F. Hanawa, 'Continuous fabrication of high silica fiber preform', *International Conference on Integrated Optics and Optical Fiber Communication*, Tokyo, July 1977, pp. 375–8.

[6] N. S. Kapany, *Fiber Optics, Principles and Applications* (New York: Academic Press, 1967).

Problems

1.1 At a given wavelength, the light in an optical waveguide can propagate in four guided modes. For one of these modes the attenuation is 2 dB km^{-1} and for the other three modes 4 dB km^{-1}. During an attenuation measurement of the fiber, the different modes are equally excited. Thus, at the transmitter end of the fiber, the input power is equally distributed over the four modes. It is assumed that there is no mode conversion, i.e. there is no power exchange between the different modes.

- Find the attenuation for 1 and 2 km lengths of the fiber.
- What can be concluded, in general, regarding the attenuation per unit of length of a multimode fiber?

2

Analysis of the slab waveguide

2.1   Theoretical models
2.2   The slab waveguide analyzed with wave optics
2.3   Transverse electric waves
2.4   Transverse magnetic waves
2.5   The propagation constant of a mode
2.6   Geometric optics interpretation
      References
      Problems

In this chapter attention will be paid to the propagation of harmonic waves in slab waveguides in order to introduce the treatment of harmonic waves in step index optical waveguides. In addition, it is worthwhile studying the properties of slab waveguides themselves, because they play an important role in the technology of integrated optics.

2.1 Theoretical models

The distances between atoms in glass are of the order of 0.1 nm, which is small compared to the wavelength of light that is used in optical communications (these wavelengths are of the order of 1000 nm). In this respect, glass is assumed to be a homogeneous medium. In Chapter 1, it was stated that the

Rayleigh scattering of light orginated from small inhomogeneities in material density or composition of the medium. Strictly speaking, therefore, glass is not homogeneous; nevertheless, any imperfections will be ignored and the glass assumed to be homogeneous. Of course, this model, like all models, is only a representation of reality. Moreoever, the medium is assumed to be isotropic, linear and time invariant. Hereafter, it will be assumed that the electrical properties of such materials can be described by the permittivity ε, the permeability μ and the conductivity σ.

Wave optics

If ε, μ and σ for the material are known, then Maxwell's equations are a good starting point for analyzing the propagation of light in glass under the name of wave optics. This powerful method can solve the problem of propagating electromagnetic waves in step index optical waveguides. If the thickness of the core of the optical waveguide is of the order of the wavelength, then the propagation can be described with a few modes that are characteristic of the optical waveguide and the wavelength of the light. If, on the other hand, the core radius is large compared to the wavelength, then many propagating modes are possible. In that instance, it will be more effective to solve the problem with the help of geometric optics.

Geometric optics

The building blocks used in optics include lenses, prisms and mirrors. The dimensions of these are usually large compared to the wavelength of the light. If this is the case, an approximating method can be used to study and describe propagation of the light. The geometric optics or ray optics method uses the language of geometry in order to formulate the laws of optics. In geometric optics, the concept of light rays is introduced in order to describe optical phenomena. The paths taken by the light rays in heterogeneous media are derived from the so-called eikonal equation that is introduced in Chapter 6, when the graded index optical waveguide is discussed.

2.2 The slab waveguide analyzed with wave optics

An insight into the relationship between the mathematical analysis and the physical interpretation of the fields and waves in glassfibers of the step index type requires, firstly, consideration of a slab waveguide. The phenomena that occur in this case are mainly analogous to those in round glassfibers and the analysis is less complex. The assumed structure is drawn in Figure 2.1, and the slab waveguide is assumed to be infinitely extended in the x- and y-directions. Medium 1, between the interfaces $x = -a$ and $x = a$, is enclosed on both sides by the optically less dense medium 2. Both media are assumed to be homogeneous, linear and isotropic. Since the materials are non-magnetic, it is assumed

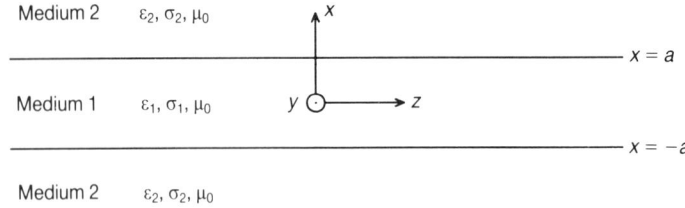

Figure 2.1 Slab waveguide.

that the magnetic permeability μ has the free space value μ_0. The permittivity, conductivity and permeability of medium 1 are denoted by ε_1, σ_1 and μ_0, respectively, and similarly for medium 2: ε_2, σ_2 and μ_0.

Maxwell's equations are taken as the starting point for the analyses:

$$\mathbf{\nabla} \times \mathbf{E} = -\frac{\partial \mathbf{B}}{\partial t} = -\mu_0 \frac{\partial \mathbf{H}}{\partial t} \tag{2.1}$$

$$\mathbf{\nabla} \times \mathbf{H} = \frac{\partial \mathbf{D}}{\partial t} + \mathbf{J} = \varepsilon \frac{\partial \mathbf{E}}{\partial t} + \sigma \mathbf{E} \tag{2.2}$$

The magnetic field vector can be eliminated from these two equations by multiplication of equation (2.1) by the vector operator $\mathbf{\nabla}$, in order to give

$$\mathbf{\nabla} \times \mathbf{\nabla} \times \mathbf{E} = -\mu_0 \frac{\partial}{\partial t} (\mathbf{\nabla} \times \mathbf{H}) \tag{2.3}$$

Developing the left-hand side of equation (2.3) and substituting equation (2.2) in the right-hand side leads to

$$\mathbf{\nabla}(\mathbf{\nabla}.\mathbf{E}) - \mathbf{\nabla}^2\mathbf{E} = -\mu_0\varepsilon \frac{\partial^2 \mathbf{E}}{\partial t^2} - \mu_0\sigma \frac{\partial \mathbf{E}}{\partial t} \tag{2.4}$$

It is assumed that there are no free charges, so that $\mathbf{\nabla}.\mathbf{D} = 0$, therefore $\mathbf{\nabla}.(\varepsilon.\mathbf{E}) = \varepsilon(\mathbf{\nabla}.\mathbf{E}) + (\mathbf{\nabla}\varepsilon)\mathbf{E} = 0$. For a homogeneous, isotropic medium, $\mathbf{\nabla}\varepsilon = 0$, and thus $\mathbf{\nabla}.\mathbf{E} = 0$, resulting in the wave equation

$$\mathbf{\nabla}^2\mathbf{E} = \mu_0\varepsilon \frac{\partial^2 \mathbf{E}}{\partial t^2} + \mu_0\sigma \frac{\partial \mathbf{E}}{\partial t} \tag{2.5}$$

2.3 Transverse electric waves

We first assume waves with vector $\mathbf{E}$ transverse to the direction of transmission; such waves are called transverse (TE) waves. To arrive at a simple solution of the wave equation for the slab waveguide in Figure 2.1, harmonic waves must be assumed that are independent of y and travel with a certain phase velocity in the positive z-direction. Every component of the electric and magnetic fields is then proportional to $\exp(j\omega t - \gamma z)$, with

$$\gamma = \alpha + j\beta \tag{2.6}$$

For TE waves polarized in the y-direction

$$E_x = E_z = 0 \tag{2.7}$$

With regard to equation (2.7), the wave equation (2.5) reduces to

$$\frac{\partial^2 E_y}{\partial x^2} + k_c^2 E_y = 0 \tag{2.8}$$

with

$$k_c^2 = k^2 + \gamma^2 \tag{2.9}$$

and

$$k^2 \triangleq \omega^2 \mu_0 \varepsilon \left(1 - \frac{j\sigma}{\omega\mu_0}\right) \tag{2.10}$$

In general, k^2 and γ^2 and thus also k_c^2 are complex quantities; however, the treatment of the general case is omitted here and attention is concentrated on simple solutions where k^2 and γ^2 have real values and the medium has no losses. In that case, $\sigma = 0$ and ε is real; γ^2 is real if $\alpha = 0$ or $\beta = 0$. For unattenuated waves travelling in the positive z-direction, the condition $\alpha = 0$ must be fulfilled. Therefore, it is assumed that

$$\gamma = j\beta \tag{2.11}$$

Simple solutions of equation (2.8) can be obtained if k_c^2 is made positive in medium 1 and negative in medium 2. The solutions thus obtained describe unattenuated waves, completely bound by the waveguide, travelling in the positive z-direction and having oscillating fields (as functions of x) in medium 1 and evanescent fields in medium 2.

Medium 1
Quantities in medium 1 are indicated by the subscript 1. $k_{c1}^2 > 0$, thus k_{c1} is real. Denote

$$k_{c1} = h_1 \tag{2.12}$$

With equations (2.10) and (2.11), it follows that

$$k_1^2 > \beta^2 \tag{2.13}$$

or

$$\omega^2 \mu_0 \varepsilon_1 > \beta^2 \tag{2.14}$$

From equation (2.14), it is clear that the phase velocity ω/β of the wave in the slab waveguide is greater than the phase velocity $c_1 = 1/\sqrt{\mu_0 \varepsilon_1}$ of a plane wave in bulk material with the same electric properties as medium 1.

The solution of equation (2.8) for medium 1 becomes

$$E_{y1} = (A \cos h_1 x + B \sin h_1 x) \exp(j\omega t - j\beta z), \qquad |x| \leq a \qquad (2.15)$$

where A and B are arbitrary constants.

Medium 2

Quantities in medium 2 are denoted by the subscript 2. $k_{c2}^2 < 0$, thus k_{c2} is imaginary. Denote

$$k_{c2} = jh_2 \qquad (2.16)$$

With equations (2.10) and (2.11), it follows that

$$k_2^2 < \beta^2 \qquad (2.17)$$

or

$$\omega^2 \mu_0 \varepsilon_2 < \beta^2 \qquad (2.18)$$

It becomes clear from equation (2.18) that the phase velocity ω/β of the wave in the slab waveguide is smaller than the phase velocity $c_2 = 1/\sqrt{(\mu_0 \varepsilon_2)}$ in bulk material with the same electric properties as medium 2. The solution of equation (2.8) for medium 2 becomes

$$E_{y2} = C \exp(-h_2 x) \exp(j\omega t - j\beta z), \qquad \text{for } x \geq a \qquad (2.19)$$

$$E_{y2} = D \exp(h_2 x) \exp(j\omega t - j\beta z), \qquad x \leq -a \qquad (2.20)$$

The components of the magnetic fields in both media, expressed in E_y, follow from equation (2.1)

$$H_x = \frac{1}{j\omega\mu_0} \frac{\partial E_y}{\partial z} \qquad (2.21)$$

$$H_z = -\frac{1}{j\omega\mu_0} \frac{\partial E_y}{\partial x} \qquad (2.22)$$

The structure symmetry of Figure 2.1 can be used to find the elementary solutions. Symmetric fields are found when $B = 0$ and $C = D$; while so-called antisymmetric fields are found when $A = 0$ and $C = -D$.

2.3.1 Symmetric TE waves ($B = 0$; $C = D$)

By putting $B = 0$ and $C = D$, the following magnitudes of the electric and magnetic field components can be found from equation (2.15) and equations (2.19)–(2.22).

$$E_{y1} = A \cos(h_1 x) \exp(j\omega t - j\beta z), \qquad |x| \leq a \qquad (2.23)$$

$$E_{y2} = C \exp(-h_2|x|) \exp(j\omega t - j\beta z), \qquad |x| \geq a \qquad (2.24)$$

$$H_{x1} = -\frac{\beta}{\omega\mu_0} A \cos(h_1 x) \exp(j\omega t - j\beta z), \qquad |x| \leq a \qquad (2.25)$$

$$H_{x2} = -\frac{\beta}{\omega\mu_0} C \exp(-h_2|x|) \exp(j\omega t - j\beta z), \qquad |x| \geq a \qquad (2.26)$$

$$H_{z1} = \frac{h_1}{j\omega\mu_0} A \sin(h_1 x) \exp(j\omega t - j\beta z), \qquad |x| \leq a \qquad (2.27)$$

$$H_{z2} = \frac{h_2}{j\omega\mu_0} C \exp(-h_2 x) \exp(j\omega t - j\beta z), \qquad x \geq a \qquad (2.28)$$

$$H_{z2} = -\frac{h_2}{j\omega\mu_0} C \exp(h_2 x) \exp(j\omega t - j\beta z), \qquad x \leq -a \qquad (2.29)$$

On the interfaces between the different media, E_y and H_z must be continuous. The continuity condition for E_y together with equations (2.23) and (2.24) gives

$$A \cos u = C \exp(-w), \qquad \text{for } x = a \qquad (2.30)$$

with

$$u = h_1 a \qquad (2.31)$$

and

$$w = h_2 a \qquad (2.32)$$

The continuity condition for H_z leads to

$$uA \sin u = wC \exp(-w) \qquad (2.33)$$

Dividing equation (2.33) by equation (2.30) gives the so-called characteristic equation for symmetric TE waves

$$w = u \tan u \qquad (2.34)$$

2.3.2 Antisymmetric TE waves

By putting $A = 0$ and $C = -D$, the components of the electric and magnetic fields of the antisymmetric waves in both media follow from equation (2.15) and equations (2.19)–(2.22):

$$E_{y1} = B \sin(h_1 x) \exp(j\omega t - j\beta z), \qquad |x| \leq a \qquad (2.35)$$

$$E_{y2} = C \exp(-h_2 x) \exp(j\omega t - j\beta z), \qquad x \geq a \qquad (2.36)$$

$$E_{y2} = -C \exp(h_2 x) \exp(j\omega t - j\beta z), \qquad x \leq -a \qquad (2.37)$$

$$H_{x1} = -\frac{\beta}{\omega\mu_0} B \sin(h_1 x) \exp(j\omega t - j\beta z), \qquad |x| \leq a \qquad (2.38)$$

$$H_{x2} = -\frac{\beta}{\omega\mu_0} C \exp(-h_2 x) \exp(j\omega t - j\beta z), \qquad x \geq a \qquad (2.39)$$

$$H_{x2} = \frac{\beta}{\omega\mu_0} C \exp(h_2 x) \exp(j\omega t - j\beta z), \qquad x \leq -a \qquad (2.40)$$

$$H_{z1} = -\frac{h_1}{j\omega\mu_0} B \cos(h_1 x) \exp(j\omega t - j\beta z), \qquad |x| \leq a \qquad (2.41)$$

$$H_{z2} = \frac{h_2}{j\omega\mu_0} C \exp(-h_2 x) \exp(j\omega t - j\beta z), \qquad x \geq a \qquad (2.42)$$

$$H_{z2} = \frac{h_2}{j\omega\mu_0} C \exp(h_2 x) \exp(j\omega t - j\beta z), \qquad x \leq -a \qquad (2.43)$$

The continuity of E_y and H_z required on the interfaces between the media leads to the characteristic equation for antisymmetric waves in the same way as with the symmetric waves

$$w = -u \cot an\, u \qquad (2.44)$$

2.3.3 TE modes

From equations (2.9), (2.10), (2.12), (2.16), (2.31) and (2.32), a relationship between u and w can be established

$$u^2 + w^2 = (h_1^2 + h_2^2)a^2 = (k_{c1}^2 - k_{c2}^2)a^2 = (k_1^2 - k_2^2)a^2$$
$$= \omega^2\mu_0\varepsilon_0 a^2(\varepsilon_{r1} - \varepsilon_{r2}) = (k_0 a)^2(\varepsilon_{r1} - \varepsilon_{r2}) \overset{\triangle}{=} v^2 \qquad (2.45)$$

with

$$k_0^2 = \omega^2\mu_0\varepsilon_0 \qquad (2.46)$$

where ε_0 is the free space permittivity and ε_{ri} is the relative permittivity of medium i ($i = 1, 2$); v is called the normalized frequency.

Given a certain frequency, and thus the normalized frequency v, the characteristic equation given by equation (2.34) or equation (2.44) requires discrete values of u and w. The corresponding wave phenomena are called modes. A graphical method for solving the characteristic equations is illustrated in Figure 2.2. In the Cartesian coordinate system illustrated, the relationship between u and w from equation (2.34) for symmetric waves is represented by a solid line, and the relationship from equation (2.44) for antisymmetric waves is represented by a broken line. The solid lines are marked S_1, S_2, etc., and the broken lines are marked A_1, A_2, etc. Equation (2.45) is represented graphically by a circle that has its center at the origin of the coordinate system and its radius is v. The coordinates of the intersections of the circle with the solid lines and the broken lines are the values of u and w that belong to the modes of the symmetric and the antisymmetric waves, respectively.

If $v < 1$, which is demonstrated in Figure 2.2 by the curve marked a, then the

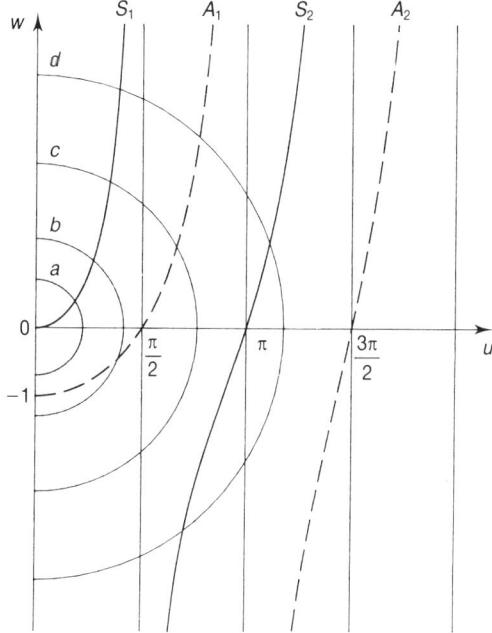

Figure 2.2 Graphical solutions for the characteristic equations.

circle has only one intersection point, namely with the curve S_1; therefore, the corresponding value of w is positive, implying that the field densities decrease exponentially in medium 2 with increasing values of $|x|$, this is clear from equations (2.19) and (2.20).

If $1 < v < \pi/2$, then the circle marked b in Figure 2.2 intersects curve A_1, at a point that corresponds to a negative value of w. Consequently, the field intensities increase exponentially in medium 2 with increasing values of $|x|$. If the electromagnetic energy remains concentrated in medium 1 and in the vicinity of medium 1, then only positive values of w are allowed.

If $w > 0$, the corresponding mode is called a proper mode. In that case, the fields in medium 2 decrease exponentially with $|x|$; the waves in this medium are then called evanescent waves. If, on the other hand, $w < 0$, the corresponding mode is called an improper mode.

If $v > \pi/2$ but is smaller than the radius of the circle that just touches the curve S_2 (marked c in Figure 2.2), then one proper mode for the symmetric wave and one proper mode for the antisymmetric wave are found. The cut-off frequency for that proper mode can be found for $v = \pi/2$. In that instance, the fields in medium 2 are independent of x. A proper mode cannot exist for a lower frequency than this cut-off frequency. Finally, a circle d has been drawn in Figure 2.2 that corresponds with two proper modes and one improper mode for the symmetric waves and one proper mode for the antisymmetric waves.

In general, the symmetric modes are denoted by TE_n, with $n = 0, 2, 4, \ldots$, and the antisymmetric modes are denoted TE_n, with $n = 1, 3, 5, \ldots$.

2.4 Transverse magnetic waves

If the magnetic field vector **H** is transverse to the direction of propagation, the wave is called a transverse magnetic (TM) wave. In a similar way to that described above for TE waves, it is possible to derive the TM waves and modes, starting by multiplying equation (2.2) with the vector operator ∇ in order to eliminate the electric field vector **E** from Maxwell's equations and to derive a wave equation for **H**. For TM waves, polarized in the y-direction, H_x and H_z are assumed to be zero. Next, the reduced wave equation needs to be solved, assuming that the oscillating fields are restricted to medium 1 and the evanescent fields to medium 2.

2.5 The propagation constant of a mode

The propagation constant plays a major role in the transmission of information. The real part of this constant, namely the attenuation constant α, can usually be assumed to be independent of frequency in the range under consideration; then the phase characteristic is equal to the phase characteristic of an ideal waveguide without attenuation. The phase constant follows from equations (2.9)–(2.12) and equations (2.16), (2.31), (2.32) and (2.45)

$$\beta(\omega) = \sqrt{\omega^2 \mu_0 \varepsilon_1 - h_1^2} = \sqrt{\omega^2 \mu_0 \varepsilon_2 + h_2^2}$$

$$= \omega \sqrt{\mu_0 \varepsilon_1} \sqrt{1 - \frac{u^2}{v^2}\left(1 - \frac{\varepsilon_2}{\varepsilon_1}\right)} \tag{2.47}$$

Since each mode has its own value of u for a certain normalized frequency v, each mode has a unique value of β. If $w = 0$ for a certain mode and thus $h_2 = 0$, then the frequency equals the cut-off frequency. Denoting ω at cut-off by ω_c allows equation (2.48) to be derived from equation (2.47)

$$\beta(\omega_c) = \omega_c \sqrt{\mu_0 \varepsilon_2} \tag{2.48}$$

Thus, at the cut-off frequency, β equals the phase constant of a plane wave in a homogeneous medium with permittivity ε_2.

If $\omega \to \infty$, then $u/v \to 0$ and it follows from equation (2.47) that

$$\beta \to \omega \sqrt{\mu_0 \varepsilon_1} \tag{2.49}$$

Thus, with increasing frequency, β approaches the phase constant of a plane wave in a homogeneous medium with permittivity ε_1.

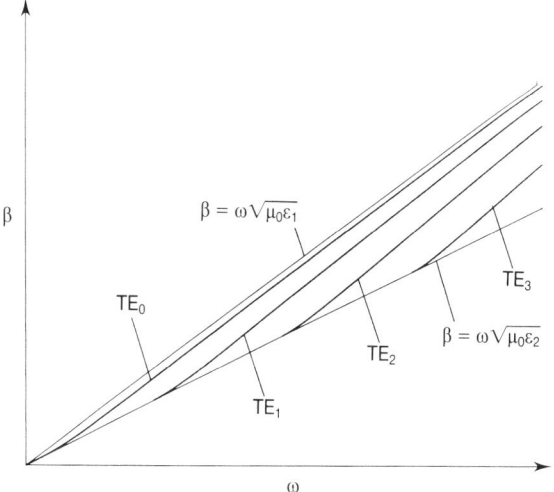

Figure 2.3 The phase constant β as a function of frequency for the modes TE_0, TE_1, TE_2 and TE_3.

The phase constant β is shown in Figure 2.3 as a function of the frequency for the modes TE_0, TE_1, TE_2 and TE_3. Unlike for the modes where $n > 0$, the TE_0 mode has no cut-off frequency. The phase characteristic of this mode touches the line described by $\beta = \omega\sqrt{(\mu_0\varepsilon_2)}$ in the origin.

2.6 Geometric optics interpretation

Equations (2.23), (2.25) and (2.27), which describe the fields for symmetric waves in medium 1, can be written in the following way:

$$E_{y1} = \frac{A}{2}\exp\left[j(\omega t + h_1 x - \beta z)\right] + \frac{A}{2}\exp\left[j(\omega t - h_1 x - \beta z)\right] \qquad (2.50)$$

$$H_{x1} = -\frac{\beta A}{2\omega\mu_0}\exp\left[j(\omega t + h_1 x - \beta z)\right] - \frac{\beta A}{2\omega\mu_0}\exp\left[j(\omega t - h_1 x - \beta z)\right]$$

$$(2.51)$$

$$H_{z1} = \frac{h_1 A}{2\omega\mu_0}\exp\left[j(\omega t + h_1 x - \beta z)\right] + \frac{h_1 A}{2\omega\mu_0}\exp\left[j(\omega t - h_1 x - \beta z)\right]$$

$$(2.52)$$

Looking at the first terms of the right-hand side of these three equations, it can be seen that, at every moment denoted by the value of t, the phase is the same on surfaces for which the following condition is fulfilled;

$$h_1x - \beta z = \omega t \tag{2.53}$$

These are plane surfaces; therefore, the first terms of the right-hand side of equations (2.50)–(2.52) describe a plane wave in medium 1. Likewise the second set of terms describes a plane wave with surfaces of equal phase given by

$$h_1x + \beta z = -\omega t \tag{2.54}$$

Thus a mode can be seen as the superposition of two plane waves in medium 1. In Figure 2.4, two surfaces of equal phase, also called wavefronts, are depicted. Note that superposition of the two components of the magnetic field gives the expected resultant magnetic field vector in the plane of the wavefront. The Poynting vector is perpendicular to the wavefront and coincides with the direction of the light rays corresponding to that wavefront. The angle between a ray and the z-axis is called θ and between a ray and the x-axis is called θ'.

From figure 2.4, it follows that

$$\cos \theta = \frac{\beta}{k_1} \tag{2.55}$$

From equation (2.47), it follows that

$$\left(\frac{\beta}{k_1}\right)^2 = 1 - \frac{u^2}{v^2}\left(1 - \frac{n_2^2}{n_1^2}\right) \tag{2.56}$$

At the cut-off frequency $u = v$, therefore

$$\frac{\beta(\omega_c)}{k_1} = \frac{n_2}{n_1} = \sin\left(\frac{\pi}{2} - \theta_c\right) = \sin \theta'_c \tag{2.57}$$

where θ_c and θ'_c are the angles θ and θ' at cut-off. From equation (2.57) it is clear that the angle of incidence at the cut-off, on the interface between the different media, equals the critical angle for total internal reflection. If the frequency is lower than the cut-off frequency, then the light rays are not totally reflected and the condition $\alpha = 0$ is not satisfied. For a proper mode, it is

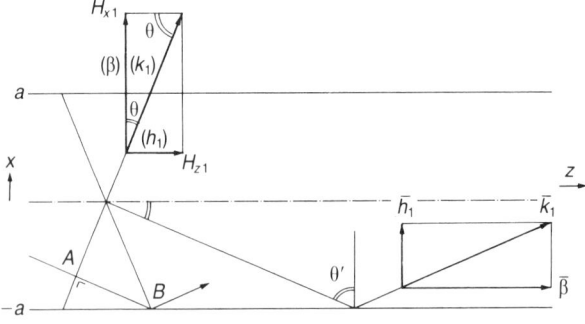

Figure 2.4 Wavefronts of two plane waves and perpendicular light rays.

necessary for the angle θ' to be greater than the critical angle for total internal reflection.

For the TE_0 mode, which has no cut-off frequency, $u/v > 1$ if $\omega > 0$. In that case the angle of incidence will approach the critical angle when the frequency decreases without limit.

2.6.1 The phase change on reflection

Since the phases of the two wavefronts must be equal, the phase change going from point A on the first wavefront to point B on the second wavefront must equal a whole multitude of 2π. The distance from A to B when measured along a light ray that is reflected between A and B is

$$\overline{AB} = \frac{2h_1 a}{k_1} \tag{2.58}$$

In medium 1, this corresponds to a phase change

$$\Delta\phi = -\overline{AB}\, k_1 = -2h_1 a = -2u \tag{2.59}$$

Therefore, the reflection of a light ray is accompanied by a phase change of $-2u$.

The slab waveguide provides an insight into optical waveguides for preparing an analysis of circular optical waveguides. In itself, the slab waveguide forms a starting point for the development of integrated optical circuits which concern mainly three or more different media and have more than two interfaces

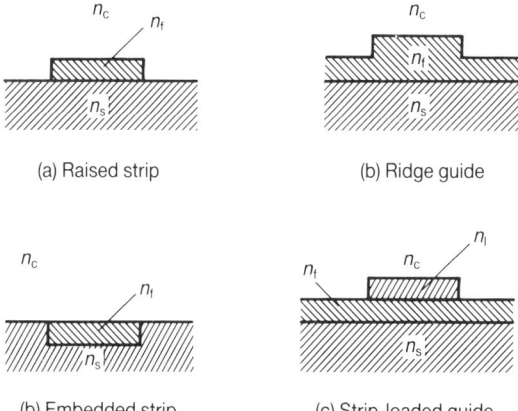

(a) Raised strip   (b) Ridge guide

(b) Embedded strip   (c) Strip-loaded guide

Figure 2.5 Some possible slab waveguide configurations; n_s is the refractive index of the substrate, n_f the refractive index of the guiding layer, n_c the refractive index of the covering layer, and n_l the refractive index of the loading strip. (Source: reproduced with permission from *Integrated Optics*, by T. Tamir, Springer-Verlag, 1979.)

between the media; for instance, with configurations like those sketched in Figure 2.5.

References

[1] N. S. Kapany and J. J. Burke, *Optical Waveguides* (New York: Academic Press, 1972).
[2] D. Marcuse, *Theory of Dielectric Optical Waveguides* (New York: Academic Press, 1974).
[3] D. Marcuse, *Light Transmission Optics* (New York: Van Nostrand Reinhold, 1972).
[4] H. G. Unger, *Planar Optical Waveguides and Fibres* (Oxford: Clarendon Press, 1977).

Problems

2.1 A slab waveguide as drawn in Figure 2.1. has $n_1 = 1.5$ and $n_2 = 1.48$. Assume monochromatic light with a wavelength such that for the TE_0 mode $u = \pi/4$.

- Find the corresponding value of w and the normalized frequency v.
- Find the ration λ_0/a.
- Sketch E_{y1} and E_{y2} for $t = 0$ and $z = 0$ as functions of x.
- Prove that

$$\left(\frac{dE_{y1}}{dx}\right)_{x=a} = \left(\frac{dE_{y2}}{dx}\right)_{x=a}$$

2.2 Find expressions for the electric and magnetic fields in the case that medium 1 of Figure 2.1 is enclosed between media of different permittivity.

2.3 Repeat the derivation of the modes as done in Section 2.3.3 for TM modes. Do the TE modes and the TM modes have the same cut-off frequencies?

2.4 A slab waveguide as given by Figure 2.1 has the following properties: $n_1 = 1.5$, $n_2 = 1.4$, $\sigma_1 = \sigma_2 = 0$, $a = 1\ \mu m$.

Which proper modes can propagate in this waveguide if the wavelength of the light is $\lambda_0 = 1\ \mu m$?

2.5 Derive an expression equivalent to the expression in equation (2.47) for the phase constant of a TM mode in a symmetric slab waveguide and compare the resulting phase characteristics with those of the TE modes as depicted in Figure 2.3.

3

Analysis of the step index fiber

3.1 The general solution of the wave equation
3.2 Unattenuated waves
3.3 Transverse and hybrid waves and modes
3.4 Transverse waves and modes
3.5 Hybrid modes
3.6 Leaky modes
3.7 Linearly polarized (LP) modes
     References
     Problems

In this chapter, as in the previous one about slab waveguides, the theory of wave optics will be used to trace the way in which harmonic waves propagate in a step index optical waveguide [1–6]. Here too, discrete solutions will be found that are called modes. We will derive the mode fields and phase constants as functions of the frequency, because these phase functions provide the key to solving dispersion problems in optical waveguides. In a later chapter, particular attention will be paid to the fundamental modes that can propagate in single-mode fibers and to the related dispersion. Finally, the last section of this chapter deals with the weakly guiding fiber with its linearly polarized (LP) modes.

33

3.1 The general solution of the wave equation

The cross-section and the refractive index profile of a step index optical waveguide are illustrated in Figure 3.1. Though the cladding of the real fiber shown in Figure 3.1 has a limited diameter, we will analyze a cladding of infinite thickness in order to simplify the analysis considerably and make it suitable for a first approximation because the fields in the cladding generally decay strongly with the distance to the core.

The core and cladding materials are assumed to be homogeneous, linear and isotropic. As in Chapter 2, the fiber material is non-magnetic; therefore, it can be assumed that the magnetic permeability μ takes the free-space value μ_0.

Our starting point for the analysis is the reduced wave equation for the electric field vector $\mathbf{E}$ that was derived in Chapter 2, equation (2.5). This equation is rewritten below:

$$\nabla^2 \mathbf{E} + k^2 \mathbf{E} = 0 \tag{3.1}$$

where

$$k^2 = \omega^2 \mu_0 \varepsilon - j\omega\mu_0 \sigma \tag{3.2}$$

The material is assumed to be lossless, thus $\sigma = 0$ and the permittivities ε_1 and ε_2 of the core and the cladding are real. Then, with

$$k_0 \triangleq \omega \sqrt{\mu_0 \varepsilon_0} = \frac{\omega}{c_0} \tag{3.3}$$

equation (3.2) becomes

$$k^2 = k_0^2 \varepsilon_r \tag{3.4}$$

where ε_r is the relative permittivity.

The cross-sectional geometry of the optical waveguide with its circular

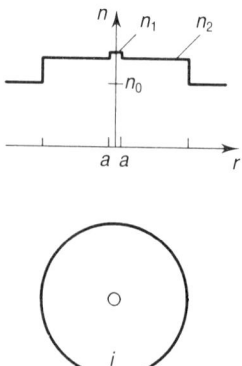

Figure 3.1 The cross-section and refractive index profile of a step index fiber.

symmetrical configuration suggests that the cylindrical polar coordinates (r, φ, z) can be used. We make the z-axis coincide with the axis of the optical waveguide.

Using cylindrical polar coordinates, equation (3.1) becomes

$$\nabla^2 \mathbf{E} = \mathbf{a}_r \left(\nabla^2 E_r - \frac{2}{r^2} \frac{\partial E_\varphi}{\partial \varphi} - \frac{E_r}{r^2} \right)$$

$$+ \mathbf{a}_\varphi \left(\nabla^2 E_\varphi + \frac{2}{r^2} \frac{\partial E_r}{\partial \varphi} - \frac{E_\varphi}{r^2} \right) + \mathbf{a}_z (\nabla^2 E_z)$$

$$= -k_0^2 \varepsilon_r (\mathbf{a}_r E_r + \mathbf{a}_\varphi E_\varphi + \mathbf{a}_z E_z) \tag{3.5}$$

where $\mathbf{a}_r$, $\mathbf{a}_\varphi$ and $\mathbf{a}_z$ are the unit vectors in the directions r, φ and z, respectively.

First we concentrate on the phasor E_z, the magnitude of the harmonic varying component $\mathbf{a}_z E_z$ of the electric field $\mathbf{E}$. From equation (3.5) we obtain

$$\nabla^2 E_z = -k_0^2 \varepsilon_r E_z \tag{3.6}$$

Writing out the Laplace operator ∇^2 with cylindrical polar coordinates, we get

$$\frac{\partial^2 E_z}{\partial r^2} + \frac{1}{r} \frac{\partial E_z}{\partial r} + \frac{1}{r^2} \frac{\partial^2 E_z}{\partial \varphi^2} + \frac{\partial^2 E_z}{\partial z^2} + k_0^2 \varepsilon_r E_z = 0 \tag{3.7}$$

Next, we use the method of separation of variables and assume that E_z can be written as a product of three functions:

$$E_z = R(r) \, \phi(\varphi) \, Z(z) \tag{3.8}$$

where $R(r)$ is a function of r only, $\phi(\varphi)$ a function of φ only and $Z(z)$ a function of z only.

Substituting equation (3.8) into the differential equation (3.7) and dividing by $R\phi Z$ gives

$$\frac{1}{Z} \frac{d^2 Z}{dz^2} = -\left(\frac{1}{R} \frac{d^2 R}{dr^2} + \frac{1}{r} \frac{1}{R} \frac{dR}{dr} + \frac{1}{r^2} \frac{1}{\phi} \frac{d^2 \phi}{d\varphi^2} + k^2 \varepsilon_r \right) \tag{3.9}$$

The right-hand side of this equation does not depend on z; therefore, varying z alone has no effect on the right-hand side of equation (3.9) and it can be replaced by a constant, γ^2, in order to get

$$\frac{1}{Z} \frac{d^2 Z}{dz^2} = \gamma^2 \tag{3.10}$$

and the solution becomes

$$Z(z) = C_1 \exp(-\gamma z) + C_2 \exp(\gamma z) \tag{3.11}$$

where C_1 and C_2 are arbitrary constants that are determined by the boundary conditions. The constant γ is usually called the propagation constant; in general, it is complex and can be written as

$$\gamma = \alpha + j\beta \tag{3.12}$$

where α is called the attenuation constant and β the phase constant.

In this chapter we shall restrict ourselves to waves that travel in the positive z-direction; this implies that $C_2 = 0$.

Substituting equation (3.10) in equation (3.9) and multiplying with r^2, results in

$$\frac{1}{\phi}\frac{d^2\phi}{d\varphi^2} = -\left(\frac{r^2}{R}\frac{d^2R}{dr^2} + \frac{r}{R}\frac{dR}{dr} + n^2\gamma^2 + n^2k^2\varepsilon_r\right) \tag{3.13}$$

The right-hand side of this equation is independent of φ and, therefore, the left-hand side must be constant. If we call the constant $-n^2$, we get

$$\phi(\varphi) = C_3 \cos n\varphi + C_4 \sin n\varphi \tag{3.14}$$

where C_3 and C_4 are arbitrary constants.

As $\phi(\varphi)$ is immune against a rotation over an angle 2π, n must be an integer.

If we introduce the two constants γ^2 and n^2 into equation (3.9), we obtain the following linear differential equation:

$$\frac{d^2R}{dr^2} + \frac{1}{r}\frac{dR}{dr} + \left(\gamma^2 + k_0^2\varepsilon_r - \frac{n^2}{r^2}\right)R = 0 \tag{3.15}$$

It is permissible to introduce the notation

$$h^2 = \gamma^2 + k_0^2\varepsilon_r \tag{3.16}$$

Equation (3.15) can then be rewritten as

$$\frac{d^2R}{dr^2} + \frac{1}{r}\frac{dR}{dr} + \left(h^2 - \frac{n^2}{r^2}\right)R = 0 \tag{3.17}$$

This equation is known as Bessel's differential equation and its solutions are Bessel functions.

3.2 Unattenuated waves

As in Chapter 2 for the slab waveguides, we restrict ourselves here to simple solutions; therefore, we assume that $\alpha = 0$ and, thus

$$\gamma = j\beta \tag{3.18}$$

so that equation (3.17) can be rewritten as

$$h^2 = -\beta^2 + k_0^2\varepsilon_r \tag{3.19}$$

Since we are interested in guided waves, completely within a waveguide, as far as the dependence on r is concerned, we need to search for solutions with oscillating fields in the core and evanescent fields in the cladding. Hence, we can assume h^2 to be positive in the core and negative in the cladding.

Furthermore, to distinguish between core and cladding material, we will use ε_{r1} and ε_{r2} to denote the relative permittivity of the core and the cladding, respectively, and write n_1 and n_2 for the refractive indices.

Now

$$n_1 = \sqrt{\varepsilon_{r1}} \quad \text{and} \quad n_2 = \sqrt{\varepsilon_{r2}} \tag{3.20}$$

3.2.1 E_z in the core

For the core, we can write

$$h = h_1 \tag{3.21}$$

where h_1 is a real.

As stated before, in the core, h^2 is positive, thus

$$h_1^2 = -\beta^2 + k_0^2 n_1^2 > 0 \tag{3.22}$$

This means that the phase velocity ω/β of the guided wave is greater than the phase velocity $\omega/k_0 n_1$ of a plane wave in the bulk core material with refractive index n_1.

For the core, equation (3.17) becomes

$$\frac{d^2 R}{dr^2} + \frac{1}{r}\frac{dR}{dr} + \left(h_1^2 - \frac{n^2}{r^2}\right)R = 0 \tag{3.23}$$

The solution of this equation is

$$R(r) = C_5 J_n(h_1 r) + C_6 N_n(h_1 r) \tag{3.24}$$

where J_n and N_n are Bessel functions of the first and second kind, respectively, and of order n. Some of these functions have been plotted in Figure 3.2. C_5 and C_6 are arbitrary constants. In the core, solutions proportional to N_n cannot exist because $N_n(0) = -\infty$; therefore, $C_6 = 0$. With the notations $A_1 = C_1 C_3 C_5$ and $B_1 = C_1 C_4 C_5$, we obtain

$$E_z = J_n(h_1 r)\,(A_1 \cos n\varphi + B_1 \sin n\varphi)\exp(j\omega t - \gamma z), \quad r < a \tag{3.25}$$

where a is the core radius and the subscript 1 for A and B denotes that they are related to the core.

3.2.2 E_z in the cladding

For the cladding, we can write

$$h = jh_2 \tag{3.26}$$

where h_2 is real.

As stated before, in the cladding, $h^2 = (jh_2)^2$, and it is negative, thus

$$-h_2^2 = -\beta^2 + k_0^2 n_2^2 < 0 \tag{3.27}$$

This means that the phase velocity ω/β of the guided wave is smaller than the phase velocity of a plane wave in the bulk cladding material with refractive index n_2.

For the cladding, equation (3.17) becomes

$$\frac{d^2 R}{dr^2} + \frac{1}{r}\frac{dR}{dr} - \left(h_2^2 + \frac{n^2}{r^2}\right)R = 0. \tag{3.28}$$

The solution of this equation is

$$R(r) = C_7 I_n(h_2 r) + C_8 K_n(h_2 r) \tag{3.29}$$

where I_n and K_n are modified Bessel functions of the first and second kind, respectively, and of order n. Some of these functions have been plotted in Figure 3.3. C_7 and C_8 are arbitrary constants.

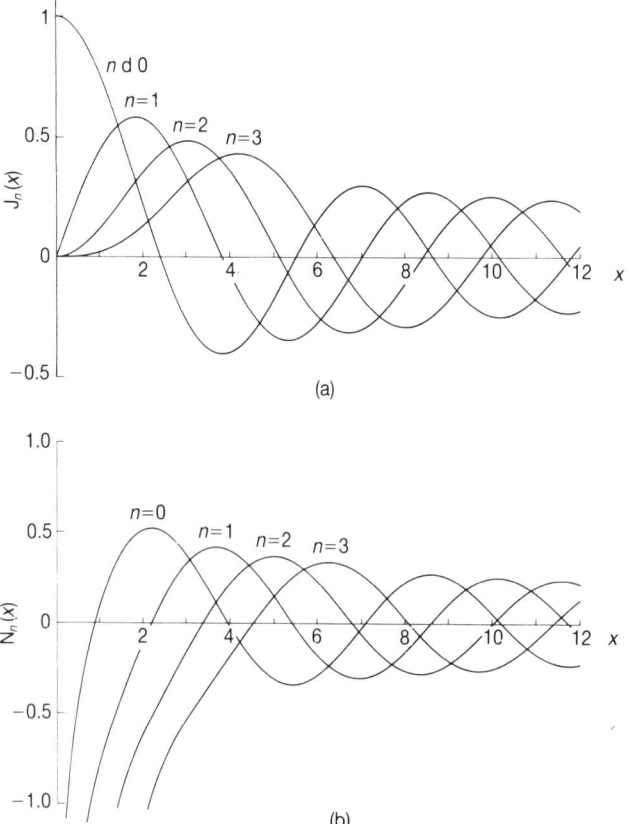

Figure 3.2 (a) Bessel functions of the first kind. (b) Bessel functions of the second kind.

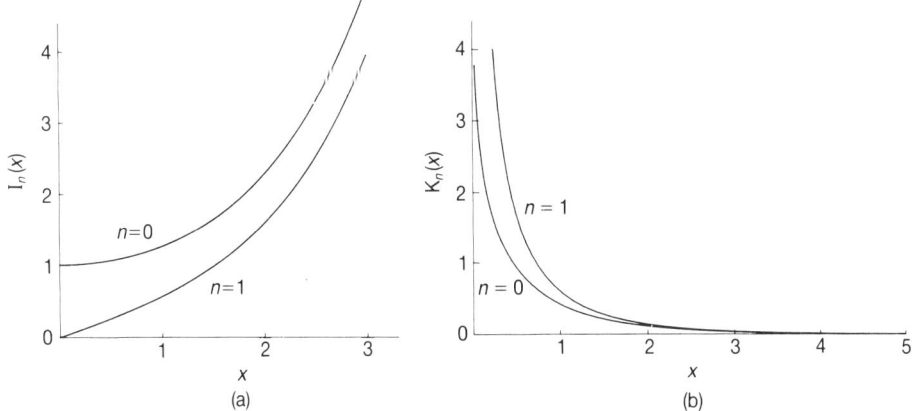

Figure 3.3 (a) Modified Bessel functions of the first kind. (b) Modified Bessel functions of the second kind.

In the cladding, solutions that are proportional to I_n are not possible, because $I_n(\infty) = \infty$; therefore, $C_7 = 0$. With the notations $A_2 = C_1 C_3 C_8$ and $B_2 = C_1 C_4 C_8$ we obtain

$$E_z = K_n(h_2 r)(A_2 \cos n\varphi + B_2 \sin n\varphi) \exp(j\omega t - \gamma z), \qquad r > 0 \qquad (3.30)$$

where the subscript 2 denotes that the term relates to the cladding.

3.2.3 The magnitude of the other field components

We can now find solutions for H_z in a similar way to those for E_z, in order to obtain

$$H_z = J_n(h_1 r)(F_1 \cos n\varphi + G_1 \sin n\varphi) \exp(j\omega t - \gamma z), \qquad r < a \qquad (3.31)$$

$$H_z = K_n(h_2 r)(F_2 \cos n\varphi + G_2 \sin n\varphi) \exp(j\omega t - \gamma z), \qquad r > a \qquad (3.32)$$

where F_1, G_1, F_2, and G_2 are arbitrary constants.

If E_z and H_z are known, we can obtain expressions for the remaining phasors E_r, E_φ, H_r and H_φ by inserting E_z and H_z into the following four equations. These equations can be derived directly from Maxwell's equations.

$$E_r = -\frac{1}{h^2}\left[\gamma \frac{\partial E_z}{\partial r} + \frac{j\omega\mu}{r}\frac{\partial H_z}{\partial \varphi}\right] \tag{3.33}$$

$$E_\varphi = \frac{1}{h^2}\left[-\frac{\gamma}{r}\frac{\partial E_z}{\partial \varphi} + j\omega\mu \frac{\partial H_z}{\partial r}\right] \tag{3.34}$$

$$H_r = \frac{1}{h^2} \left[\frac{j\omega\varepsilon}{r} \frac{\partial E_z}{\partial\varphi} - \gamma \frac{\partial H_z}{\partial r} \right] \tag{3.35}$$

$$H_\varphi = -\frac{1}{h^2} \left[j\omega\varepsilon \frac{\partial E_z}{\partial r} + \frac{\gamma}{r} \frac{\partial H_z}{\partial\varphi} \right] \tag{3.36}$$

3.3 Transverse and hybrid waves and modes

The waves in a step index optical waveguide can be divided into three distinct categories:

- Transverse electric (TE) waves, which are sometimes designated as magnetic (H) waves. These waves are characterized by $E_z = 0$ and $H_z \neq 0$.
- Transverse magnetic (TM) waves, which are sometimes designated as electric (E) waves. These waves are characterized by $E_z \neq 0$ and $H_z = 0$.
- Hybrid waves. The hybrid waves are characterized by $E_z \neq 0$ and $H_z \neq 0$.

Although it is possible to consider the first two categories as special cases of the third, for the sake of clarity we start with an analysis of the transverse waves.

3.4 Transverse waves and modes

3.4.1 TE or H waves

Excluding the factors that depend on z and t, we can rewrite equations (3.31) and (3.32) as

$$H_z = J_n(h_1 r)\,(F_1 \cos n\varphi + G_1 \sin n\varphi), \qquad r < a \tag{3.37}$$

$$H_z = K_n(h_2 r)\,(F_2 \cos n\varphi + G_2 \sin n\varphi), \qquad r > a \tag{3.38}$$

As previously stated, $E_z = 0$ holds for TE and H waves and introducing it into equation (3.36), and including equations (3.37) and (3.38), gives

$$H_\varphi = \frac{n\gamma}{h_1^2 r}\, J_n(h_1 r)\,(F_1 \sin n\varphi - G_1 \cos n\varphi), \qquad r < a \tag{3.39}$$

$$H_\varphi = -\frac{n\gamma}{h_2^2 r}\, K_n(h_2 r)\,(F_2 \sin n\varphi - G_2 \cos n\varphi), \qquad r > a \tag{3.40}$$

At the interface between the core and the cladding, the size of the field for the core must match the field for the cladding. One of the boundary conditions is the continuity of H_z for $r = a$. Introducing this condition in equations (3.37) and (3.38), we see that

$$F_1 J_n(u) = F_2 K_n(w) \tag{3.41}$$

Again, as in Chapter 2, u and w are defined as

$$u \triangleq h_1 a \tag{3.42}$$

$$w \triangleq h_2 a \tag{3.43}$$

Another boundary condition is the continuity of H_φ for $r = a$ and this condition leads to the requirement of

$$-\frac{F_1}{h_1^2} J_n(u) = \frac{F_2}{h_2^2} K_n(w) \tag{3.44}$$

Since h_1^2 and h_2^2 are both positive, equations (3.41) and (3.44) are incompatible; therefore, the boundary conditions can only be satisfied when $n = 0$. In that case H_φ is zero and the second condition becomes meaningless.

With $n = 0$, equations (3.37) and (3.38) become

$$H_z = F_1 J_0(h_1 r), \qquad r < a \tag{3.45}$$

$$H_z = F_2 K_0(h_2 r), \qquad r > a \tag{3.46}$$

and equation (3.41) becomes

$$F_1 J_0(u) = F_2 K_0(w) \tag{3.47}$$

Since H_z is independent of φ, therefore $E_r = 0$, according to equation (3.33).

Rewriting equation (3.35) with equations (3.45) and (3.46), successively, gives

$$H_r = -\frac{\gamma}{h_1} F_1 J_0'(h_1 r), \qquad r < a \tag{3.48}$$

$$H_r = \frac{\gamma}{h_1} F_2 K_0'(h_2 r), \qquad r > a \tag{3.49}$$

Substituting equations (3.45) and (3.46) successively in equation (3.34) finally gives the last phasors, as follows

$$E_\varphi = \frac{1}{h_1} F_1 j\omega\mu_0 J_0'(h_1 r), \qquad r < a \tag{3.50}$$

$$E_\varphi = -\frac{1}{h_2} F_2 j\omega\mu_0 K_0'(h_2 r), \qquad r > a \tag{3.51}$$

Due to the need for continuity of H_r and E_φ at the interface between the core and cladding, we obtain the condition that

$$\frac{1}{h_1} F_1 J_0'(u) = -\frac{1}{h_2} F_2 K_0'(w). \tag{3.52}$$

The two conditions given by equations (3.47) and (3.52) can be combined to produce the characteristic equation or eigen equation

$$\frac{1}{u} \frac{J_0'(u)}{J_0(u)} = -\frac{1}{\omega} \frac{K_0'(w)}{K_0(w)} \tag{3.53}$$

We still need a relationship between u, w and the frequency ω or the wavelength λ_0 in order to solve this equation. Then we can define the normalized frequency as

$$v \triangleq \frac{\omega a}{c_0} \sqrt{n_1^2 - n_2^2} = \frac{2\pi a}{\lambda_0} \sqrt{n_1^2 - n_2^2} \qquad (3.54)$$

Subtracting equation (3.27) from equation (3.22) and multiplying with a produces a second relationship between u, w and v, namely:

$$v^2 = u^2 + w^2 \qquad (3.55)$$

For a given value of the parameter v, the characteristic equation can be solved with the help of equation (3.55). In general, there are a limited number of discrete solutions – the eigenvalues of the system. The values of u and w for a particular solution can be substituted into the equations that describe E_φ, H_z and H_r as functions of r in order to determine the shape of the electromagnetic fields in either the core or cladding. The particular electromagnetic wave that belongs to a discrete solution is called a transmission mode, or just mode for short.

It is not possible to solve the characteristic equation analytically and we will use a graphical method to illustrate its solution.

3.4.2 Graphical solution of the characteristic equation

The curves for $J_0(u)$ and $J_0'(u)$ are shown in Figure 3.4 and the behaviour of the left-hand side of equation (3.53) appears in Figure 3.5. as a function of u.

The curves for $K_0(w)$ and $-K_0'(w)$ are shown in Figure 3.6. The right-hand side of equation (3.53) as a function of w is shown in Figure 3.7. Figure 3.8 presents the curves for each side of the characteristic equation superimposed on

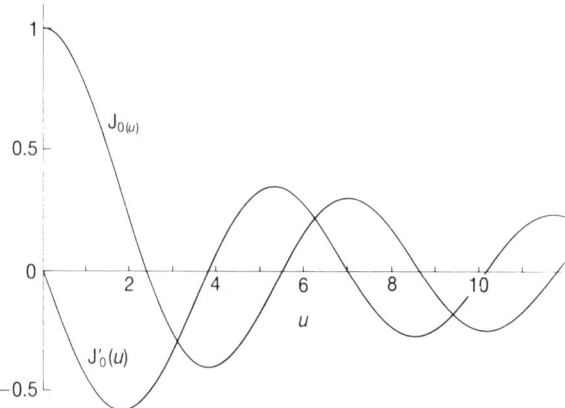

Figure 3.4 Curves for the functions $J_0(u)$ and $J_0'(u)$.

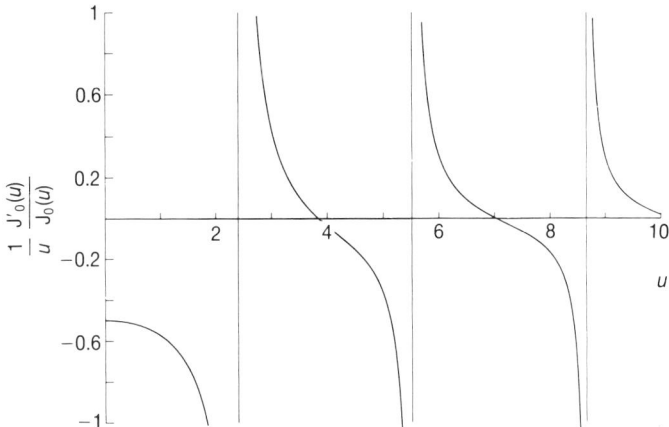

Figure 3.5 The curves for the left-hand side of equation (3.53) as a function of u.

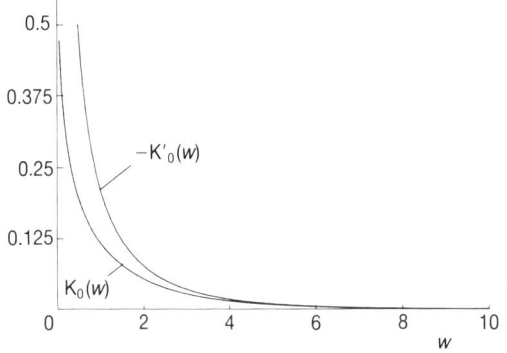

Figure 3.6 Curves for the functions $K_0(w)$ and $-K_0'(w)$.

one another. The normalized frequency has been assumed to have the value $v = 8$. In this case, the two curves intersect at two points where the u values at these intersections are the roots of the characteristic equation. The corresponding values of w can be obtained from equation (3.55). The reader is advised to trace the behaviour of the intersection points and the corresponding values of u and w when the frequency is changed.

The mode that corresponds to the intersection between the mth and the $(m + 1)$th zero crossing of $J_0(u)$ is called the H_{0m} mode or the TE_{0m} mode. If $u = 2.405 \ldots$ we have the first zero crossing of $J_0(u)$. If $u < 2.405 \ldots$, then there is no intersection point and thus no H_{0m} mode can exist. The value of v at

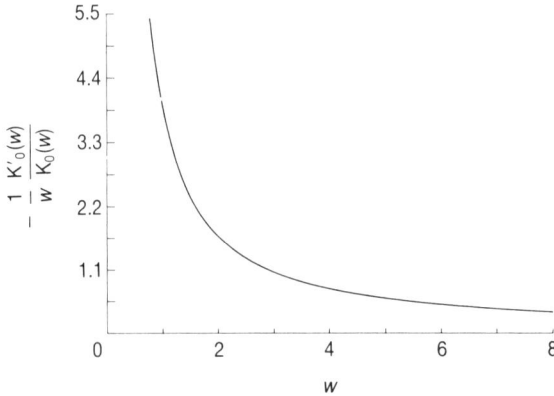

Figure 3.7 The curve for the right-hand side of equation (3.53) as a function of w.

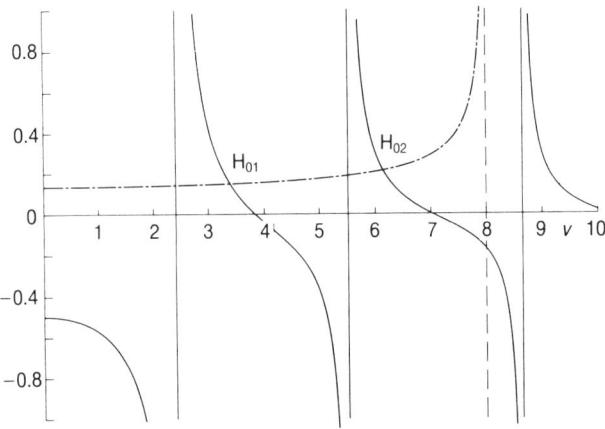

Figure 3.8 The curves for both sides of equation (3.53) superimposed. The normalized frequency is assumed to have a value of $v = 8$.

the mth zero crossing of $J_0(u)$ is called the cut-off frequency of the H_{0m} mode. In that case $u = v$ and $w = 0$. If v increases without limit, the value of u that corresponds to the H_{0m} mode approaches the value of the mth zero crossing of $J_0'(u)$ and w increases without limit.

For each mode, a certain value of the normalized frequency v is accompanied by values of u and w; thus, all the non-zero components, namely $\mathbf{E}_\varphi$, $\mathbf{H}_r$ and $\mathbf{H}_z$, are known.

3.4.3. TM or E waves

The treatment of TM or E waves, where $H_z = 0$, is analogous to TE or H waves, the only difference being the appearance of the permittivities, ε_1 and ε_2, in the characteristic equation

$$\frac{\varepsilon_1}{u} \frac{J_0'(u)}{J_0(u)} = - \frac{\varepsilon_2}{w} \frac{K_0'(w)}{K_0(w)} \tag{3.56}$$

This is due to the continuity of the normal component of $\varepsilon\mathbf{E}$ needed on the interface between the core and cladding.

For values of $u < 2.405 \ldots$, E_{0m} modes do not exist, nor do H_{0m} modes.

3.5 Hybrid modes

3.5.1 The characteristic equation

Here we repeat the general expressions that are given in equations (3.25), (3.30), (3.31) and (3.32) for the magnitude of the z components of $\mathbf{E}$ and $\mathbf{H}$ in the core and in the cladding, excluding the factors that depend on z and t.

$$E_z = J_n(h_1 r) (A_1 \cos n\varphi + B_1 \sin n\varphi), \qquad r < a \tag{3.57}$$

$$E_z = K_n(h_2 r) (A_2 \cos n\varphi + B_2 \sin n\varphi), \qquad r > a \tag{3.58}$$

$$H_z = J_n(h_1 r) (F_1 \cos n\varphi + G_1 \sin n\varphi), \qquad r < a \tag{3.59}$$

$$H_z = K_n(h_2 r) (F_2 \cos n\varphi + G_2 \sin n\varphi), \qquad r > a \tag{3.60}$$

The phasor E_φ can be derived from equation (3.34) as follows:

$$E_\varphi = \frac{1}{h_1^2} \left[-\frac{\gamma}{r} J_n(h_1 r) (-A_1 n \sin n\varphi + B_1 n \cos n\varphi) \right.$$
$$\left. + j\omega\mu_0 h_1 J_n'(h_1 r) (F_1 \cos n\varphi + G_1 \sin n\varphi) \right], \qquad r < a \tag{3.61}$$

$$E_\varphi = -\frac{1}{h_2^2} \left[-\frac{\gamma}{r} K_n(h_1 r) (-A_2 n \sin n\varphi + B_2 n \cos n\varphi) \right.$$
$$\left. + j\omega\mu_0 h_2 K_n'(h_2 r) (F_2 \cos n\varphi + G_2 \sin n\varphi) \right], \qquad r > a \tag{3.62}$$

The continuity of E_φ at the interface $r = a$ leads to the following equation:

$$\left\{ \frac{1}{h_1^2} \left[\frac{n\gamma}{a} A_1 J_n(u) + j\omega\mu_0 h_1 G_1 J_n'(u) \right] \right.$$
$$\left. + \frac{1}{h_2^2} \left[\frac{n\gamma}{a} A_2 K_n(w) + j\omega\mu_0 h_2 G_2 K_n'(w) \right] \right\} \sin n\varphi$$

$$+ \left\{ \frac{1}{h_1^2} \left[-\frac{n\gamma}{a} B_1 J_n(u) + j\omega\mu_0 h_1 F_1 J_n'(u) \right] \right.$$

$$+ \frac{1}{h_2^2} \left[-\frac{n\gamma}{a} B_2 K_n(w) + j\omega\mu_0 h_2 F_2 K_n'(w) \right] \Bigg\} \cos n\varphi = 0 \qquad (3.63)$$

Since the requirements of equation (3.63) must be satisfied for each value of φ, the factors between the braces vanish.

If we assume that B_1, B_2, F_1 and F_2 are equal to zero, then we can find non-trivial solutions with finite values for A_1, A_2, G_1 and G_2. If we assume the aforesaid constants to be zero, then we can find non-trivial solutions with finite values for B_1, B_2, F_1 and F_2. The latter solutions are orthogonal with the first solutions.

Starting with an analysis of the first case, it follows that A_1, A_2, G_1 and G_2 are not zero.

Equations (3.57)–(3.60) now reduce to the following equations:

$$E_z = A_1 J_n(h_1 r) \cos n\varphi, \qquad\qquad r < a \qquad (3.64)$$

$$E_z = A_2 K_n(h_2 r) \cos n\varphi, \qquad\qquad r > a \qquad (3.65)$$

$$H_z = G_1 J_n(h_1 r) \sin n\varphi, \qquad\qquad r < a \qquad (3.66)$$

$$H_z = G_2 K_n(h_2 r) \sin n\varphi, \qquad\qquad r > a \qquad (3.67)$$

While equations (3.61) and (3.62) become

$$E_\varphi = \frac{1}{h_1^2} \left[\frac{n\gamma}{r} A_1 J_n(h_1 r) + j\omega\mu_0 h_1 G_1 J_n'(h_1 r) \right] \sin n\varphi, \qquad r < 0 \quad (3.68)$$

$$E_\varphi = -\frac{1}{h_2^2} \left[\frac{n\gamma}{r} A_2 K_n(h_2 r) + j\omega\mu_0 h_2 G_2 K_n'(h_2 r) \right] \sin n\varphi, \quad r < 0 \quad (3.69)$$

In order to derive the characteristic equation, in this case, we need one extra phasor, which means, therefore, finding E_r from equation (3.33), as follows:

$$E_r = -\frac{1}{h_1^2} \left[\gamma h_1 A_1 J_n'(h_1 r) + \frac{j\omega\mu n}{r} G_1 J_n(h_1 r) \right] \cos n\varphi, \qquad r < a \qquad (3.70)$$

$$E_r = \frac{1}{h_2^2} \left[\gamma h_2 A_2 K_n'(h_2 r) + \frac{j\omega\mu n}{r} G_2 K_n(h_2 r) \right] \cos n\varphi, \qquad r > a \qquad (3.71)$$

The boundary conditions that require E_z and H_z to be continuous are

$$A_2 = A_1 \frac{J_n(u)}{K_n(w)} \qquad (3.72)$$

and

$$G_2 = G_1 \frac{J_n(u)}{K_n(w)} \qquad (3.73)$$

The boundary condition that requires E_φ to be continuous is

$$\frac{1}{h_1^2}\left[\frac{n\gamma}{a}A_1J_n(u) + j\omega\mu_0h_1\acute{G}_1J'_n(u)\right]$$
$$= -\frac{1}{h_2^2}\left[\frac{n\gamma}{a}A_2K_n(w) + j\omega\mu_0h_2G_2K'_n(w)\right] \tag{3.74}$$

The boundary condition that requires εE_r to be continuous is

$$-\frac{\varepsilon_1}{h_1^2}\left[\gamma h_1A_1J'_n(u) + \frac{j\omega\mu n}{a}G_1J_n(u)\right]$$
$$= \frac{\varepsilon_2}{h_2^2}\left[\gamma h_2A_2K'_n(w) + \frac{j\omega\mu n}{a}G_2K_n(w)\right] \tag{3.75}$$

Replacing A_2 and G_2 from equations (3.72) and (3.73) in the last two equations and putting $\gamma = j\beta$, $h_1a = u$ and $h_2a = w$, we have the following statements after rearranging the terms:

$$\left(\frac{1}{u^2} + \frac{1}{w^2}\right)n\beta A_1 + \left[\frac{1}{u}\frac{J'_n(u)}{J_n(u)} + \frac{1}{w}\frac{K'_n(w)}{K_n(w)}\right]\omega\mu_0G_1 = 0 \tag{3.76}$$

$$\left[\frac{\varepsilon_1}{u}\frac{J'_n(u)}{J_n(u)} + \frac{\varepsilon_2}{w}\frac{K'_n(w)}{K_n(w)}\right]\beta A_1 + \left(\frac{\varepsilon_1}{u^2} + \frac{\varepsilon_2}{w^2}\right)\omega\mu_0nG_1 = 0 \tag{3.77}$$

For convenience, we introduce

$$Y_n(u) \triangleq \frac{1}{u}\frac{J'_n(n)}{J_n(u)} \tag{3.78}$$

$$X_n(w) \triangleq \frac{1}{w}\frac{K'_n(w)}{K_n(w)} \tag{3.79}$$

When solving for finite constants A_1 and G_1, the determinant of the system of coefficients in equations (3.76) and (3.77) must vanish, thus

$$\begin{vmatrix} \left(\dfrac{1}{u^2} + \dfrac{1}{w^2}\right)n\beta & [Y_n(u) + X_n(w)]\omega\mu_0 \\ [\varepsilon_1Y_n(u) + \varepsilon_2X_n(\omega)]\beta & \left(\dfrac{\varepsilon_1}{u^2} + \dfrac{\varepsilon_2}{w^2}\right)\omega\mu_0n \end{vmatrix} = 0 \tag{3.80}$$

This leads to the characteristic equation

$$n^2\left(\frac{1}{u^2} + \frac{1}{w^2}\right)\left(\frac{1}{u^2} + \frac{p}{w^2}\right) = [Y_n(u) + X_n(w)][Y_n(u) + pX_n(w)] \tag{3.81}$$

where

$$p = \frac{\varepsilon_2}{\varepsilon_1} \tag{3.82}$$

Either equation (3.76) or (3.77) can be used to obtain the ratio between the constants A_1 and G_1, as follows:

$$G_1 = -A_1 \frac{n\beta}{\omega\mu_0} \frac{(1/u^2) + (1/w^2)}{Y_n(u) + X_n(w)} = -A_1 \frac{\beta}{\omega\mu_0 n} \frac{\varepsilon_1 Y_n(u) + \varepsilon_2 X(w)}{(\varepsilon_1/u^2) + (\varepsilon_2/w^2)}$$

(3.83)

With equation (3.83) and equations (3.72) and (3.73), the ratios between all four constants can be determined. For the orthogonal modes, when B_1, B_2, F_1 and F_2 are finite, we can obtain the same characteristic equation and similar relationships between the constants.

3.5.2 Solution of the characteristic equation

The characteristic equation cannot be solved analytically, therefore it must be solved with a numerical method. For the sake of clarity, we shall resort again to a graphical method, by developing the right-hand side of equation (3.81) and finding the following quadratic equation, after rearranging the terms:

$$Y_n^2(u) + (1 + p) X_n(w) Y_n(u) + pX_n^2(w)$$

$$- n^2 \left(\frac{1}{u^2} + \frac{1}{w^2}\right)\left(\frac{1}{u^2} + \frac{p}{w^2}\right) = 0$$

(3.84)

The roots of this quadratic are

$$Y_n(u) = -\frac{1 + p}{2} X_n(w)$$

$$\pm \sqrt{\left[\frac{1}{4} (1 - p)^2 X_n^2(w) + n^2 \left(\frac{1}{u^2} + \frac{1}{w^2}\right)\left(\frac{1}{u^2} + \frac{p}{w^2}\right)\right]}$$

(3.85)

This equation can be solved graphically by superimposing the left-hand and right-hand sides on each other, as a function of u; this is shown in Figure 3.9 for $n = 1$ and in Figure 3.10 for $n = 2$.

The right-hand side of equation (3.85) is not only a function of u but also depends on the parameters p and v. The value of w is determined by v and u according to the relationship given in equation (3.55).

In the examples presented in Figures 3.9 and 3.10, the following values were assigned to the parameters above:

$$p = 0.99 \quad \text{and} \quad v = 9$$

The curves in Figures 3.9 and 3.10 labelled a and b refer to the right-hand side of equation (3.85) with the plus and the minus signs, respectively. The intersections of the curves for $Y_n(u)$ with the curve for the right-hand side with a plus sign (curve a) correspond to the modes that are called EH_{nm} modes. Otherwise, the intersections with the curve resulting from the minus sign correspond to the modes that are called HE_{nm} modes.

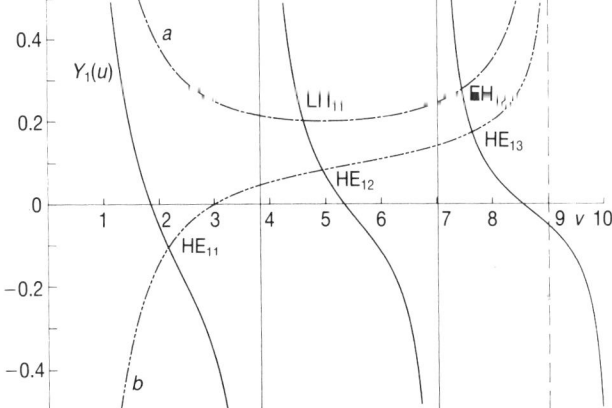

Figure 3.9 Graphical solution of the characteristic equation when $n = 1$. $Y_1(u)$ is the curve for the left-hand side of equation (3.85). If $v = 9$, then the curves marked a and b represent the right-hand side of equation (3.85) with a plus sign or minus sign, respectively.

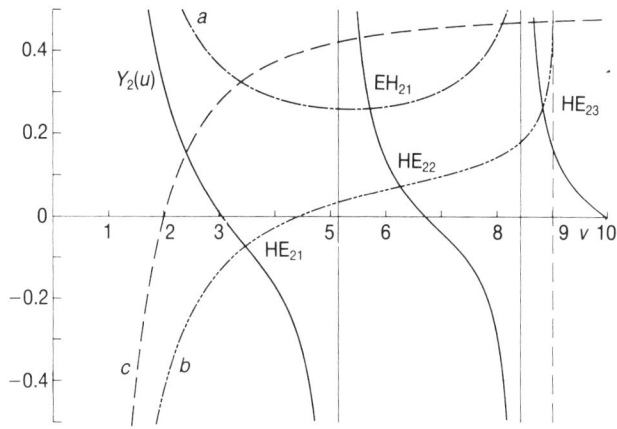

Figure 3.10 Graphical solution of the characteristic equation when $n = 2$. $Y_2(u)$ gives the curve for the left-hand side of equation (3.85). If $v = 9$, then the curves a and b represent the right-hand side of equation (3.85) with a plus or minus sign, respectively. Curve c represents the set of end-points for the curves that represent the right-hand side of equation (3.85) with minus signs for different values of v.

The HE_{11} mode is the only mode that has no cut-off frequency. If v decreases, then the intersection that corresponds to this mode is always present. There is typically a difference between the curve b for $n = 1$ and the curve b

for $n > 1$ in the vicinity of $u = v$. To analyze this difference we will derive functions that describe the curves a and b in the vicinity of $u = v$. $K_n(w)$ and $K'_n(w)$ can be expressed as functions of $K_{n-1}(w)$ and $K_{n+1}(w)$ [7]

$$-2K'_n(w) = K_{n-1}(w) + K_{n+1}(w) \tag{3.86}$$

$$-\frac{2n}{w}K_n(w) = K_{n-1}(w) - K_{n+1}(w) \tag{3.87}$$

Combining these equations gives

$$K'_n(w) = -K_{n-1}(w) - \frac{n}{w}K_n(w) \tag{3.88}$$

Inserting equation (3.88) into equation (3.79) results in

$$X_n(w) = -\frac{1}{w}\frac{K_{n-1}(w)}{K_n(w)} - \frac{n}{w^2} \tag{3.89}$$

We can now examine the behaviour of $X_n(w)$ for small values of w. If $0 < w \ll 1$, then [7]

$$K_0(w) \approx \ln\frac{2}{w} - C \tag{3.90}$$

where C is Euler's constant, which is defined as

$$C \overset{\triangle}{=} \lim_{m \to \infty}\left(1 + \frac{1}{2} + \frac{1}{3} + \cdots + \frac{1}{m} - \ln m\right) = 0.577215665 \ldots \tag{3.90a}$$

$$K_n(w) \approx \frac{1}{2}(n-1)!\left(\frac{2}{w}\right)^2, \qquad n = 1, 2, 3, \ldots \tag{3.91}$$

Inserting equations (3.90) and (3.91) successively in equation (3.89) gives

$$X_1(w) = -\ln\frac{2}{\gamma w} - \frac{1}{w^2} \tag{3.92}$$

and

$$X_n(w) = -\frac{n}{w^2} - \frac{1}{2(n-1)}, \qquad n = 2, 3, 4, \ldots \tag{3.93}$$

When these expressions for $X_n(w)$ are put into equation (3.84) and the constant terms containing w^2 in the denominator are neglected, then we get the two quadratic equations below for $n = 1$ and $n > 1$, respectively:

$$Y_1^2(u) - (1 + p)\left[\frac{1}{w^2} + \ln\left(\frac{2}{\gamma w}\right)\right]Y_1(u)$$

$$+ \left[\frac{2p}{w^2}\ln\left(\frac{2}{\gamma w}\right) + p\ln^2\left(\frac{2}{\gamma w}\right) - \frac{p+1}{u^2 w^2}\right] = 0 \tag{3.94}$$

and

$$Y''_n(u) - \frac{n}{w^2}(1 + p)Y_n(u) + \frac{pu^2 n - n^2(n-1)(p+1)}{(n-1)u^2 w^2} = 0,$$

$$\text{for } n = 2, 3, 4, \ldots \tag{3.95}$$

If $x^2 - bx + c = 0$ and if $b^2 \gg 4c$ then the roots of this denominator can be denoted as

$$x_{1,2} = \frac{b \pm \sqrt{b^2 - 4c}}{2} \approx \frac{b \pm \left(b - \dfrac{2c}{b}\right)}{2} \tag{3.96}$$

$$x_1 = b \qquad (+ \text{ sign}) \tag{3.97}$$

$$x_2 = \frac{c}{b} \qquad (- \text{ sign}) \tag{3.98}$$

We apply this approximation of the roots of a quadratic to equations (3.94) and (3.95) with the following results:

If $0 < w \ll 1$ then

$$Y_1(u) \approx \frac{1+p}{w^2}, \qquad\qquad (+ \text{ sign}) \tag{3.99}$$

$$Y_1(u) \approx \frac{2p}{1+p}\ln\left(\frac{2}{\gamma w}\right), \qquad (- \text{ sign}) \tag{3.100}$$

$$Y_n(u) \approx \frac{n(1+p)}{w^2}, \qquad\qquad (+ \text{ sign}) \tag{3.101}$$

$$Y_n(u) \approx \frac{p}{(n-1)(1+p)} - \frac{n}{u^2}, \quad (- \text{ sign}) \tag{3.102}$$

for $n = 2, 3, 4, \ldots$

The curves that correspond to the right-hand side of equations (3.99) and (3.100) are labelled a and b in Figure 3.9; both have the line $u = v$ as their asymptote. The curve that corresponds to the right-hand side of equation (3.101), namely the curve labelled a in Figure 3.10, also has as its asymptote the line $u = v$, unlike the curve for equation (3.102), which is labelled b in Figure 3.10. As u approaches v this latter curve approaches a finite limit that is a function of u according to equation (3.102); it is plotted in Figure 3.10 and labelled c. The intersections of this curve c with the curve of $Y_2(u)$ correspond to the cut-off frequencies for the HE_{2m} modes. Thus, it is clear that the HE_{n1} mode, with $n > 1$, must have a cut-off frequency.

3.6 Leaky modes

The solutions that we have considered so far described unattenuated waves in an ideal step index fiber, with the restriction that $\alpha = 0$. The modes that were derived in this way are called guided modes. It is possible to find another class of solutions where $\alpha \neq 0$ – its modes are called leaky modes. The leaky modes, especially those with small α values, can play a role in the transmission of light through an optical waveguide [2, 3].

3.7 Linearly polarized (LP) modes

As demonstrated in the preceding sections, Maxwell's equations can lead us to exact solutions for electromagnetic waves in a dielectric waveguide; however, these solutions are rather complex, even assuming an infinitely thick cladding and real values of h^2. However, if we are concerned with weakly guiding fibers, i.e. when the relative difference between the refractive indices of the core and cladding is small, then we can find approximate solutions that are much simpler than the exact solutions.

As will be shown, solving the characteristic equation leads to similar eigenvalues for both the $EH_{n-1,m}$ modes and the $HE_{n+1,m}$ modes. If the fields of these modes are composed, the resulting transversal field will be linearly polarized. The linearly polarized modes are labelled $LP_{n,m}$ modes and are composed in the following way:

$$LP_{0m} = HE_{1m}$$

$$LP_{1m} = HE_{2m} + E_{0m} + H_{0m}$$

$$LP_{nm} = HE_{n+1,m} + EH_{n-1,m} \qquad n > 2 \tag{3.103}$$

For each m there are two LP_{0m} modes, polarized perpendicular to each other, while for each m and each $n > 0$ there are four LP_{nm} modes.

Making the approximation $p = 1$, the characteristic equation (3.81) simplifies to

$$n\left(\frac{1}{u^2} + \frac{1}{w^2}\right) = \pm[Y_n(u) + X_n(u)] \tag{3.104}$$

Far away from the cut-off frequency, if $w \to \infty$, this equation reduces to

$$\frac{n}{u} = \pm \frac{J_n'(u)}{J_n(u)} \tag{3.105}$$

In this new characteristic equation, again the plus sign corresponds to the EH modes, while the minus sign denotes HE modes.

To solve the characteristic equation we use the following relations between Bessel functions (see Appendix 4, equation A4.16):

$$u J_n'(u) = +n J_n(u) - u J_{n+1}(u) \tag{3.106}$$

$$u J_n'(u) = u J_n(u) + u J_{n-1}(u) \tag{3.107}$$

With equation (3.107) we can write equation (3.105) with the minus sign

$$\frac{n+1}{u} = -\frac{J_{n+1}'(u)}{J_{n+1}(u)} = \frac{n+1}{u} - \frac{J_n(u)}{J_{n+1}(u)} \tag{3.109}$$

Thus, the eigenvalue u for the $EH_{n-1,m}$ mode equals the value of the argument u at the mth zero crossing of $J_n(u)$.

With equation (3.107) we can write equation (3.105) with the minus sign

$$\frac{n+1}{u} = -\frac{J_{n+1}'(u)}{J_{n+1}(u)} = \frac{n+1}{u} - \frac{J_n(u)}{J_{n+1}(u)} \tag{3.109}$$

Thus, the eigenvalue u for the HE_{n+1} mode also equals the value of the argument u at the mth zero crossing of $J_n(u)$.

As far as the phase constant β is concerned, from equations (3.22) and (3.27) it follows that

$$k_0 n_1 > \beta > k_0 n_2. \tag{3.110}$$

And when $n_1 \rightarrow n_2$, we get

$$\beta = \omega \sqrt{\varepsilon \mu_o} \tag{3.111}$$

where

$$\varepsilon = \frac{\varepsilon_1 + \varepsilon_2}{2} \tag{3.112}$$

Inserting equations (3.104) and (3.111) into equation (3.76) gives the ratio between G_1 and A_1

$$G_1 = \pm \sqrt{\frac{\varepsilon}{\mu}} A_1 \tag{3.113}$$

The plus sign corresponds to the HE modes (the minus sign in equation 3.104) and the minus sign corresponds to the EH modes (the plus sign in equation 3.104).

In the core of the fiber, E_z and H_z can be determined from equations (3.64), (3.66) and (3.113)

$$E_z = A_1 J_n(h_1 r) \cos n\varphi \tag{3.114}$$

$$H_z = \pm A_1 \sqrt{\frac{\varepsilon}{\mu}} J_n(h_1 r) \sin n\varphi \tag{3.115}$$

For the orthogonal mode, we find that the ratio between F_1 and B_1 is

$$F_1 = \pm \sqrt{\frac{\varepsilon}{\mu}} B_1 \tag{3.116}$$

In this case, the minus sign signifies the HE modes and the plus sign the EH modes.

From equations (3.57) and (3.59), where $A_1 = G_1 = 0$, and equation (3.116), we can find E_z and H_z for the orthogonal mode

$$E_z = B_1 J_n(h_1 r) \sin n\varphi \tag{3.117}$$

$$H_z = \pm B_1 \sqrt{\frac{\varepsilon}{\mu}} J_n(h_1 r) \cos n\varphi \tag{3.118}$$

The two $EH_{n-1,m}$ modes and the two HE_{n+1} modes that can exist for any value of n and m may combine in four different ways to give an LP_{nm} mode; we choose equal values for the amplitude A_1 and B_1.

To calculate the transverse fields we change the coordinate system to a Cartesian system; in that case, we have the following four expressions:

$$E_x = -\frac{1}{h^2} \left(\gamma \frac{\partial E_z}{\partial x} + j\omega\mu \frac{\partial H_z}{\partial y} \right) \tag{3.119}$$

$$E_y = \frac{1}{h^2} \left(-\gamma \frac{\partial E_z}{\partial y} + j\omega\mu \frac{\partial H_z}{\partial x} \right) \tag{3.120}$$

$$H_x = \frac{1}{h^2} \left(j\omega\varepsilon \frac{\partial E_z}{\partial y} - \gamma \frac{\partial H_z}{\partial x} \right) \tag{3.121}$$

$$H_y = -\frac{1}{h^2} \left(j\omega\varepsilon \frac{\partial E_z}{\partial x} + \gamma \frac{\partial H_z}{\partial y} \right) \tag{3.122}$$

Since the phasors E_z and H_z are denoted with cylindrical polar coordinates, we must use the following relationships:

$$\frac{\partial r}{\partial x} = \cos\varphi, \quad \frac{\partial r}{\partial y} = \sin\varphi, \quad \frac{\partial \varphi}{\partial x} = -\frac{1}{r}\sin\varphi, \quad \frac{\partial \varphi}{\partial y} = \frac{1}{r}\cos\varphi$$

that are derived from

$$r = \sqrt{x^2 + y^2} \quad \text{and} \quad \varphi = \arctan\frac{y}{x}$$

Writing out equations (3.119)–(3.122) for all the possible combinations of plus and minus signs and using equations (3.106) and (3.107) leads to the following set of formulae for the fields of the LP_{nm} modes, where E_n is the maximum value of the electric field on the interface between core and cladding.

First combination

$$E_y = -H_x \sqrt{\frac{\mu}{\varepsilon_1}} = E_n \frac{J_n\left(u \frac{r}{a}\right)}{J_n(u)} \cos n\varphi, \quad r < a \tag{3.123a}$$

$$E_y = -H_x \sqrt{\frac{\mu}{\varepsilon_2}} = E_n \frac{K_n\left(w\dfrac{r}{a}\right)}{K_n(w)} \cos n\varphi, \qquad r > a \tag{3.123b}$$

$$E_x = 0, \quad H_y = 0 \tag{3.123c}$$

Second combination

$$E_y = -H_x \sqrt{\frac{\mu}{\varepsilon_1}} = E_n \frac{J_n\left(u\dfrac{r}{a}\right)}{J_n(u)} \sin n\varphi, \qquad r < a \tag{3.124a}$$

$$E_y = -H_x \sqrt{\frac{\mu}{\varepsilon_2}} = E_n \frac{K_n\left(w\dfrac{r}{a}\right)}{K_n(w)} \sin n\varphi, \qquad r > a \tag{3.124b}$$

$$E_x = 0, \quad H_y = 0 \tag{3.124c}$$

Third combination

$$E_x = H_y \sqrt{\frac{\mu}{\varepsilon_1}} = E_n \frac{J_n\left(u\dfrac{r}{a}\right)}{J_n(u)} \cos n\varphi, \qquad r < s \tag{3.125a}$$

$$E_x = H_y \sqrt{\frac{\mu}{\varepsilon_2}} = E_n \frac{K_n\left(w\dfrac{r}{a}\right)}{K_n(w)} \cos n\varphi, \qquad r > a \tag{3.125b}$$

$$E_y = 0, \quad H_x = 0 \tag{3.125c}$$

Fourth combination

$$E_x = H_y \sqrt{\frac{\mu}{\varepsilon_1}} = E_n \frac{J_n\left(u\dfrac{r}{a}\right)}{J_n(u)} \sin n\varphi, \qquad r < a \tag{3.126a}$$

$$E_x = H_y \sqrt{\frac{\mu}{\varepsilon_2}} = E_n \frac{K_n\left(w\dfrac{r}{a}\right)}{K_n(w)} \sin n\varphi, \qquad r > a \tag{3.126b}$$

$$E_y = 0, \quad H_x = 0 \tag{3.126c}$$

The longitudinal phasors can be derived directly from Maxwell's equations, as follows:

$$E_z = \frac{j}{k_0\varepsilon_{r1}} \sqrt{\frac{\mu}{\varepsilon_0}} \left(\frac{\partial H_x}{\partial y} - \frac{\partial H_y}{\partial x}\right) \tag{3.127}$$

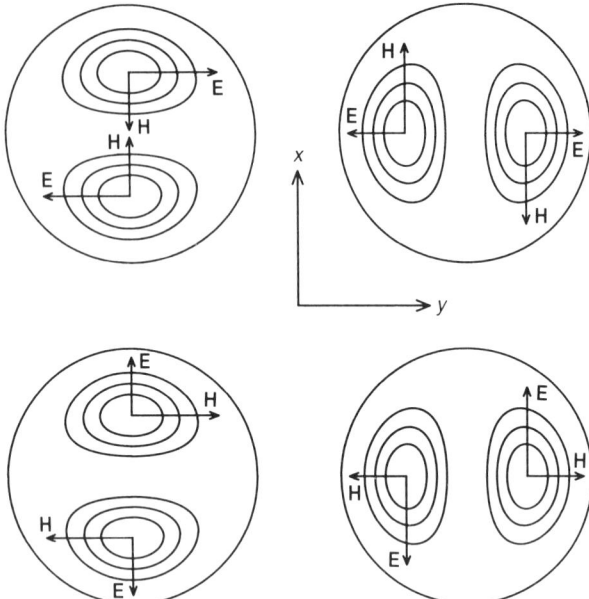

Figure 3.11 The four orthogonal LP_{11} modes according to equations (3.123)–(3.126).

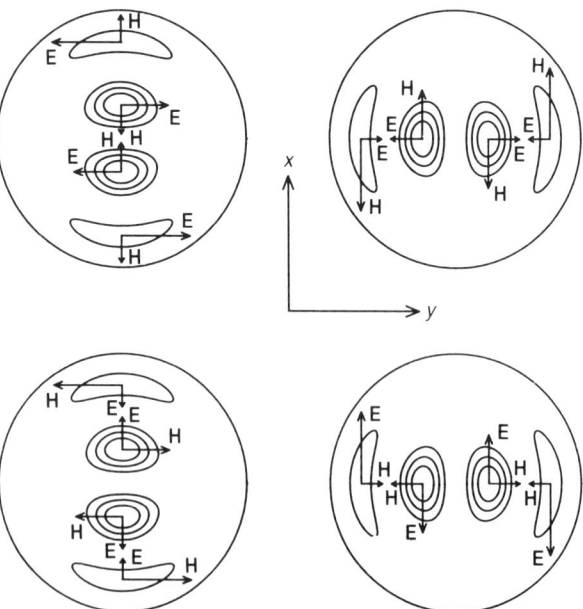

Figure 3.12 The four orthogonal LP_{12} modes according to equations (3.123)–(3.126).

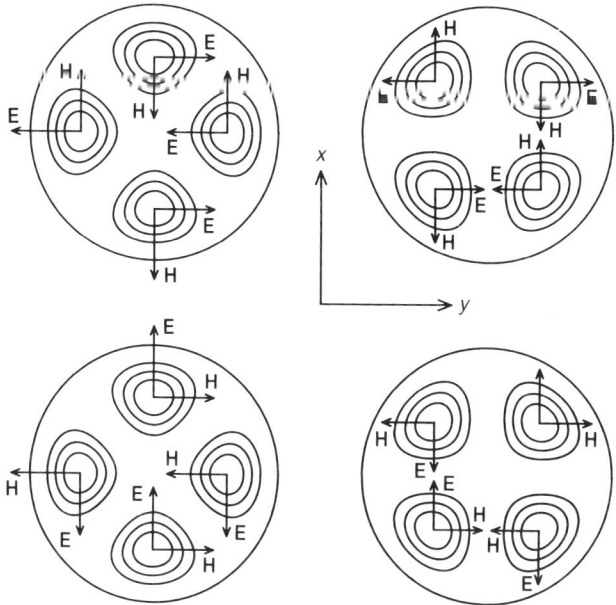

Figure 3.13 The four orthogonal $LP_{21'}$ modes according to equations (3.123)–(3.126).

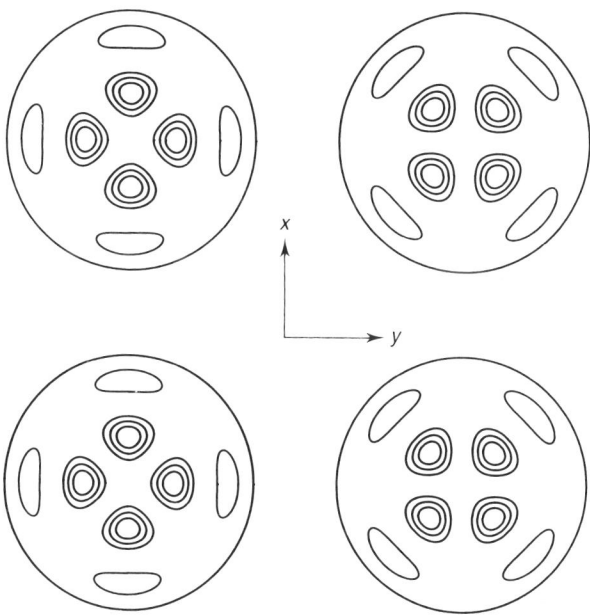

Figure 3.14 The four orthogonal LP_{22} modes according to equations (3.123)–(3.126).

$$H_z = \frac{j}{k_0} \sqrt{\frac{\varepsilon_0}{\mu}} \left(\frac{\partial E_y}{\partial x} - \frac{\partial E_x}{\partial y} \right) \tag{3.128}$$

For small values of Δ, the longitudinal components are small compared to the transversal components and their ratios, are, respectively, u/ak and w/ak. These factors are both of the order $\sqrt{\Delta}$.

In Figures 3.11–3.14 the power distribution of the light in the core of the fiber is depicted by lines of constant intensity, i.e. $|\mathbf{E} \times \mathbf{H}|$ = constant. The constants have been taken as 0.25, 0.5 and 0.75 times the maximum intensity.

References

[1] S. Ramo, J. R. Whinnery and Th. van Duzer, *Fields and Waves in Communication Electronics* (New York: Wiley, 1967).

[2] N. S. Kapani and J. J. Burke, *Optical Waveguides* (New York: Academic Press, 1972).

[3] D, Marcuse, *Theory of Dielectric Optical Waveguides* (New York: Academic Press, 1974).

[4] F. P. Kapron, 'Maximum information capacity of fibre-optic waveguides', *Electronic Letters*, vol. 13, February 1977, no 4, pp. 96–7.

[5] D. Gloge, 'Weakly guiding fibres', *Applied Optics*, vol. 10, October 1971, pp. 2252–8.

[6] L. G. Cohen and Chinlon Lin, 'Pulse delay measurements in the zero material dispersion wavelength region for optical fibers', *Applied Optics*, vol. 16, no. 12. December 1977, pp. 3136–9.

[7] E. Jahnke, F. Emde and F. Lösch, *Taflen höheren Funktionen*, 7th edn (Stuttgart: Teubner Verlagsgesellschaft, 1966).

Problems

3.1 Assume an arbitrary vector $\mathbf{A}$. The components of the vector $\mathbf{\nabla} \times \mathbf{A}$ are described in circular cylindrical coordinates as follows:

$$(\nabla \times A)_r = \frac{1}{r} \frac{\partial A_z}{\partial \varphi} - \frac{\partial A_\varphi}{\partial z}$$

$$(\nabla \times A)_\varphi = \frac{\partial A_r}{\partial z} - \frac{\partial A_z}{\partial r}$$

$$(\nabla \times A)_z = \frac{1}{r} A_\varphi + \frac{\partial A_\varphi}{\partial r} - \frac{1}{r} \frac{\partial A_r}{\partial \varphi}$$

Use these equalities to derive equations (3.33)–(3.36) from Maxwell's equations.

3.2 The light of a single-mode laser with a wavelength $\lambda_0 = 1.3 \ \mu$m is coupled into a step index fiber with a core diameter of 9 μm. The refractive index of the cladding is $n_2 = 1.446$. The normalized frequency $v = 5.136$.

• What is the value of n_1?

- Which guided modes can propagate in this fiber?
- How can the cut-off frequencies of these modes be estimated?
- Which LP modes can propagate in this fiber?
- What is typical for the field of the EH_{21} mode in the cladding?

3.3 Assume a weakly guiding step index fiber and a single-mode semiconductor laser. The wavelength of the light in vacuum is $\lambda_0 = 0.85 \, \mu m$. In the wavelength region considered, the refractive indices of the core and cladding can be approximated by the formulae

$$n_1 = 1.25 + 0.2\lambda_0^{-1} \quad \text{and} \quad n_2 = 1.24 + 0.2\lambda_0^{-1},$$

where λ_0 is the wavelength in vacuum, expressed in μm.

The cut-off frequency of the E_{01}-mode in the fiber is $0.8 \, \mu m$.

- Calculate the diameter of the fiber core.
- How many LP modes can exist in the fiber using the laser concerned?

3.4 Find the expressions for the electric and magnetic fields of the LP_{0m} mode in the core and the cladding of a weakly guiding fiber.

3.5 Combine an $HE_{n+1,m}$ mode with an $EH_{n-1,m}$ mode of equal power in a weakly guiding fiber and prove by calculating the resulting fields that the combination gives a linearly polarized wave.

4

Dispersion in the step index fiber

4.1   Phase characteristic and modulation bandwidth
4.2   The propagation constant of a mode in a step index fiber
4.3   Waveguide dispersion
4.4   Material dispersion
4.5   Waveguide and material dispersion in a step index fiber
4.6   Multimode dispersion
4.7   Dispersion caused by a non-monochromatic source
      References
      Problems

In telecommunications, an optical waveguide belongs to the general class of transmission lines, the purpose of which is to transport signals from one place to another. The telecommunications engineer is interested in the terminal properties of transmission lines, i.e. in the relationship between an input signal and the resulting output signal. Therefore, we can describe glassfibers as systems with input and output. The electromagnetic fields that have been treated in the preceding chapters were derived from linear differential equations with constant coefficients; therefore, we can model the transmission line as a linear, time-invariant system. If the input to such a system is a harmonic signal, then the output is also a harmonic signal with the same frequency as the input signal, but, in general, with a different amplitude and a different phase. We can assume that a harmonic wave travels in a certain mode through an optical waveguide of length l without a power coupling to other modes or a change in the direction of polarization. The field at the input end of the fiber, where $z = 0$, is related to the field at the output end of the fiber, where $z = l$. For instance, the electric field component $E_z(r, \varphi, 0)$ of the input is related to the electric field component $E_z(r, \varphi, l)$ of the output. The complex ratio between the harmonic output signal and the harmonic input signal is denoted as the system function or the transfer function $H(\omega)$ of the mode concerned. This system function is specified by the attenuation $a(\omega)$ and the phase shift $b(\omega)$

$$H(\omega) = \frac{E_z(r, \varphi, l)}{E_z(r, \varphi, 0)} = \exp[-a(\omega) - jb(\omega)] \tag{4.1}$$

Naturally, this relation holds for each component of the electric and the magnetic fields. The attenuation $a(\omega)$ and the phase shift $b(\omega)$ of the mode concerned are simply related to the specific attenuation $\alpha(\omega)$ and the specific phase shift $\beta(\omega)$ of that mode

$$a(\omega) = \alpha(\omega) l \tag{4.2}$$

$$b(\omega) = \beta(\omega) l \tag{4.3}$$

For each mode in a fiber, the propagation constant γ is a separate function of the frequency. The unique value of the propagation constant γ that corresponds to a mode determines the z-dependence of the electromagnetic field of that particular mode according to equation (3.11). The attenuation constant, i.e. the real part α of the propagation constant, describes the z-dependence of the amplitudes of the fields; the phase constant, i.e. the imaginary part β of the propagation constant, describes field phases.

The attenuation constant α of a guided mode depends, in the first place, on

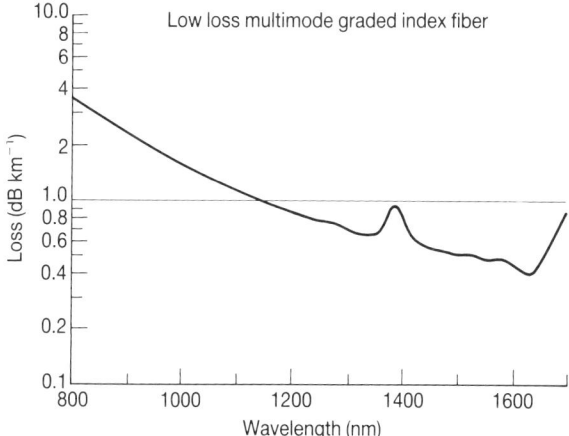

Figure 4.1 The attenuation of a multimode silica fiber as function of the wavelength. (Source: reproduced with permission from *Optical Communication Systems*, by John Gowar, Prentice Hall, 1984.)

the material chosen, and α shows a minimum for silica, at a wavelength of 1500 nm. In Figure 4.1 an experimental attenuation curve of a multimode silica glassfiber is displayed and the attenuation is expressed in dB km^{-1}. If we consider a 5 km length of this fiber, then the frequency range between the 3 dB points found from this curve is roughly 10^{14} Hz. This is a very large bandwidth that can, theoretically, support millions of television channels. However, when bandwidth is mentioned in glassfiber communications, another type of band-width is usually meant, namely the modulation bandwidth, which is closely related to the dispersion of the fiber. This will be explained later.

4.1 Phase characteristic and modulation bandwidth

Generally, the attenuation constant α is assumed to be independent of fre-quency in the relatively small band of frequencies that a modulated carrier occupies. If we focus our attention on the mechanics of wave propagation, then attenuation plays a minor role. We assume that $\alpha = $ constant in the model so that there is no amplitude distortion. The transmission channel associated with a mode can therefore be seen as a cascade of an attenuator and a waveguide without losses.

Each mode has a unique value of the phase constant β, which is sometimes called the phase propagation coefficient and is a function of frequency. It plays a major role in communications, because it determines the dispersion and the pulse dispersion and, to a great extent, the capacity of a communication link. The phase shift that a harmonic signal experiences when it travels in a certain mode through an optical waveguide is a function of frequency. As an illustra-tion, an arbitrary phase characteristic is drawn in Figure 4.2.

The phase velocity and the group velocity are defined as follows:

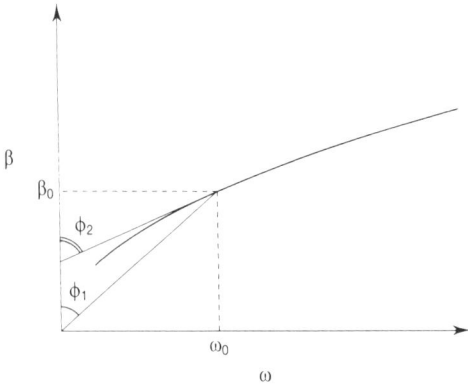

Figure 4.2 An arbitrary phase characteristic.

Phase velocity: $\quad v_p \overset{\triangle}{=} \dfrac{\omega}{\beta}$ $\hspace{4cm}$ (4.4)

Group velocity: $\quad v_g \overset{\triangle}{=} \dfrac{d\omega}{d\beta}$ $\hspace{4cm}$ (4.5)

In Figure 4.2, a specific phase shift β_0 corresponds to a frequency ω_0. The phase velocity at a frequency ω_0 is then equal to the tangent of the angle ϕ_1 and the group velocity equals the tangent of the angle ϕ_2.

Often the specific phase delay, i.e. the phase delay per unit of length, is introduced instead of the phase velocity. We denote

$$\tau_p = \frac{1}{v_p} = \frac{\beta}{\omega} \hspace{4cm} (4.6)$$

The phase delay after a distance z follows from

$$t_p = \tau_p z = \frac{\beta z}{\omega} \hspace{4cm} (4.7)$$

Instead of the group velocity, the specific group delay (the group delay per unit of length), is often used. We denote

$$\tau_g = \frac{1}{v_g} = \frac{d\beta}{d\omega} \hspace{4cm} (4.8)$$

The group delay after a distance z is given by

$$t_g = \tau_g z = \frac{d\beta}{d\omega} z \hspace{4cm} (4.9)$$

In this section, the physical meaning that must be attached to these names will be explained.

Consider that part of the phase characteristic that is in the frequency range

$$\omega_0 - \mu < \omega < \omega_0 + \mu, \qquad \mu \ll \omega_0 \hspace{2cm} (4.10)$$

Assume that β_0 corresponds to ω_0 and that $\beta' = d\beta/d\omega$ is constant in the frequency range concerned. Assume the attenuation in this frequency is also constant. In Figure 4.3 the assumed phase characteristic is depicted as a straight line. Next we consider successively different signals that travel along the transmission line. First a monochromatic carrier of frequency ω_0 that is amplitude modulated by a harmonic signal of frequency μ, followed by a carrier modulated by a more general, pulse-like signal that is considered to be composed of an infinite number of harmonic components.

Assuming an amplitude-modulated component of the electric field at the input of the fiber

$$E_i(t) = (1 + m \cos \mu t) \sin \omega_0 t \hspace{3cm} (4.11)$$

Equation (4.11) can be written as the sum of three frequency components

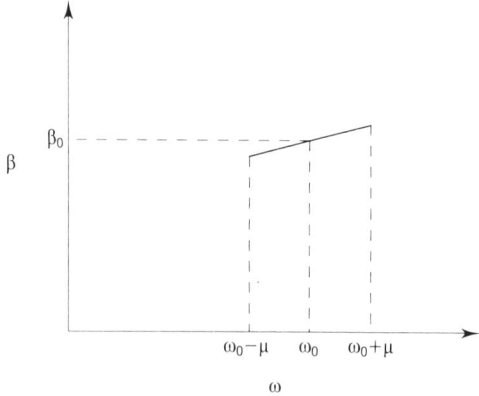

Figure 4.3 A linear phase characteristic in the frequency range of interest.

$$E_i(t) = \sin \omega_0 t + \frac{m}{2} \sin (\omega_0 - \mu)t + \frac{m}{2} \sin (\omega_0 + \mu)t \tag{4.12}$$

In the fiber, three harmonic waves occur, each corresponding to one of the components of equation (4.12)

$$E_1(t, z) = \sin (\omega_0 t - \beta_0 z) \tag{4.13}$$

$$E_2(t, z) = \frac{m}{2} \sin [(\omega_0 - \mu)t - (\beta_0 - \beta' \mu)z] \tag{4.14}$$

$$E_3(t, z) = \frac{m}{2} \sin [(\omega_0 + \mu)t - (\beta_0 + \beta' \mu)z] \tag{4.15}$$

After a distance z along the fiber, the phases of the three components are shifted by $\beta_0 z$, $(\beta_0 - \beta' \mu)z$ and $(\beta_0 + \beta' \mu)z$ respectively. The corresponding component of the electric field at a distance z from the input becomes the sum of the three components E_1, E_2 and E_3; they can be written as

$$E(t, z) = [1 + m \cos \mu(t - \beta' z)] \sin (\omega_0 t - \beta_0 z) \tag{4.16}$$

The resulting wave is shown in Figure 4.4.

We regard the sum $E(t, z)$ as a high-frequency harmonic wave

$$\sin (\omega_0 t - \beta_0 z) \tag{4.17}$$

that is modulated in amplitude with the low-frequency wave

$$\cos \mu(t - \beta' z) \tag{4.18}$$

If we keep the argument of the sine constant so that

$$\omega_0 t - \beta_0 z = \text{constant} \tag{4.19}$$

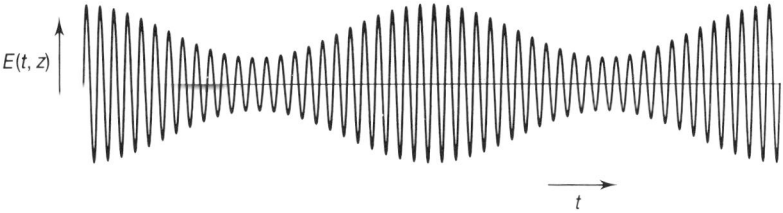

Figure 4.4 The wave $E(t, z)$ as a function of t.

then we find the velocity of the high-frequency wave

$$v_0 = \frac{\mathrm{d}z}{\mathrm{d}t} = \frac{\omega_0}{\beta_0} \tag{4.20}$$

This velocity is equal to the phase velocity defined in equation (4.4). If we keep the argument of the cosine constant, then

$$t - \beta' z = \text{constant} \tag{4.21}$$

and the velocity of the envelope is

$$v = \frac{\mathrm{d}z}{\mathrm{d}t} = \frac{\mathrm{d}\omega}{\mathrm{d}\beta} \tag{4.22}$$

This velocity is equal to the group velocity defined in equation (4.5). Equation (4.16) can then be written as

$$E(t, z) = [1 + \mathrm{m} \cos \mu(t - t_\mathrm{g})] \cos \omega_0(t - t_\mathrm{p}) \tag{4.23}$$

From this expression, it is clear that the envelope is delayed by t_g and the carrier is delayed by t_p.

Next, we consider a more general case rather than the simple example with only three harmonic waves. We assume that an input signal can be written as an amplitude-modulated signal

$$f(t) = f_\mathrm{m}(t) \cos \omega_0 t \tag{4.24}$$

As an illustration of a pulse-modulated carrier, in Figure 4.5 a carrier modulated with a Gaussian envelope is presented.

Assume that the Fourier transform $F_\mathrm{m}(\omega)$ of the modulating signal $f_\mathrm{m}(t)$ is practically band limited

$$|F_\mathrm{m}(\omega)| = 0, \qquad \text{for } |\omega| > \mu \tag{4.25}$$

Like the preceding example, the attenuation and the derivative of the phase characteristic with respect to ω are assumed to be constant in the frequency interval

$$\omega_0 - \mu < |\omega| < \omega_0 + \mu \tag{4.26}$$

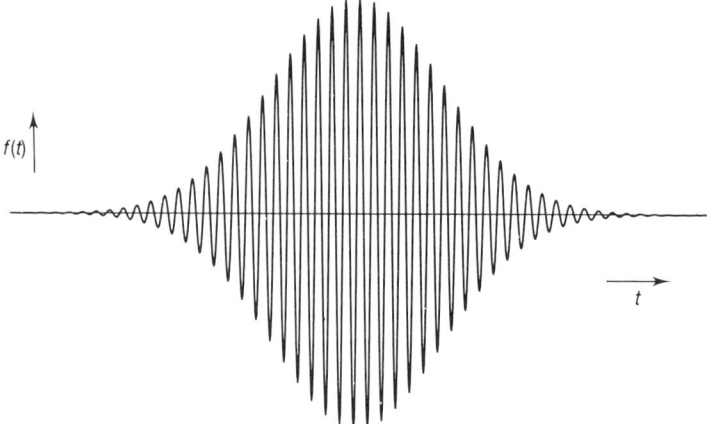

Figure 4.5 A pulse-modulated carrier.

Thus

$$\beta(\omega) = \beta_0 + \beta'(\omega - \omega_0), \qquad \text{for } \omega > 0 \tag{4.27}$$

$$\beta(\omega) = -\beta_0 + \beta'(\omega + \omega_0), \quad \text{for } \omega < 0 \tag{4.28}$$

In the assumed frequency band, the system behaves as an ideal bandpass filter and its response on the input signal is

$$g(t) = f_{\mathrm{m}}(t - t_{\mathrm{g}}) \cos \omega_0(t - t_{\mathrm{p}}) \tag{4.29}$$

The modulating signal is only delayed and attenuated and the pulse shape remains unaltered. In this case, a definition of the pulse delay gives no problem. Here, it will also be found that the travel time of the envelope equals the group delay. From the foregoing, it should be clear that the signal is not distorted if the phase characteristic is a straight line. In general, however, the phase characteristic of an optical waveguide is not linear. The influence of a non-linear phase characteristic can be analyzed when we expand the phase characteristic into a power series. As a first approximation, we will restrict this series to its first three terms

$$\beta = \beta_0 + \beta_0'(\omega - \omega_0) + \frac{\beta_0''}{2}(\omega - \omega_0)^2 \tag{4.30}$$

Again assume that the signal is amplitude modulated as in equation (4.11); then the phase shifts of the three components of this signal are $\beta_0 z$, $[\beta_0 - \beta_0'\mu + (\beta_0''/2)\mu^2]z$ and $[\beta_0 + \beta_0'\mu + (\beta_0''/2)\mu^2]z$, respectively, and the three harmonic waves that result are

$$E_1(t, z) = \sin(\omega_0 t - \beta_0 z) \tag{4.31}$$

$$E_2(t, z) = \frac{m}{2} \sin\left[(\omega_0 - \mu)t - \left(\beta_0 - \beta_0'\mu + \frac{\beta_0''}{2}\mu^2\right)z\right]$$ (4.32)

$$E_3(t, z) = \frac{m}{2} \sin\left[(\omega_0 + \mu)t - \left(\beta_0 + \beta_0'\mu + \frac{\beta_0''}{2}\mu^2\right)z\right]$$ (4.33)

The sum of these waves can be written as

$$E(t, z) = A(t, z) \sin[\omega_0 t + \psi(t, z)]$$ (4.34)

where

$$A(t, z) = \sqrt{\left\{\left[1 + m\cos\left(\frac{\beta_0''}{2}\mu^2 z\right)\cos(\mu t - \beta_0'\mu z)\right]^2\right.}$$

$$\left. + \left[m\sin\left(\frac{\beta_0''}{2}\mu^2 z\right)\cos(\mu t - \beta_0'\mu z)\right]^2\right\}$$ (4.35)

and

$$\psi(t, z) = -\arctan\frac{m\cos(\mu t - \beta_0'\mu z)\sin\frac{\beta_0''}{2}\mu^2 z}{1 + m\cos(\mu t - \beta_0'\mu z)\cos\frac{\beta_0''}{2}\mu^2 z} - \beta_0 z$$ (4.36)

From equation (4.36) it is now clear that a non-linear phase characteristic causes phase modulation. If $\beta_0'' = 0$, then equation (4.35) reduces to

$$A(t, z) = 1 + m\cos(\mu t - \beta_0'\mu z)$$ (4.37)

and equation (4.36) to

$$\psi(t, z) = -\beta_0 z$$ (4.38)

In reality, there is always a deviation from the linear phase. This deviation is caused by the waveguide and the material dispersion, as will be described later. If $m \ll 1$, then equation (4.35) can be approximated by

$$A(t) = 1 + m\cos\left(\frac{\beta_0''}{2}\mu^2 z\right)\cos(\mu t - \beta'\mu z)$$ (4.39)

With $m\cos\mu t$ as the input signal and $m\cos(\beta_0''\mu^2 z/2)\cos(\mu t - \beta_0'\mu z)$ as the output signal, a transfer function can be formed with an amplitude characteristic

$$|H(\mu)| = \left|\cos\left(\frac{\beta_0''}{2}\mu^2 z\right)\right|$$ (4.40)

A graph of this transfer function is depicted in Figure 4.6. For low modulation frequencies, the envelope of the output signal is the same as the envelope of the input signal. With an increasing modulation frequency, the ratio between the output and input amplitudes of the envelope decreases. If the modulation

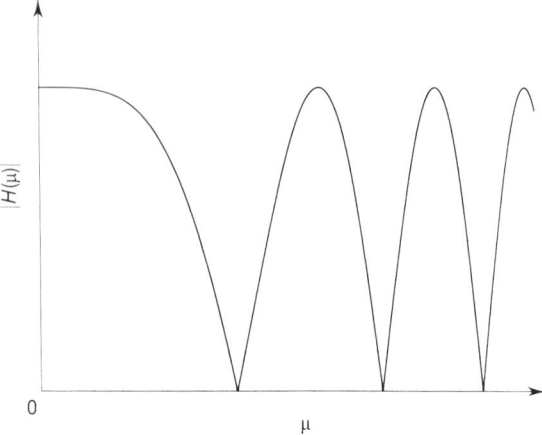

Figure 4.6 The ratio between the amplitudes of the envelopes of the output and input signal of a modulated carrier for $m \ll 1$.

frequency is increased to a value where $\beta_0'' \mu^2 z = \pi$, then the amplitude of the output envelope becomes zero. The modulation bandwidth therefore depends on the second derivative of the phase constant with respect to the frequency, i.e. on the dispersion of the fiber. If $\beta_0'' \mu^2 z = 2\pi/3$, then the amplitude characteristic is half the maximum value. The 3 dB bandwidth for small signals is then given by

$$B = \sqrt{\frac{2\pi}{3\beta_0'' z}} \tag{4.41}$$

If the phase characteristic is not a linear function of the frequency, then the output signal does not have the same shape as the input signal. If there is a big difference in the shape, it may become difficult to determine the travel time of the signal. In that case, the simple relationship between the group delay and the travel time of the envelope is lost.

4.2 The propagation constant of a mode in a step index fiber

For glassfiber communications the general aim is to create a transmission channel that has an attenuation as small as possible and a modulation bandwidth as large as possible. As was shown in the preceding section, the bandwidth depends mainly on the phase characteristics of the modes. The phase constant β of a mode in a step index fiber can be deduced by starting from equation (3.22) and introducing the variable u as defined in equation (3.42), or by starting with equation (3.27) and introducing the variable w as defined in equation (3.43). After some manipulation, we obtain

$$\beta = \sqrt{k_0^2 n_1^2 - \frac{u^2}{a^2}} \tag{4.42}$$

and

$$\beta = \sqrt{k_0^2 n_2^2 + \frac{w^2}{a^2}} \tag{4.43}$$

If the normalized frequency v, as defined by equation (3.54), is also introduced, then these expressions can be written as

$$\beta = k_0 n_1 \sqrt{1 - \frac{u^2}{v^2}(1 - p)} = k_0 n_1 \sqrt{1 - \frac{u^2}{v^2} 2\Delta} \tag{4.44}$$

and

$$\beta = k_0 n_2 \sqrt{1 + \frac{w^2}{v^2}\left(\frac{1}{p} - 1\right)} = k_0 n_2 \sqrt{1 + \frac{w^2}{v^2} 2\Delta} \tag{4.45}$$

with

$$2\Delta = 1 - p = \frac{\varepsilon_1 - \varepsilon_2}{\varepsilon_1} = \frac{n_1^2 - n_2^2}{n_1^2} \tag{4.46}$$

If the relative difference $(n_1 - n_2)/n_1$ between the refractive indices n_1 of the core and the refractive index n_2 of the cladding is small, then

$$\Delta \approx \frac{n_1 - n_2}{n_1} \tag{4.47}$$

The refractive indices n_1 and n_2 are functions of the frequency; therefore, the material dispersions, which are described by the functions $n_1(\omega)$ and $n_2(\omega)$, determine the function $\beta(\omega)$ as well as the functions $u(\omega)$, $w(\omega)$ and $v(\omega)$ for the particular mode that is being considered. Nevertheless, we can assume that n_1 and n_2 are constant, in order to introduce the concept of waveguide dispersion. Thereafter, the influence of the frequency dependence on the refractive index of the materials is included.

4.3 Waveguide dispersion

Figure 4.7 illustrates the curves for β as a function of ω for some of the modes described in equations (4.44) or (4.45). From equation (4.44), it becomes clear that β must be smaller than $k_0 n_1$ and, from equation (4.45), that β must be greater than $k_0 n_2$. Therefore in Figure 4.7 the curves that represent the phase characteristics of the modes must be situated between the two straight lines that correspond to $k_0 n_1$ and $k_0 n_2$, respectively.

In general, each mode has a cut-off frequency that can be denoted as ω_c. At the cut-off frequency $u = v$, thus

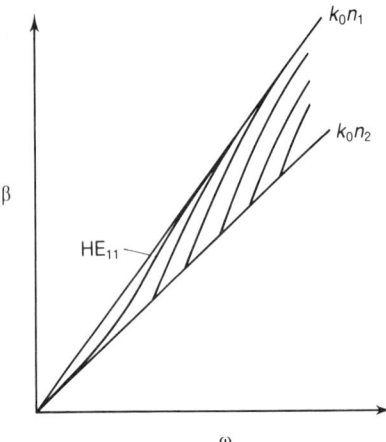

Figure 4.7 The phase constant β as a function of the frequency ω drawn for the HE_{11} mode and for several higher-order modes.

$$\beta(\omega_c) = k_0 n_2 = \frac{\omega_c n_2}{c_0} \tag{4.48}$$

The only exception is the HE_{11} mode, which has no cut-off frequency. For this mode, we can derive

$$\lim_{\omega \to 0} \beta(\omega) = k_0 n_2 = \frac{\omega n_2}{c_0} \tag{4.49}$$

All the curves start on the line $k_0 n_2$. The curve that describes $\beta(\omega)$ for the HE_{11} mode starts at $\omega = 0$ and is tangential to this line for $\omega = 0$. The other curves start at the cut-off frequency of each particular mode. The phase velocity v_p at these points equals the phase velocity c_0/n_2 of a plane wave in a homogeneous medium with a refractive index of n_2. At the cut-off frequency $w = 0$ in which case the guided wave degenerates into a plane wave that propagates in the cladding. On the other hand, if ω (or v) increases, then w increases accordingly and u approaches a certain limit. Therefore, with equation (4.44), we can obtain

$$\lim_{\omega \to \infty} \beta(\omega) = k_0 n_1 = \frac{\omega n_1}{c_0} \tag{4.50}$$

In this instance, the phase velocity approaches the phase velocity of a plane wave in a homogeneous medium with refractive index n_1. Instead of a curve of β as a function of ω, we often see that β/k_0 is plotted as a function of ω, or as a function of v. In Figure 4.8, for instance, β/k_0 is plotted as a function of v for a

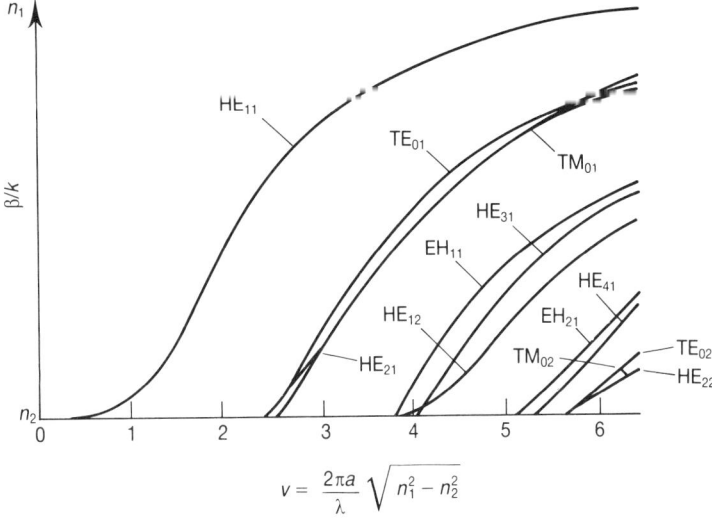

$$v = \frac{2\pi a}{\lambda} \sqrt{n_1^2 - n_2^2}$$

Figure 4.8 The normalized phase constant β/k_0 as a function of the normalized frequency v for a number of lower-order modes. (Source: reproduced with permission from *Optical Communication Systems*, by John Gowar, Prentice Hall, 1984.)

number of lower-order modes. The lower-order modes are denoted by the subscripts that are small numbers.

More important than $\beta(\omega)$ itself is the first derivative τ_g of $\beta(\omega)$ with respect to ω. When multiplied by the length of the fiber, the group delay t_g becomes

$$t_g = l\tau_g = l \frac{d\beta}{d\omega} \tag{4.51}$$

where l = the length of the fiber.

With some restrictions, the group delay t_g is the delay time for the envelope of a modulated harmonic carrier, and thus is roughly equal to the delay time for a pulse that passes through the fiber in the particular mode concerned.

Differentiating β with respect to ω we find the specific group delay as

$$\tau_g = \frac{d\beta}{d\omega} = \frac{d\beta}{dv} \frac{dv}{d\omega} \tag{4.52}$$

with

$$\frac{dv}{d\omega} = \frac{a}{c_0} \sqrt{n_1^2 - n_2^2} \tag{4.53}$$

With u and w, respectively, as parameters, equation (4.52) can be rewritten as

$$\tau_{\mathrm{g}} = \frac{1}{c_0} \frac{n_1^2 - (n_1^2 - n_2^2) \dfrac{u}{v} \dfrac{\mathrm{d}u}{\mathrm{d}v}}{\sqrt{n_1^2 - (n_1^2 - n_2^2) \dfrac{u^2}{v^2}}} \tag{4.54}$$

or

$$\tau_{\mathrm{g}} = \frac{1}{c_0} \frac{n_2^2 + (n_1^2 - n_2^2) \dfrac{w}{v} \dfrac{\mathrm{d}w}{\mathrm{d}v}}{\sqrt{n_2^2 + (n_1^2 - n_2^2) \dfrac{w^2}{v^2}}} \tag{4.55}$$

In summary, the group delay for a particular mode at a given normalized frequency v can be established after first finding the corresponding values of u or w from the characteristic equation, see Chapter 3, and the derivative of u or w with respect to v.

4.4 Material dispersion

Before analyzing the phase constant of a harmonic wave that propagates in an optical waveguide with refractive indices that are functions of frequency, some definitions will be given on the basis of a harmonic plane wave in bulk material. The refractive index of the material equals the ratio between the phase velocities of a plane harmonic wave in vacuum and in the material, respectively; it is also the same ratio as that between the wavelengths in vacuum and in the material:

$$n = \frac{\lambda_0}{\lambda} = \frac{\beta c_0}{\omega} \tag{4.56}$$

where β is the specific phase constant in the direction of propagation. For example, the refractive index n of pure silica as a function of the wavelength λ_0 is shown in Figure 4.9.

For a plane harmonic wave with frequency ω the specific phase delay in bulk material with refractive index n is

$$\frac{\beta}{\omega} = \frac{n}{c_0} \tag{4.57}$$

Taking the derivative of β with respect to ω and introducing the group index N for a plane harmonic wave in bulk material, defined as

$$N \triangleq n + \omega \frac{\mathrm{d}n}{\mathrm{d}\omega} = n - \lambda_0 \frac{\mathrm{d}n}{\mathrm{d}\lambda_0} \tag{4.58}$$

we obtain the specific group delay for a plane harmonic wave

$$\frac{\mathrm{d}\beta}{\mathrm{d}\omega} = \frac{N}{c_0} \tag{4.59}$$

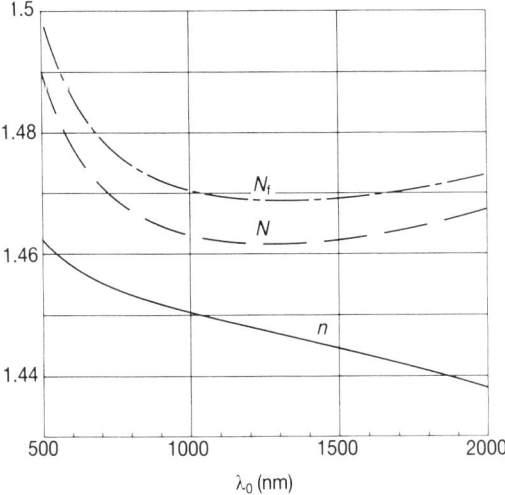

Figure 4.9 The refractive index n and the group index N of pure silica as a function of the wavelength λ_0 and the group index N_f for the HE_{11} mode of a step index fiber.

Note the similarity of the formulae in equations (4.57) and (4.59); the refractive index n determines the phase delay of a plane wave and, likewise, the group index N determines the group delay of the plane wave. The group index of pure silica as a function of λ_0 is also shown in Figure 4.9. That curve has a minimum at a wavelength of $\lambda_0 = 1.27 \, \mu\text{m}$. The material dispersion D for a plane harmonic wave is defined as the derivative of the specific group delay τ_g with respect to the wavelength λ_0

$$D \triangleq \frac{d\tau_g}{d\lambda_0}. \tag{4.60}$$

The material dispersion for SiO_2 and SiO_2 doped with 4.1 percent GeO_2 is displayed in Figure 4.10 as a function of the wavelength λ_0.

4.5 Waveguide and material dispersion in a step index fiber

Next we analyze β of the wave that propagates in the glassfiber, as a function of frequency, and take into account the frequency dependence of the refractive indices of the materials used. The specific group delay of a particular mode can be found by differentiating β with respect to ω, e.g. as given in equation (4.44). The result is

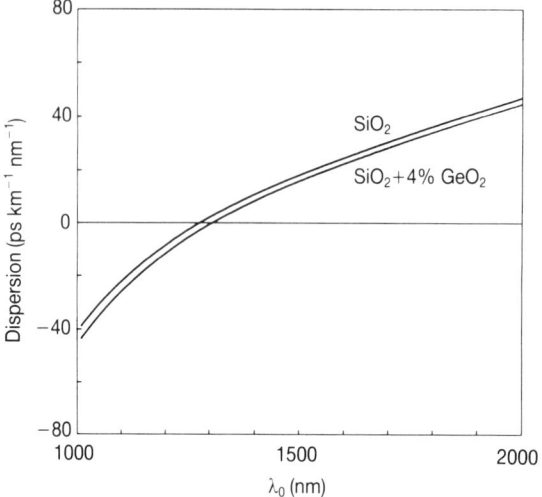

Figure 4.10 The material dispersion of SiO_2 and for SiO_2 doped with 4.1 percent GeO_2.

$$\frac{d\beta}{d\omega} = \frac{N_f}{c_0} \qquad (4.61)$$

where N_f denotes the group index of the fiber mode and is given by

$$N_f = \left\{ N_1 - \frac{n_1}{1 - 2\Delta(u/v)^2} \; \omega \; \frac{d[\Delta(u/v)^2]}{d\omega} \right\} \sqrt{1 - 2\Delta(u/v)^2} \qquad (4.62)$$

where N_1 is the group index of a plane wave in bulk material with refractive index n_1. As an example, the group index for the HE_{11} mode of a step index fiber is the top curve in Figure 4.9. This curve is for a step index fiber with a core (radius = 4 μm) of SiO_2 doped with 4.1 percent GeO_2 and a cladding of pure silica. For step index monomode fibers, more dispersion curves are depicted in Chapter 5.

It is clear from equation (4.62) that the dispersion in a fiber is caused partly by the variable refractive indices. This kind of dispersion is called material dispersion. Furthermore, equation (4.62) shows that the waveguide dispersion and the material dispersion are closely interweaved; it is not possible to separate these two different kinds of dispersion mathematically.

In digital transmission, the light source – such as a semiconductor laser – is modulated with short pulses. If the group delay is constant in the frequency interval of the modulated pulse, the pulse is not disturbed and the pulse duration remains the same; therefore, the wavelength where N_f has its minimum is a favourable wavelength for high-speed digital transmission with a monomode fiber.

4.6 Multimode dispersion

Assume a monochromatic source with a frequency ω_0 that is modulated with a pulse-like input current $i_i(t)$, also, assume that the input power $p_i(t)$ launched into the fiber is proportional to the input signal

$$p_i(t) = ai_i(t) \tag{4.63}$$

where a is a constant.

The power $p_{in}(t)$ denotes the power that is coupled to the nth mode of the fiber

$$p_{in}(t) = a_n i_i(t) \tag{4.64}$$

where a_n is a factor that depends on the way in which the light from the source is coupled into the fiber. Then we can associate every mode in the fiber with a transmission channel with its own characteristic attenuation and a group delay as depicted in Figure 4.11. The output power $p_{on}(t)$ of the nth channel depends upon the attenuation, the group delay τ_{gn} and the material and waveguide pulse dispersion. The output power $p_o(t)$ is the sum of the power outputs of the individual channels, and the output current $i_o(t)$ from the photodetector is proportional to this summation. Thus we get

$$i_o(t) = bp_o(t) = b\sum_{n=1}^{N} p_{on}(t) \tag{4.65}$$

where b is a constant and N the total number of modes involved.

The transfer from $i_i(t)$ to $i_o(t)$ for a step index multimode fiber is determined largely by the ratios of the input powers $p_{in}(t)$ and by the differences in the group delays τ_{gn} of the individual channels. The pulse dispersion that is caused by these differences in delay times is called multimode dispersion, or mode dispersion for short.

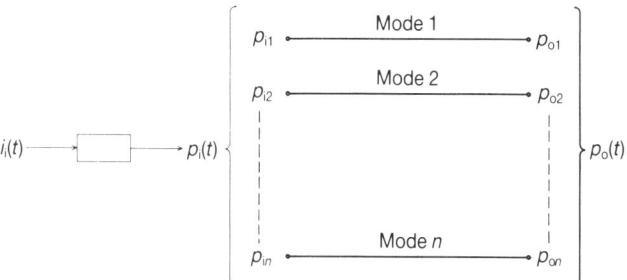

Figure 4.11 The model of a multimode fiber with source and detector.

4.7 Dispersion caused by a non-monochromatic source

If a non-monochromatic source is used, then each mode or each transmission channel in this model causes a change in the pulse duration, if $d\tau_g/d\lambda_0 \neq 0$. If the difference in wavelength between two carriers equals $\Delta\lambda_0$, then a difference in the time delay equal to $\Delta\lambda_0(d\tau_g/d\lambda_0)$ results.

The transfer is thus determined by the following factors:

- The attenuation per mode.
- The waveguide dispersion per mode.
- The material dispersion.
- The (multi-)mode dispersion.
- The power spectrum of the source.
- The distribution of power over the modes.

However, the mode conversion and the influence of joints have not been included and these effects will be treated in Chapters 10 and 11, respectively.

References

[1] J. A. Arnaud, *Beam and Fiber Optics* (New York: Academic Press, 1976).
[2] H.-G. Unger, *Planar Optical Waveguides and Fibers* (Oxford: Clarendon Press, 1977).
[3] D. Marcuse, *Principles of Optical Fiber Measurements* (New York: Academic Press, 1981).
[4] T. Okoshi, *Optical Fibers* (New York: Academic Press, 1982).
[5] A. W. Snyder and J. D. Love, *Optical Waveguide Theory* (London: Chapman and Hall, 1983).
[6] L. B. Jeunhomme, *Single-mode Fiber Optics* (New York: Marcel Dekker, 1983).
[7] C. K. Kao, *Optical Fibre* (London: Peter Peregrinus, 1988).
[8] E.-G. Neumann, *Single-model Fibers* (Berlin: Springer Verlag, 1988).

Problems

4.1 Describe the origin and the behaviour of the various kinds of dispersion that can be distinguished in glassfibers.

4.2 Assume a weakly guiding step index fiber with a length of 10 km and a single-mode semiconductor laser. The wavelength of the light in vacuum is $\lambda_0 = 0.85\ \mu$m. In the wavelength region considered, the refractive indices of the core and cladding can be approximated by the formulae $n_1 = 1.25 + 0.2\lambda_0^{-1}$ and $n_2 = 1.24 + 0.2\lambda_0^{-1}$, where λ_0 is the wavelength in vacuum, expressed in μm. The cut-off wavelength of the E_{01} mode in the fiber is 0.8 μm.

- Find the diameter of the fibre core.
- Using this laser, how many LP modes can exist in the fiber?
- Find the phase delay and the group delay of the LP modes in the fiber.

4.3 Assume a step index fiber with refractive indices of core and cladding, respectively, $n_1 = 1.50$ and $n_2 = 1.49$. The diameter of the core is 50 μm. A single-mode laser with $\lambda_0 = 840$ nm is used as a light source.

1. Derive the characteristic equation for the TM or E waves in the fiber.
2. Find the cut-off wavelength for the E_{01} mode.
3. Show that for the E_{01} mode β is smaller than for the H_{01} mode.
4. Sketch E_r and H_φ as functions of r for the E_{01} mode.

4.4 Assume a step index fiber with $n_1 = 1.455$ and $n_2 = 1.445$. The radius of the core is $a = 3000$ nm. The light source used is monochromatic and the wavelength is $\lambda_0 = 1350$ nm. In the wavelength region concerned, the value of u for the HE_{11} mode can be approximated by

$$u = 2.7484 \frac{\lambda_c}{\lambda_0} - 0.9960$$

where λ_c is the cut-off wavelength of the E_{01} mode.

1. Is the fiber a single-mode fiber?
2. Find the cut-off wavelength λ_c for the E_{01} mode.
3. Find the specific phase shift and the specific phase delay at the wavelength $\lambda_0 = 3000$ nm.

4.5 The cladding of a step index fiber is made of pure silica. The diameter of the core is 5500 nm and $\varepsilon_1/\varepsilon_2 = 1.01$.

- Use Figure 4.9 to find the wavelength above which the fiber is single mode.
- Find the cut-off wavelength of the E_{02} mode.
- Sketch the graphs of E_z and H_φ as functions of r for the E_{02} mode at the cut-off wavelength of this mode and for a frequency much larger than the cut-off frequency.

5

The monomode fiber

5.1   The electromagnetic field in a monomode step index fiber
5.2   Power flow in the z-direction
5.3   The mode field diameter
5.4   Waveguide dispersion in monomode fibers
5.5   Waveguide and material dispersion in monomode fibers
5.6   Dispersion-shifted and dispersion-flattened fibers
         References
         Problems

The monomode fiber, or single-mode fiber, plays a major role in modern glassfiber communication systems. It is clearly the most promising optical waveguide for long-distance transmission systems, and it is also generally regarded as the most suitable for future local networks. Monomode fibers have low attenuation and large bandwidths. In Chapter 3, a general method for solving the wave equation in a step index optical waveguide was developed; in this chapter, the results of Chapter 3 are used to describe the fields in a monomode or single-mode fiber. To that end, the general formulae for the fields are applied to the special case of the HE_{11} mode to gain some insight into the field distribution for this important case. A link is made between these fields and the mode field diameter, sometimes also called the spot size. The mode field diameter plays a major role in coupling light sources with monomode fibers, in mutually coupling monomode fibers and calculating microbending losses.

The results from Chapter 4 are then used to describe the dispersion in monomode fibers. In this connection, dispersion-shifted and dispersion-flattened fibers are treated.

5.1 The electromagnetic field in a monomode step index fiber

As should be clear from the previous chapters, the electromagnetic field that belongs to a particular mode of a step index fiber depends on the value of the normalized frequency v. Since a step index fiber is a monomode fiber if the value of the normalized frequency is less than 2.405, the value of v that corresponds to the cut-off frequency of the E_{01} and the H_{01} modes, namely $v = 2.405$, seems to be a good starting point for analyzing any particular field of the HE_{11} mode. In later sections, we will see how varying v influences the field properties.

If a value of v has been chosen, then the values of u and w follow from equation (3.55) and from the characteristic equation, for instance as given by equation (3.85), for the HE_{11} mode with the minus sign and with $n = 1$. With $p = 0.99$, u and w for the HE_{11} mode can be calculated as functions of v and are plotted in Figure 5.1.

The field for a particular case can be described when v is taken as 2.405 and the value of p is taken as 0.99. With these values, the numerical evaluation of the characteristic equation gives u and w.

With u and w known, $J_0(u)$, $J_1(u)$, $K_0(w)$ and $K_1(w)$ can be calculated, and

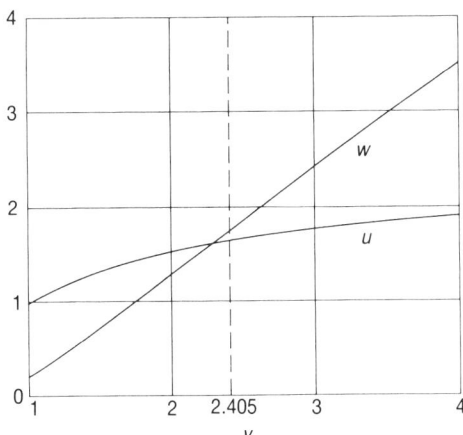

Figure 5.1 The variables u and w as functions of the normalized frequency v for a step index fiber and for $p = 0.99$.

the derivatives $J_1'(u)$ and $K_1'(w)$ follow from equations (A4.18) and (A4.30) of Appendix 4.

First we must determine the electric field as a function of r. If $\varphi = 0$, then E_z follows from equations (3.64) and (3.65)

$$E_z = A_1 J_1(h_1 r), \qquad\qquad r < a \qquad\qquad (5.1)$$

$$E_z = A_2 K_1(h_2 r) = R_1 A_1 K_1(h_1 r), \quad r > a \qquad (5.2)$$

where R_1 is the ratio between A_1 and A_2; it follows from equation (3.72) that

$$R_1 = \frac{A_2}{A_1} = \frac{J_1(u)}{K_1(w)} \qquad\qquad (5.3)$$

From equation (3.70) with $\gamma = j\beta$ and, again, for $\varphi = 0$

$$E_r = -\frac{1}{h_1^2}\left[j\beta h_1 A_1 J_1'(h_1 r) + \frac{j\omega\mu G_1}{r} J_1(h_1 r) \right]$$

$$= -j\frac{\beta A_1}{h_1}\left[J_1'(h_1 r) + R_2 \frac{1}{h_1 r} J_1(h_1 r) \right], \qquad r < a \qquad (5.4)$$

where, according to equations (3.76), (3.78) and (3.79)

$$R_2 = \frac{\omega\mu G_1}{\beta A_1} = -\frac{(1/u^2) + (1/w^2)}{Y_1(u) + X_1(w)} \qquad\qquad (5.5)$$

Introducing the ratio R_3 as follows

$$R_3 = \frac{\beta}{h_1} = \frac{\beta a}{u} = \sqrt{\frac{v^2}{u^2(1-p)} - 1} \qquad\qquad (5.6)$$

gives then, in the core

$$E_r = -jA_1 R_3\left[(R_2 - 1)\frac{1}{h_1 r} J_1(h_1 r) + J_0(h_1 r) \right], \qquad r < a \qquad (5.7)$$

From equation (3.71), we can derive E_r in the cladding as

$$E_r = \frac{1}{h_2^2}\left[j\beta h_2 A_2 K_1'(h_2 r) + \frac{j\omega\mu G_2}{r} K_1(h_2 r) \right]$$

$$= j\frac{\beta A_2}{h_2}\left[K_1'(h_2 r) + R_2 \frac{1}{h_2 r} K_1(h_2 r) \right], \qquad r > a \qquad (5.8)$$

Introducing the ratio R_4

$$R_4 = R_3 \frac{u}{w} = \frac{\beta a}{w} = \frac{\beta}{h_2} \qquad\qquad (5.9)$$

gives

$$E_r = jA_1 R_1 R_4\left[(R_2 - 1)\frac{1}{h_2 r} K_1(h_2 r) - K_0(h_2 r) \right], \qquad r > a \qquad (5.10)$$

If $\varphi = \pi/2$, then from equation (3.68) it follows that

$$
E_\varphi = \frac{1}{h_1^2}\left[\frac{j\beta A_1}{r}J_1(h_1r) + j\omega\mu_1 G_1 J_1'(h_1r)\right]
$$

$$
= j\frac{\beta A_1}{h_1}\left[\frac{1}{h_1 r}J_1(h_1r) + R_2 J_1'(h_1r)\right]
$$

$$
= jA_1 R_3\left[(1 - R_2)\frac{1}{h_1 r}J_1(h_1r) + R_2 J_0(h_1r)\right], \qquad r < a \qquad (5.11)
$$

$$
E_\varphi = -\frac{1}{h_2^2}\left[\frac{j\beta A_2}{r}K_1(h_2r) + j\omega\mu_2 G_2 K_1'(h_2r)\right]
$$

$$
= -j\frac{\beta A_2}{h_2}\left[\frac{1}{h_2 r}K_1(h_2r) + R_2 K_1'(h_2r)\right]
$$

$$
= -jA_1 R_1 R_4\left[(1 - R_2)\frac{1}{h_2 r}K_1(h_2r) - R_2 K_0(h_2r)\right], \qquad r > a \qquad (5.12)
$$

In Figure 5.2 the amplitudes of E_z, E_r and E_φ of the electric field that are obtained from equations (5.1), (5.2), (5.7), (5.10), (5.11) and (5.12) have been plotted as functions of r, in the correct proportions, given the values $p = 0.99$ and $v = 2.405$. Similar curves can be found for the amplitudes of H_z, H_r and H_φ of the magnetic field. Only for E_r is there a small discontinuity at the interface between the core and cladding. This discontinuity is very small if $p = 0.99$, so that it is hardly apparent in Figure 5.2.

The components E_r and E_φ differ by π in phase and have a phase difference of $\pi/2$ with the component E_z. If we consider a plane perpendicular to the z-axis, then periodically, at certain moments, in every point of that plane

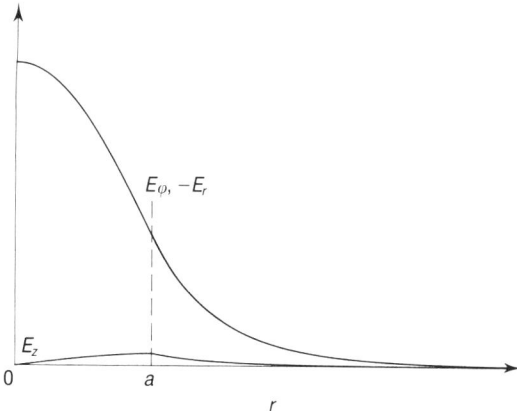

Figure 5.2 The amplitudes E_z, E_r and E_φ as functions of r.

$E_z = 0$, and at these moments the components E_r and E_φ have their maximum or minimum values. Also, if we have to deal with a weakly guiding fiber, the amplitudes of E_r and E_φ are almost equal. For the ratio E_r/E_φ at the interface between the core and the cladding, we can see from equations (5.7) and (5.11) that

$$\frac{E_r}{E_\varphi} = -\frac{(R_2 - 1)\dfrac{1}{u}J_1(u) + J_0(u)}{(1 - R_2)\dfrac{1}{u}J_1(u) + R_2J_0(u)} \tag{5.13}$$

With the values given in the previous example, this ratio becomes 0.998, which means that the transversal resultant $\mathbf{E}_t$ of the components $\mathbf{E}_r$ and $\mathbf{E}_\varphi$ where $E_r = -E_{r\max}\cos\varphi$ and $E_\varphi = E_{\varphi\max}\sin\varphi$, in the plane perpendicular to the axis of the fiber, points almost in the direction $\varphi = \pi$, and is independent of r and φ. This is shown in Figure 5.3, and agrees with the results derived from the LP modes.

The electric field lines in planes perpendicular to the z-axis are thus nearly parallel. Combining the components $\mathbf{H}_r$ and $\mathbf{H}_\varphi$ of the magnetic field results in a transverse component $\mathbf{H}_t$ perpendicular to $\mathbf{E}_t$ that points in the direction $\varphi = -\pi/2$. The magnetic field lines are nearly parallel too, and perpendicular to the electric field lines. The electric and magnetic field patterns in the plane $\varphi = 0$, through the axis of the fiber, are drawn in Figure 5.4, again with $p = 0.99$ and $v = 2.405$.

The contribution of the Bessel functions J_1 and K_1 in the equations that describe E_r and E_φ are shown in Figure 5.5, where the electric field pattern has been plotted for the extreme case of a fiber with a core refractive index of $n_1 = 1.5$ and a cladding refractive index of $n_2 = 1$.

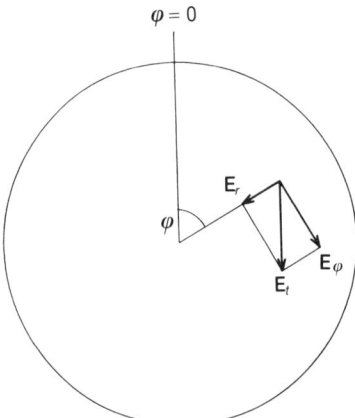

Figure 5.3 The composition of the vector $\mathbf{E}_r$ and $\mathbf{E}_\varphi$ in a plane perpendicular to the z-axis, where $E_z = 0$.

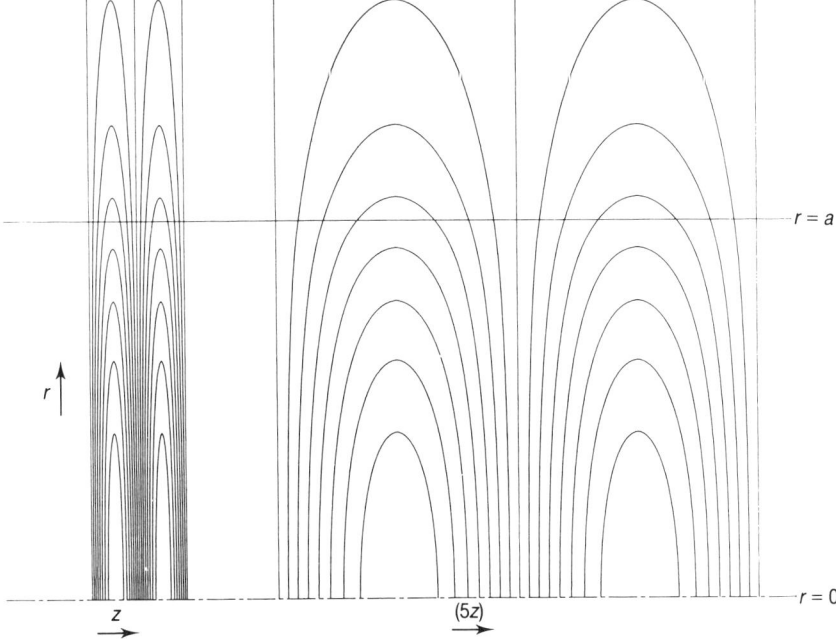

Figure 5.4 The electric field line pattern in the plane $\varphi = 0$.

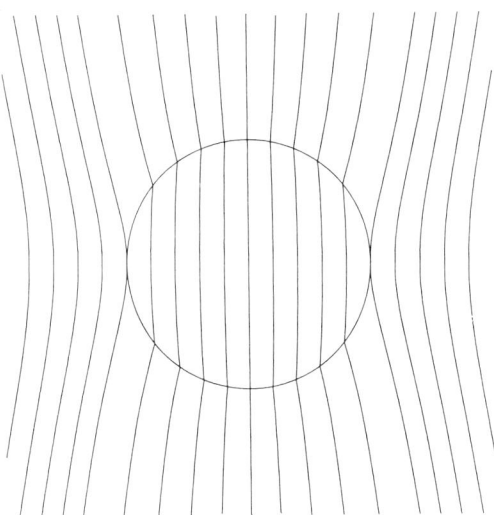

Figure 5.5 The electric field pattern for a fiber with $p = 0.44$ and $v = 2.405$.

5.2 Power flow in the *z*-direction

In this section we determine the density of the power and the power flow in the direction of the axis of the fiber. In the previous section it was shown that the transversal components $\mathbf{E}_t$ and $\mathbf{H}_t$ of the electric and the magnetic field are practically independent of the coordinate φ. For practical values of the parameter p, the amplitudes can be approximated by

$$E_t = H_t \sqrt{\frac{\mu_0}{\varepsilon}} = E_0 \frac{J_0(ur/a)}{J_0(u)}, \qquad r < a \tag{5.14}$$

$$E_t = H_t \sqrt{\frac{\mu_0}{\varepsilon}} = E_0 \frac{K_0(wr/a)}{K_0(w)}, \qquad r > a \tag{5.15}$$

The intensity I in the direction of the axis, i.e. the mean density of the power flow in the z-direction per unit of surface area, is

$$I = \tfrac{1}{2} E_t H_t \tag{5.16}$$

For equations (5.14) and (5.15) the intensity in the core and in the cladding, respectively, becomes

$$I_1 = \frac{1}{2} \sqrt{\frac{\varepsilon}{\mu_0}} \left[E_0 \frac{J_0(ur/a)}{J_0(u)} \right]^2, \qquad r < a \tag{5.17}$$

$$I_2 = \frac{1}{2} \sqrt{\frac{\varepsilon}{\mu_0}} \left[E_0 \frac{K_0(wr/a)}{K_0(w)} \right]^2, \qquad r > a \tag{5.18}$$

The power flow in the core is

$$P_1 = \int_0^{2\pi} \int_0^a I_1 r \, dr \, d\varphi \tag{5.19}$$

Substituting equation (5.17) in equation (5.19) and using equation (A4.45) of Appendix 4 as follows

$$\int z J_0^2(az) \, dz = \frac{z^2}{2} \left[J_0^2(az) + J_1^2(az) \right] \tag{A.45}$$

leads to

$$P_1 = \frac{na^2}{240} E_0^2 \left[1 + \frac{J_1^2(u)}{J_0^2(u)} \right] \tag{5.20}$$

where $n \simeq n_1 \simeq n_2$.

The power flow in the cladding can be found, using equation (A.46) to give

$$P_2 = \int_0^{2\pi} \int_a^\infty I_2 r \, dr \, d\varphi = \frac{na^2}{240} E_0^2 \left[\frac{K_1^2(w)}{K_0^2(w)} - 1 \right] \tag{5.21}$$

The total power flow in the z-direction is

$$P_t = \frac{na^2}{240} E_0^2 \left[\frac{J_1^2(u)}{J_0^2(u)} + \frac{K_1^2(w)}{K_0^2(w)} \right] \tag{5.22}$$

Using the characteristic equation (3.105) for the HE_{11} mode, the total power flow can be written as

$$P_t = \frac{na^2}{240} \left[E_0 \frac{v}{w} \frac{J_1(u)}{J_0(u)} \right]^2 = \frac{na^2}{240} \left[E_0 \frac{v}{u} \frac{K_1(w)}{K_0(w)} \right]^2 \tag{5.23}$$

If the total power is normalized to $P_t = 1$, then

$$E_0 = \frac{1}{a} \sqrt{\frac{240}{n}} \frac{w}{v} \frac{J_0(u)}{J_1(u)} = \frac{1}{a} \sqrt{\frac{240}{n}} \frac{u}{v} \frac{K_0(w)}{K_1(w)} \tag{5.24}$$

With equation (5.24) the normalized intensity of the light in the core and in the cladding is

$$I_{n1} = \frac{1}{\pi a^2} \frac{w^2}{v^2} \frac{J_0^2[u(r/a)]}{J_1^2(u)}, \qquad r < a \tag{5.25}$$

$$I_{n2} = \frac{1}{\pi a^2} \frac{u^2}{v^2} \frac{K_0^2[w(r/a)]}{K_1^2(w)}, \qquad r > a \tag{5.26}$$

In Figure 5.6 the normalized intensity for the HE_{11} mode in a step index fiber and as a function of r can be plotted for different values of the normalized frequency v. From these plots, it becomes clear that a substantial part of the power flows in the cladding if the fiber is monomode, i.e. if $v < 2.405$. In Figure 5.7 the radii that enclose various percentages of the total power flow are plotted as functions of v for the HE_{11} mode and $p = 0.99$.

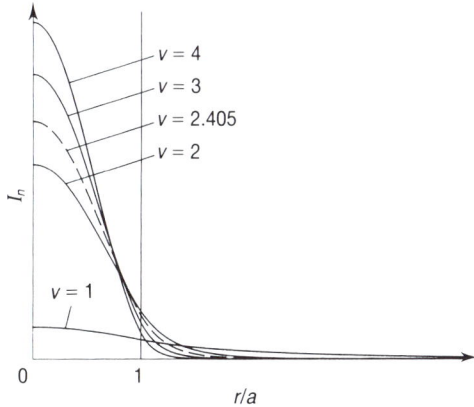

Figure 5.6 The normalized intensity of the HE_{11} mode in a step index fiber as a function of r and for different values of the frequency v.

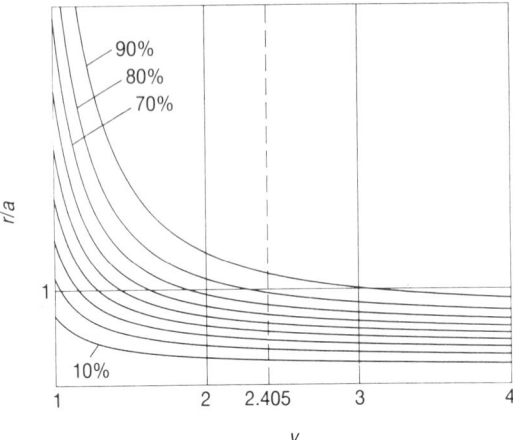

Figure 5.7 The normalized radii r/a that enclose various percentages of the total power flow for the HE_{11} mode of a step index fiber; $p = 0.99$.

5.3 The mode field diameter

The electromagnetic field of a monomode step index fiber and the power distribution of the light propagating over the cross-section of the fiber is represented in detail by the formulae from the previous section. In practice, it is more common to express certain properties of the field by a single parameter. If well chosen, such a parameter will be appropriate in most cases. The mode field diameter (MFD) is a simple numerical representation assigned to the relative complex electromagnetic field according to a fixed rule. For instance, it can be used to establish an efficient coupling between a laser diode and a fiber, or to describe the coupling between two fibers or the microbending losses. The rules by which the MFD is assigned to the electromagnetic field of the fiber can vary according to the situation. The three commonest rules will be described.

5.3.1 MFD or spot size related to Gaussian beam

Assume that the light of a Gaussian beam is launched into a fiber, as depicted in Figure 5.8, in such a way that the waist or narrowest part of the beam coincides

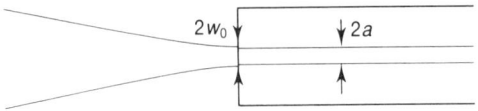

Figure 5.8 A Gaussian beam coupled into a step index fiber.

with the front face of the fiber. The amplitude of the magnetic field at the waist of a power normalized Gaussian beam is

$$H = \frac{2}{w_0\pi} \sqrt{\frac{n}{120}} \exp\left[-(r/w_0)^2\right]$$

(5.27)

If this amplitude is multiplied by half the normalized amplitudes of the electric field given in equations (5.14) and (5.15), with the E_0 as given in equation (5.24), then the density of the resulting power flow of the HE_{11} mode in the fiber is found and the efficiency η of the coupling follows from integrating this density over the surface of the coupling.

$$\eta = \frac{2\sqrt{2}}{w_0 a} \left\{ \frac{w}{vJ_1(u)} \int_0^a J_0\left(u\frac{r}{a}\right) \exp\left[-(r/w_0)^2\right] r\, dr \right.$$
$$\left. + \frac{u}{vK_1(w)} \int_a^\infty K_0\left(w\frac{r}{a}\right) \exp\left[-(r/w_0)^2\right] r\, dr \right\}$$

(5.28)

The efficiency as a function of the ratio for half the width w_0 and the core radius a is plotted in Figure 5.9 for $v = 2.405$. The width $2w_0$ of the waist of the beam with the maximum efficiency is by definition the MFD or spot size of the HE_{11} mode of the fiber.

In Figure 5.10 the maximum efficiency is plotted as a function of the normalized frequency v together with the corresponding normalized half-width w_0/a. In the literature [1], a formula has been given that approximates to this MFD, within 1 percent, over the range $v = 1.2$–3, namely

$$\frac{w_0}{a} \approx 0.65 + 1.619v^{-3/2} + 2.879v^{-6}$$

(5.29)

A curve that corresponds to this approximation is also shown in Figure 5.10.

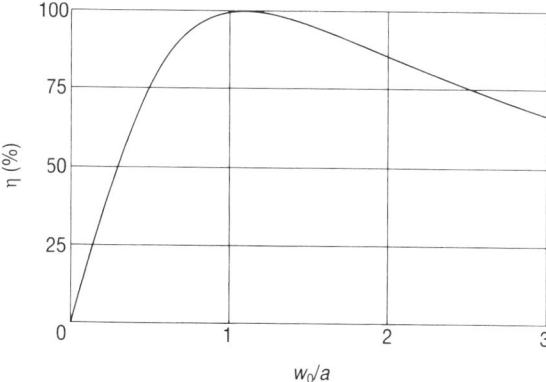

Figure 5.9 The efficiency η of the coupling of a Gaussian beam with the HE_{11} mode of a step index fiber or $v = 2.405$.

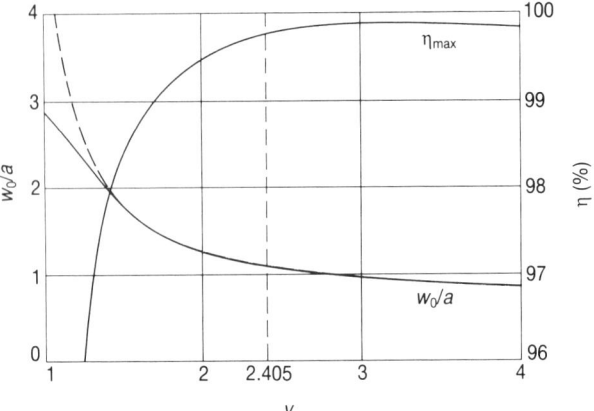

Figure 5.10 The maximum efficiency of coupling a Gaussian beam with the HE_{11} mode of a step index fiber as a function of v, together with the corresponding normalized half-width w_0/a of the Gaussian beam (solid line) and the approximation of w_0/a from equation (5.29).

5.3.2 The mode field diameters d_m and d_j

Microbending losses and joint losses due to small tilts [2] can be described with the help of the functional d_m that is defined as follows:

$$d_m = 2 \left[\frac{2\int_0^\infty E^2(r)r^3 \, dr}{\int_0^\infty E^2(r) \, dr} \right]^{1/2} \tag{5.30}$$

Losses due to small lateral offsets [3] and waveguide dispersion [4] depend on the functional d_j that is defined in the following way:

$$d_j = 2 \left[\frac{2\int_0^\infty E^2(r)r \, dr}{\int_0^\infty \left(\frac{dE}{dr}\right)^2 r \, dr} \right]^{1/2} \tag{5.31}$$

The mode field diameters d_m and d_j for the HE_{11} mode of a step index fiber, normalized to the core diameter $2a$, are depicted in Figure 5.11 as a function of v, and compared with the normalized half-width w_0/a described in the previous section.

5.4 Waveguide dispersion in monomode fibers

In Chapter 4, the general formulae for waveguide dispersion of step index fibers with constant refractive indices were derived. As an example, $c_0(d\beta/d\omega) = c_0\tau_g$ has been calculated for the HE_{11} mode, using equation (4.13) or equation

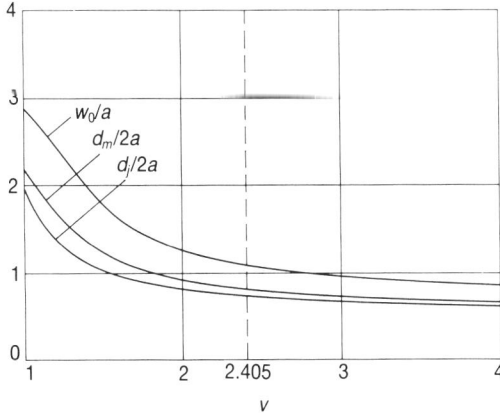

Figure 5.11 The normalized mode field radii $d_m/2a$ and $d_j/2a$ and the normalized half-width w_0/a of the HE_{11} mode in a step index fiber as a function of the normalized frequency v.

(4.14), and the results are plotted in Figure 5.12. It was assumed that the cladding had a constant refractive index, namely $n_2 = 1.44$, and different constants were assumed for the parameter p, namely 0.99, 0.98 and 0.97.

In general, the specific group delay τ_g is also a function of the frequency ω or the wavelength λ_0. If $d\tau_g/d\lambda_0 = 0$, then the envelope is not distorted and the transmission channel behaves as a definite delay line. However, in general,

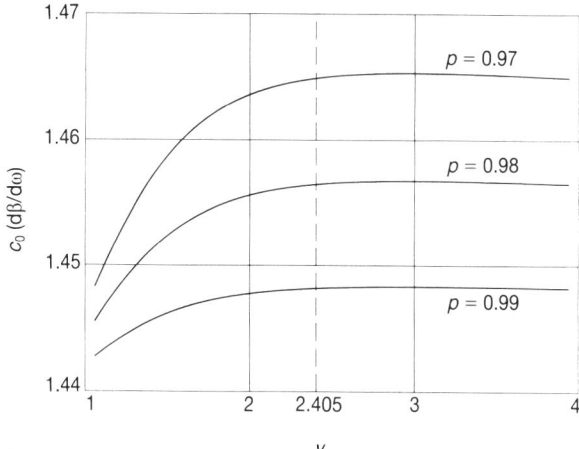

Figure 5.12 The normalized specific group delay $c_0(d\beta/d\omega)$ as a function of v for the HE_{11} mode of a step index fiber with $n_2 = 1.44$ and different values of the parameter p.

$d\tau_g/d\lambda_0 \neq 0$ and the envelope is distorted. In general, any power pulse that is transmitted along an individual transmission channel increases the pulse duration. This is called pulse dispersion. As an example the dispersion $D = d\tau_g/d\lambda_0$ for the HE_{11} mode of the fiber shown in Figure 5.12 has been plotted in Figure 5.13 as a function of v. Usually, the dispersion is expressed in ps $km^{-1} nm^{-1}$, which means that it is the difference in specific group delay (in ps km^{-1}) per nm variation in the wavelength of the light. From this figure, it is clear that the dispersion becomes zero when $v \approx 3$; the dispersion is independent of the value of the parameter p. For that value of v, the corresponding fibers are multimode. As will become clear, if the material dispersion is taken into account, then the value of v is such that the fiber is monomode if the HE_{11} mode has no dispersion.

5.5 Waveguide and material dispersion in monomode fibers

As stated in Chapter 4, it is impossible to separate the waveguide and material dispersion of a fiber mathematically. If the core radius a of a step index fiber is known, then the v value can be calculated from the equation

$$v = \frac{2\pi a}{\lambda_0} \sqrt{n_1^2(\lambda_0) - n_2^2(\lambda_0)} \tag{5.32}$$

If the dependence of the refractive indices of the core and cladding on the wavelength are known, as described in Chapter 3, the characteristic equation can be solved and the specific phase constant obtained. For example, this has been done for fibers with core radii of 2, 3 and 4 μm, respectively. The core

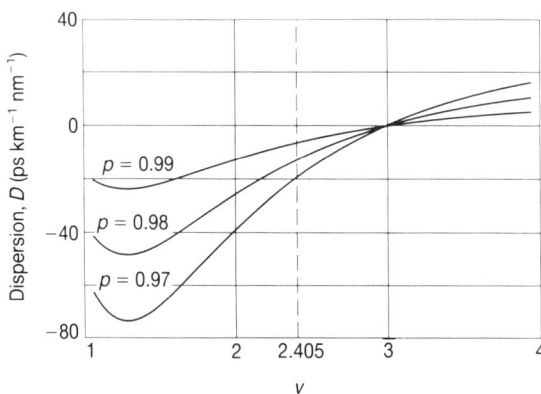

Figure 5.13 The dispersion $D = d\tau_g/d\lambda_0$ as a function of v for the HE_{11} mode of a step index fiber with $n_2 = 1.44$ and different values of p.

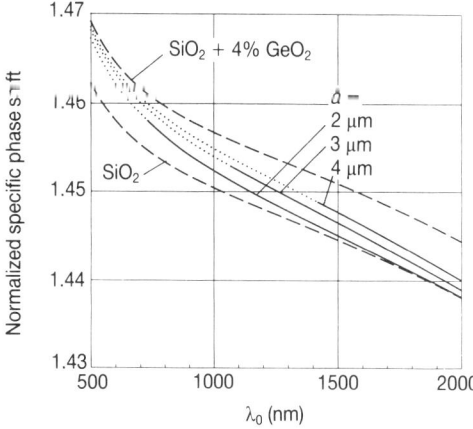

Figure 5.14 The refractive indices n_1 and n_2 for the core and the cladding of a step index fiber as functions of the wavelength λ_0 (dashed lines), together with the normalized specific phase constants of three fibers with core radii of 2, 3 and 4 μm, respectively (dotted curves). These latter three curves have been drawn with a solid line in the region where the fiber is monomodal.

material is SiO_2 doped with 4 percent GeO_2 and the cladding material is SiO_2. In Figure 5.14 the dashed lines represent the graphs for $n_1(\lambda_0)$ and $n_2(\lambda_0)$.

As was described in Chapter 4, the curve for a normalized phase constant β/k_0 must be located between the two curves of the refractive indices. The curves for the normalized specific phase constants for the three fibers are shown in Figure 5.14. For short wavelengths, the curves approach the refractive index curve of the core, and for long wavelengths, they approach the refractive index curve of the cladding. The parts of the phase curves that have been drawn with a solid line correspond to monomode behaviour and the other parts with multimode behaviour. In Figure 5.15 the group indices $N_1(\lambda_0)$ and $N_2(\lambda_0)$ are shown as dashed lines and the normalized specific group delay for the three fibers of different core radii are shown as solid lines in the monomode region and dotted lines in the multimode region.

For short wavelengths, the electromagnetic field is concentrated mainly in the core, and the wave tends to behave like a wave in bulk core material; therefore, the curves that describe the normalized specific group delay approach the curve that represents the group index of the core material with decreasing wavelength. On the other hand, for relatively long wavelengths, the curves for the specific group delay approach the curve that represents the group index of the cladding material.

The minima of the specific group delay curves of the fibers occur at longer wavelengths than the minima of the group index curves.

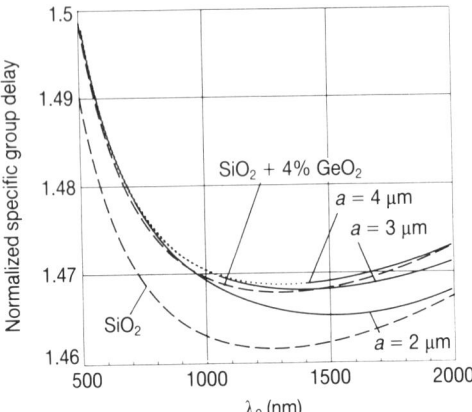

Figure 5.15 The group indices N_1 and N_2 of the core and the cladding material respectively of a step index fiber shown as functions of the wavelength λ_0 (dashed curves), together with the normalized specific group delays of three fibers with core radii of 2, 3 and 4 μm, respectively (dotted curves). These latter three curves have been drawn with a solid line in the region where the fiber is monomodal.

5.6 Dispersion-shifted and dispersion-flattened fibers

From Figure 5.15 it is clear that the wavelength that corresponds to the minimum delay increases when the core radius decreases. The smaller the core, the longer the wavelength of the corresponding minimum group delay.

In Figure 5.16 the dispersion in ps km^{-1} nm^{-1} for the three fibers with different core radii is depicted, together with the material dispersion of the core and cladding material. From these curves, it is clear that the wavelength of zero dispersion moves to higher values of λ_0 if the core radius is reduced. It is therefore possible to choose the core radius so that the wavelength of zero dispersion coincides with the wavelength of minimum attenuation. Fibers showing this property are called dispersion-shifted fibers.

Another way to shift the wavelength of zero dispersion is to increase the difference between the refractive indices of the core and cladding. This is illustrated in Figure 5.17, where the doping of the core material has been increased from 4 to 7 percent GeO$_2$. A great difference between the refractive indices of the core and cladding produces high mechanical tension in the fiber that results in increased attenuation. This effect can be prevented by using different profiles for the step profile [5]. By choosing the proper materials with their specific material dispersions and the proper refractive index profile, a fiber can be realized that has minimum dispersion in the 1300 nm region as well as in the 1550 nm region. Such a fiber is called a dispersion-flattened fiber.

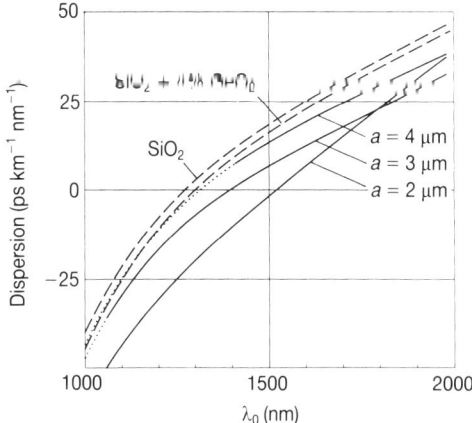

Figure 5.16 Dispersion in the core ($SiO_2 + 4$ percent GeO_2) and cladding (SiO_2) material of a step index fiber (dashed lines) and dispersion of the HE_{11} mode for three fibers of differing core radii.

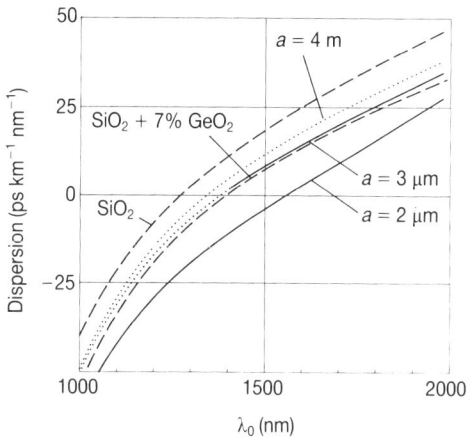

Figure 5.17 Material dispersion for the core ($SiO_2 + 7$ percent GeO_2) and cladding (SiO_2) of a step index fiber (dashed lines) and the dispersion of the HE_{11} mode for three fibers with different core radii.

The modulation bandwidth of a transmission channel, described in Chapter 4, that can be associated with a mode in a fiber depends, in the first place, on the dispersion. Equation (4.41) gives a first approximation of the modulation bandwidth using a monochromatic source; this equation is repeated here, for convenience

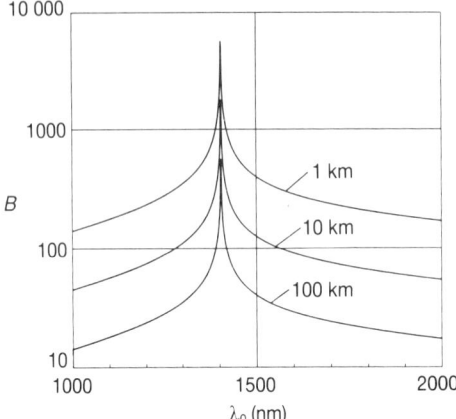

Figure 5.18 The modulation bandwidth of a step index fiber with a core of SiO_2 doped with 4 percent GeO_2 and a cladding of SiO_2. Core radius = 3 μm.

$$B = \sqrt{\frac{2\pi}{3\beta_0'' z}} \tag{5.33}$$

If this equation is applied to a fiber with a core radius of 3 μm, the dispersion of which is shown in Figure 5.16, then the resulting bandwidth is depicted in Figure 5.18 as a function of the wavelength of the carrier, with the length of the fiber as a parameter.

References

[1] D. Marcuse, 'Loss analysis of single-mode fiber splices', *Bell System Technical Journal*, vol. 56, no. 5, May–June 1977, pp. 703–18.

[2] Revised version of Recommendation G652, 'Characteristics of a single-mode optical fibre cable', *CCITT document COM.XV/TD 46-E*, May 1984.

[3] K. Petermann, 'Constraints for fundamental-mode spotsize for broadband dispersion-compensated single-mode fibres', *Electronics Letters*, vol. 19, no. 18, September 1983, pp. 712–14.

[4] G. Coppa *et al.*, 'Near-field measurements in monomode fibres: determination of chromatic dispersion', *Electronics Letters*, vol. 19, no. 18, 1 September 1983, pp. 731–3.

[5] Luc B. Jeunhomme, *Single-mode Fiber optics, Principles and Applications* (New York and Basel: Marcel Dekker, 1983).

[6] E.-G. Neumann, *Single-mode Fibers* (Berlin: Springer Verlag, 1988).

Problems

5.1 Assume a semiconductor laser and a step index fiber. The wavelength in vacuum of the monochromatic laser light is 830 nm. The length of the fiber is 20 km. In the wavelength region considered the refractive indices n_1 and n_2 of the core and cladding material, respectively, can be approximated by the formula $n_1 = 1.46 + 4000\lambda_0^{-2}$ and $n_2 = 1.45 + 4000\lambda_0^{-2}$, where λ_0 is the wavelength of the light in vacuum, expressed in nm.

- Find the maximum core diameter to make the fiber a single-mode fiber if the given laser diode is used as light source.
- Find the maximum core diameter for the fiber such that it guides both the LP_{01} and the LP_{11} modes.

For the single-mode fiber it can be assumed that the group index for the HE_{11} mode equals the group index of the bulk material in the wavelength region concerned. The laser is modulated directly by a random binary signal with a bit rate of 100 Mbit s^{-1}.

- Show that the pulse broadening that a single pulse out of the random signal undergoes is small compared with the width of the pulse itself and no serious intersymbol interference results.
- Is it possible to use in this system a multimode laser with spectral lines that are mutually separated by 4 nm in wavelength?

5.2 The light from a monochromatic source travels through a step index fiber in the HE_{11} mode. The v-value is 2.4. At the end of the fiber the core is gradually tapered to one-half of its original diameter. Sketch how the power is divided in the fiber and in the taper.

5.3 The core radius of a step index fiber $a = 2.5\ \mu m$. At a wavelength $\lambda_0 = 1300$ nm the refractive index of the core is $n_1 = 1.45$ and of the cladding $n_2 = 1.44$. Find the waist w_0 of a Gaussian beam for maximum coupling efficiency between this beam and the HE_{11} mode at the wavelength mentioned.

5.4 Sketch the magnetic field line pattern in the plane $\pi/2$ analogously to the electric field line pattern of Figure 5.4.

5.5 Sketch the magnetic field line pattern for Figure 5.5, or write a computer program to calculate the magnetic field lines and make a plot of the magnetic field line pattern.

6

Propagation of light rays in multimode graded index fibers

6.1 The eikonal equation

6.2 Solutions of the eikonal equation for a cylindrical symmetrical fiber and the resulting ray equations

6.3 Some analytical solutions

6.4 Numerical solutions

6.5 Ray congruencies, the h,g-coordinate system

6.6 The local numerical aperture

6.7 The relationship between ray congruences and modes

6.8 The WKB method

References

Problems

In the previous chapters, the step index fiber was described as a dielectric waveguide and it was analyzed with the help of Maxwell's equations. This wave-optical approach to the problem was used to obtain rigorous solutions that describe both the oscillating fields in the core and the evanescent fields in the cladding. The quantum character of the phenomena shows up very clearly. However, this rigorous method is less useful for describing the propagating of light through a multimode graded index fiber, because the solutions become very complicated and provide relatively little physical insight. Moreover, numerical solutions of certain problems, with their inherent inaccuracies, can reduce the precision of the model. Therefore, in this chapter, the problem of light propagation will be approached from the viewpoint of geometric optics and the phenomena will be explained with the help of light rays. This method can be used with success if the dimensions of the cross-section of the fiber are large compared with the wavelength; but its major advantage is simplicity. However, one disadvantage is that certain phenomena, such as losses in the cladding and tunnelling of light due to leaky rays, cannot be explained with geometric optics.

6.1 The eikonal equation

The reduced wave equation that was derived by substituting $\sigma = 0$ in equation (2.5) and assuming harmonic waves, gives

$$\nabla^2 \mathbf{E} + \omega^2 \mu_0 \varepsilon \mathbf{E} = 0 \tag{6.1}$$

Strictly speaking, such a reduced wave equation holds only if ε is constant and not when $\nabla \varepsilon \neq 0$. However, in a first approximation, equation (6.1) holds if ε changes over distances in the order of a wavelength are relatively small. The wave equation holds for every component V of $\mathbf{E}$ and $\mathbf{H}$, i.e. each component satisfies the scalar reduced wave equation

$$\nabla^2 V + \omega^2 \mu_0 \varepsilon V = 0 \tag{6.2}$$

Substituting the complex scalar

$$V(x,\ y,\ z,\ t) = V(x,\ y,\ z) \exp(j\omega t) \tag{6.3}$$

in equation (6.2) produces the following reduced wave equation:

$$\nabla^2 V(x,\ y,\ z) + k^2 V(x,\ y,\ z) = 0 \tag{6.4}$$

where

$$k = \omega \sqrt{\varepsilon \mu_0} \tag{6.5}$$

The complex $V(x,\ y,\ z)$ is written as

$$V(x,\ y,\ z) = V_0(x,\ y,\ z) \exp[-jk_0 S(x,\ y,\ z)] \tag{6.6}$$

where

$$k_0 = \omega \sqrt{\varepsilon_0 \mu_0} = \frac{k}{n} \tag{6.7}$$

$V_0(x,\ y,\ z)$ is the amplitude and $-k_0 S(x,\ y,\ z)$ is the phase at time $t = 0$ of the wave phenomenon as a function of its position in space. Equation (6.6) gives a snapshot view of the field at $t = 0$. This is drawn diagrammatically in Figure 6.1. The lines represent the surfaces of constant $S(x,\ y,\ z)$. Thus, on these surfaces, the phase $-k_0 S(x,\ y,\ z)$ is also constant. Therefore, these surfaces are called equiphase surfaces or wavefronts. Proceeding from the surface where $S = 0$ to

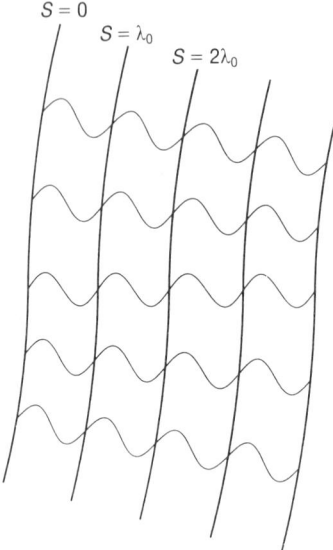

Figure 6.1 Surfaces of constant $S(x, y, z)$; also called equiphase surfaces or wavefronts.

the surface where $S = \lambda_0$, the phase decreases by 2π. Hence, each wavefront depicted in Figure 6.1 is separated by a wavelength. Substituting equation (6.6) into equation (6.4) and expanding ∇^2 leads to the following equation:

$$ n^2 - \nabla S.\nabla S = \mathrm{j}\frac{2\nabla S\nabla V_0 + V_0\nabla^2 S}{k_0 V_0} - \frac{\nabla^2 V_0}{k_0^2 V_0} \tag{6.8} $$

where $k/k_0 = n$, the refractive index.

If the relative changes in the amplitude V_0 and refractive index n over a wavelength are minute, then the right-hand side of equation (6.8) can be neglected compared to n^2 and the so-called eikonal equation results

$$ \nabla S.\nabla S = n^2 \tag{6.9} $$

This equation is the starting point for ray tracing in graded index optical waveguides. By definition, the light rays intersect the wavefronts perpendicularly; thus, the direction of the light ray at the point of intersection must coincide with the direction of the vector

$$ \nabla S = n\mathbf{t} \tag{6.10} $$

where $\mathbf{t}$ is the unit vector perpendicular to the wavefront at the point of intersection.

The line integral of the refractive index along a curve is called the optical length. The optical length measured along a light ray from a point P_1 on a wavefront where $S = S_1$ to a point P_2 on a wavefront where $S = S_2$ equals

$$S_2 - S_1 = \int_{P_1}^{P_2} n \ ds \tag{6.11}$$

6.2 Solutions of the eikonal equation for a cylindrical symmetrical fiber and the resulting ray equations

The refractive index $n(r)$ is independent of z and φ and depends on the distance r to the axis of the fiber. Solutions to equation (6.9) for this special case, when expressed in circular cylindrical coordinates, are of the form

$$S(r, \varphi, z) = \int p(r) \ dr + ha\varphi + gz \tag{6.12}$$

where

$$p(r) = \pm \sqrt{n^2(r) - g^2 - h^2 \ (a/r)^2} \tag{6.13}$$

and a is the radius of the core.

The symbols g and h are dimensionless constants which will be explained later. It can be verified that equation (6.12) is a solution of the eikonal equation (6.9) by substituting equation (6.12) into equation (6.9). In circular cylindrical coordinates, the gradient of S becomes

$$\nabla S = \mathbf{a}_r \frac{\partial S}{\partial r} + \mathbf{a}_\varphi \frac{1}{r} \frac{\partial S}{\partial \varphi} + \mathbf{a}_z \frac{\partial S}{\partial z} \tag{6.14}$$

Substituting equation (6.12) into equation (6.14) gives

$$\nabla S = \mathbf{a}_r[p(r)] + \mathbf{a}_\varphi h \frac{a}{r} + \mathbf{a}_z g \tag{6.15}$$

and

$$\nabla S . \nabla S = p^2(r) + h^2 \left(\frac{a}{r}\right)^2 + g^2 \tag{6.16}$$

And with equation (6.13) the eikonal equation follows.

Within certain limits, the constants h and g can vary continually, but a set of values for them, together with the refractive index profile $n(r)$ of the fiber, determines the corresponding set of wavefronts described by equations (6.12) and (6.13), and also the path of the light rays that intersect these wavefronts perpendicularly. The set of light rays that correspond to a certain set of values for g and h is called a ray congruence. A light ray that belongs to the congruence determined by g and h and that passes through a point $P(r, \varphi, z)$, has at that point a direction that coincides with the direction of ∇S. The vector components of ∇S in equation (6.15) with $p(r) > 0$ are depicted in Figure 6.2. The angle between the light ray and the z-axis is denoted by $\theta(r)$; the angle between the projection of ∇S on the plane of z = constant and the unit vector $\mathbf{a}_r$ is denoted by $\psi(r)$.

The relationship between the constants g and h on one side and the angles θ and ψ on the other can be deduced from Figure 6.2:

$$g = n(r)\cos\theta(r) \tag{6.17}$$

$$h = n(r)\,\frac{r}{a}\sin\theta(r)\sin\psi(r) \tag{6.18}$$

Thus, if the direction of a light ray at a certain point is known from the angles θ and ψ, then the values of the constants g and h of the ray congruence concerned follow from equations (6.17) and (6.18).

If the vectors depicted in Figure 6.2 are multiplied by the factor $k_0 = \omega\sqrt{\varepsilon_0\mu_0}$, then the vector $n(r)\mathbf{t}$ in the direction of the light ray changes to the wave vector

$$\mathbf{k} = k_0 n(r)\mathbf{t} \tag{6.19}$$

This wave vector $\mathbf{k}$ has a z-component that depends on the value of g in the corresponding ray congruence

$$\mathbf{k}_z = k_0 g \mathbf{a}_z = \beta \mathbf{a}_z \tag{6.20}$$

and the components

$$\mathbf{k}_\varphi = k_0 h\,\frac{a}{r}\,\mathbf{a}_\varphi \tag{6.21}$$

and

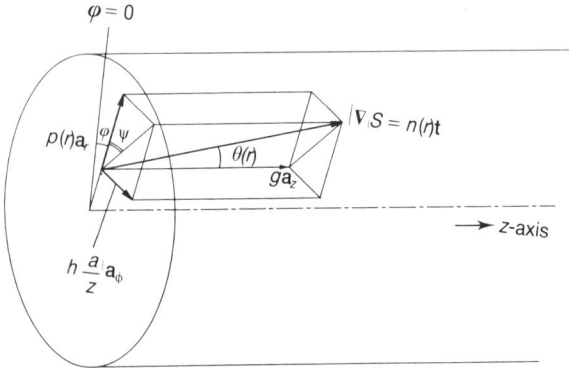

Figure 6.2 The vector ∇S and the component vectors.

$$\mathbf{k}_r = k_0 p(r)\mathbf{a}_r \tag{6.22}$$

In general, the path of a light ray can be described by a parametric expression of the coordinates of a point on the path. Often, the distance s measured along the light ray is chosen as the parameter. Here, we have chosen the z-coordinate as the parameter and the path of a light ray can be described by the following coordinates

$$r = r(z) \tag{6.23}$$

$$\varphi = \varphi(z) \tag{6.24}$$

Assuming that the light ray covers a distance ds in the direction $\mathbf{t}$, then the components of ds.$\mathbf{t}$ will compare with their respective components of ∇S, so that

$$\mathrm{d}r : r\mathrm{d}\varphi : \mathrm{d}z \doteq p(r) : h\,\frac{a}{r} : g \tag{6.25}$$

From equation (6.25), the differential equations for the path of a light ray are

$$\frac{\mathrm{d}r}{\mathrm{d}z} = \frac{p(r)}{g} \tag{6.26}$$

$$\frac{\mathrm{d}\varphi}{\mathrm{d}z} = \frac{ha}{r^2 g} \tag{6.27}$$

If a single point of a light ray is known – e.g. the intersection of a light ray with the face $z = 0$, which is $r(0)$ and $\varphi(0)$ – then the constants g and h determine the path of the ray. In general, there are two possible solutions taking into account the plus or minus sign in equation (6.13). All rays with the same g and h can be made to coincide by a move in the z-direction and a rotation around the z-axis. A set of light rays that has g and h in common is called a ray congruence.

6.3 Some analytical solutions

In general, the differential equations given in equations (6.26) and (6.27) to describe the path of a light ray cannot be solved analytically; however, in some special cases, an analytical solution is possible.

6.3.1 Meridional rays in a fiber with a parabolic refractive index profile

Suppose that the refractive index, as a function of the distance from the axis of the fiber, is given by

$$n(r) = n(0)\sqrt{1 - 2\Delta(r/a)^2}, \qquad 0 \le r \le a \tag{6.28a}$$

$$n(r) = n(0)\sqrt{1 - 2\Delta}, \qquad r \ge a \tag{6.28b}$$

$$\Delta = \frac{n^2(0) - n^2(a)}{2n^2(0)} \approx \frac{n(0) - n(a)}{n(0)} \tag{6.28c}$$

where Δ is the relative refractive index difference between the center of the core and the cladding, so that substituting equation (6.28a) into equation (6.13), and equation (6.13) into equation (6.12) gives

$$S = \pm \int \sqrt{n^2(0)[1 - 2\Delta(r/a)^2)] - g^2 - h^2(a/r)^2} \ \mathrm{d}r + ha\varphi + gz,$$

$$r \leq a \quad (6.29)$$

If a restriction is made to meridional rays, which intersect the axis of the fiber and are characterized by $h = 0$, then integrating equation (6.29) with the plus sign leads to

$$S_1(r, z) = \frac{gbc^2}{2}[(r/c)\sqrt{1 - (r/c)^2} + \arcsin(r/c)] + gz + C_1, r \leq a \tag{6.30a}$$

$$S_2(r, z) = -\frac{gbc^2}{2}[(r/c)\sqrt{1 - (r/c)^2} + \arcsin(r/c)] + gz + C_2,$$

$$r \leq a \quad (6.30b)$$

where

$$c = \frac{a}{n(0)\sqrt{2\Delta}}\sqrt{n^2(0) - g^2} = a\sqrt{\frac{n^2(0) - g^2}{n^2(0) - n^2(a)}} \tag{6.31}$$

and

$$b = \frac{n(0)}{ag}\sqrt{2\Delta} \tag{6.32}$$

C_1 and C_2 are integration constants.

Real values for the wavefronts $S_1(r, z)$ and $S_2(r, z)$ can be found if $r/c < 1$. To obtain the path equation for a meridional ray requires the solution of the differential equation that results from equation (6.13) and equation (6.28a) when they are substituted in equation (6.26) as follows:

$$\frac{\mathrm{d}r}{\mathrm{d}z} = \pm \frac{1}{g}\sqrt{n^2(0) - n^2(0).2\Delta.(r/a)^2 - g^2} \tag{6.33}$$

The solution being

$$r = c\sin(bz + \alpha), \qquad r \leq a \tag{6.34}$$

where c and b are the same as in equations (6.31) and (6.32) and α is arbitrary.

Figure 6.3. shows a number of intersections of the wavefronts with a plane where φ is constant (the plane of drawing) and a corresponding light ray. A

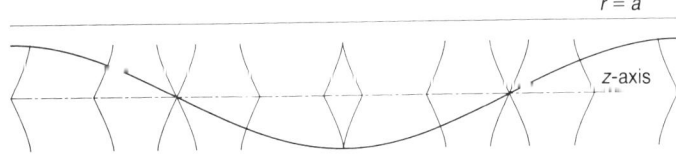

Figure 6.3 A meridional ray according to equation (6.34) and the intersections of a number of wavefronts according to equation (6.29) with a meridional plane (plane of drawing). $n(0) = 2$, $\Delta = 0.18$ and $g = 1.8$.

plane through the axis of the fiber is called a meridional plane and *a* light ray in that plane a meridional ray. For the sake of clarity, the ratio between the refractive index at the center of the core and of the cladding is much greater than would be the case for real fibers.

The solutions from equations (6.30) and (6.34) hold only for the core of a fiber. If $c < a$, then the amplitude of the sinusoidal path of the light ray is smaller than the core radius, which is the case when $g > n(a)$. The light ray stays in the core and is therefore guided. If $g < n(a)$, then $c > a$ and the path of the ray is sinusoidal in the core but, if it passes through the interface between core and cladding, where $r = a$, then it continues in a straight line because the refractive index of the cladding is constant. Such a ray is called a radiated ray. This is illustrated in Figure 6.4, where g^2 and $n^2(r)$ are given as functions of r.

The point of intersection between g^2 and $n^2(r)$ gives the amplitude of the ray path. The plus sign in equation (6.33) corresponds to those parts of the rays that proceed in the positive z-direction and away from the axis. The wavefronts that are given by equation (6.30a) are intersected perpendicularly by them. The minus sign in equation (6.33) corresponds to those parts of the rays that proceed in the positive z-direction and towards the axis. The wavefronts that are given by equation (6.30b) are intersected perpendicularly by them. Figure 6.5 shows in perspective a wavefront that belongs to rays that proceed in the positive z-direction and away from the axis.

, Each point on this surface of equal phase is an intersection of a light ray of the ray congruence that intersects the wavefront perpendicularly. Consider four neighbouring rays of a congruence that intersect the wavefront at the points (r, φ), $(r + dr, \varphi)$, $(r, \varphi + d\varphi)$ and $(r + dr, \varphi + d\varphi)$. These points are the angular points of an elementary rectangle on the surface of equal phase which has sides of $dr/\cos\theta$ and $r\,d\varphi$. All rays that intersect this elementary rectangle perpendicularly will comprise an elementary bundle of light rays, the cross-section of which varies along the path of the ray. This rectangle degenerates into a line when the bundle reaches its greatest distance from the axis (where $r = c$), or if the bundle cuts the axis of the fiber (if $r = 0$): in these cases the bundle degenerates into a focal line.

Normally, the phase of a light ray or of an elementary bundle changes by an

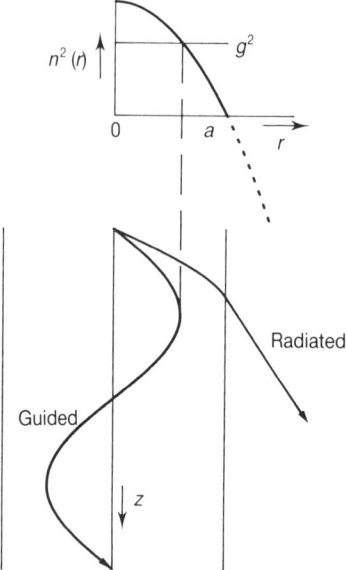

Figure 6.4 The paths of a guided meridional ray and a radiated meridional ray in graded index fiber with a parabolic refractive index profile such as that given by equation (6.28).

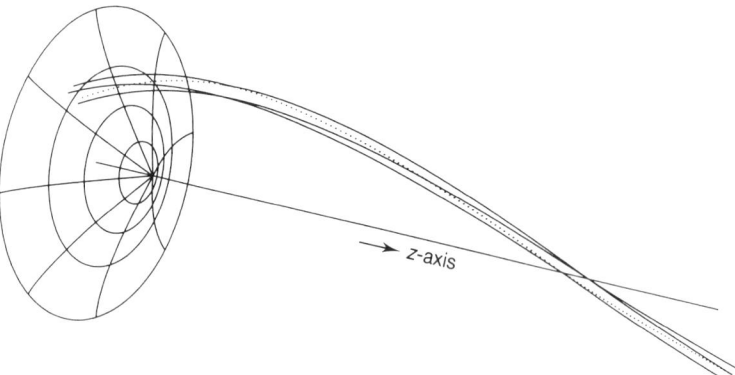

Figure 6.5 A wavefront and four neighbouring light rays that intersect it perpendicularly. $n(0) = 2$, $\Delta = 0.18$ and $g = 1.7$.

amount of $k\Delta z$ if it traverses a distance Δz; from wave theory it can be deduced that there is an extra phase change of $-\pi/2$ if the bundle passes through a focal line. Hence, if the elementary bundle passes a focal line, the phase changes by an amount equal to $(k\Delta z - \pi/2)$. The infinite number of focal lines that result from the infinite number of elementary bundles of a congruence form a surface that is called a caustic surface. A caustic line is formed in the special case of intersection with the axis. If there is no attenuation in a stationary situation, then the power in an elementary bundle remains unchanged and the intensity in the bundle is inversely proportional to the area of the cross-section. According to the model used in geometric optics, the intensity increases infinitely if the bundle passes through a focal line, and this indicates that the model is imperfect.

6.3.2 Ray congruencies and modes

Two meridional rays belonging to the same congruence and lying in the same meridional plane and separated in the z-direction intersect each other periodically; one ray moves away from the axis and the other approaches the axis. These two rays interfere with each other so that the whole bundle of rays in a congruence moving away from the axis interferes with the bundle of rays approaching the axis. In planes perpendicular to the axis, standing wave patterns will appear if g and h fulfil certain constraints so that a ray congruence forms a mode. The constraints that have to be fulfilled can be studied when we concentrate on the three wavefronts shown in Figure 6.6.

Assume that the first wavefront, $S_1(r, z) = \text{constant} = S_1(0, z_1)$ and it intersects the axis at $z = z_1$. From equation (6.30a) it can be deduced that

$$S_1(0, z_1) = gz_1 + C_1 \tag{6.35}$$

The relationship between r and z on this wavefront can be obtained from equation (6.30a) and is given by

$$z = z_1 - \frac{bc^2}{2}[(r/c)\sqrt{1 - (r/c)^2} + \arcsin(r/c)] \tag{6.36}$$

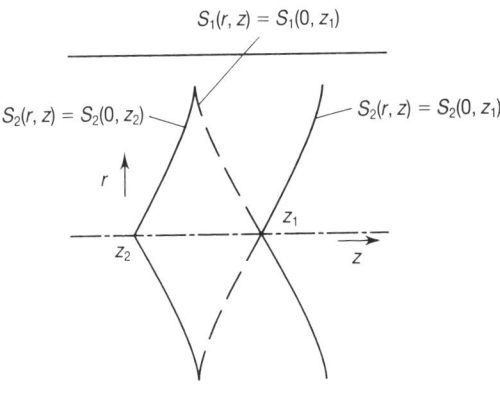

Figure 6.6 Wavefronts used for the derivation of equation (6.45).

The second wavefront, $S_2(r, z) = S_2(0, z_1)$, also cuts the axis at $z = z_1$. From equation (6.30b) it can be deduced that

$$S_2(0, z_1) = gz_1 + C_2 \tag{6.37}$$

The relationship between r and z on this wavefront can be obtained from equation (6.30b) and is given by

$$z = z_1 + \frac{bc^2}{2} [(r/c)\sqrt{1 - (r/c)^2} + \arcsin{(r/c)}] \tag{6.38}$$

The third wavefront, $S_2(r, z) = S_2(0, z_2)$, cuts the axis at $z = z_2$. From equation (6.30b) it can be deduced that

$$S_2(0, z_2) = gz_2 + C_2. \tag{6.39}$$

The relationship between r and z on this wavefront can be obtained from equation (6.30b) and is given by

$$z = z_2 + \frac{bc^2}{2} [(r/c)\sqrt{1 - (r/c)^2} + \arcsin{(r/c)}] \tag{6.40}$$

Assume that the first wavefront touches the third wavefront on the line $r = c$; then the value of z from equation (6.36) with $r = c$ must be equal to the value of z from equation (6.40) with $r = c$. This leads to

$$z_2 = z_1 - \frac{\pi bc^2}{2} \tag{6.41}$$

As there is a $-\pi/2$ phase change when a ray crosses the axis, as well as when it turns at the caustic surface, there must be a phase difference of $-\pi - m2\pi$ between the first and the third wavefronts. Thus

$$S_2(0, z_1) - S_2(0, z_2) = \frac{\lambda_0}{2} + m\lambda_0, \qquad m = 0, 1, 2, \ldots \tag{6.42}$$

Substitute equations (6.37) and (6.39) into equation (6.42) in order to obtain

$$gz_1 + C_2 - gz_2 - C_2 = \frac{\lambda_0}{2} + m\lambda_0 \tag{6.43}$$

With equation (6.41) this leads to

$$g\pi bc^2 = \lambda_0(2m + 1) \tag{6.44}$$

Finally, after substituting the expressions for b and c given by equations (6.31) and (6.32) into equation (6.44), the constraint for g to describe a mode is obtained

$$g^2 = n^2(0) - (2m + 1) \frac{\lambda_0}{a} \frac{n(0)\sqrt{2\Delta}}{\pi} \tag{6.45}$$

Constructive interference occurs and a corresponding wave pattern is found only for the discrete values of g that are obtained from equation (6.45). In

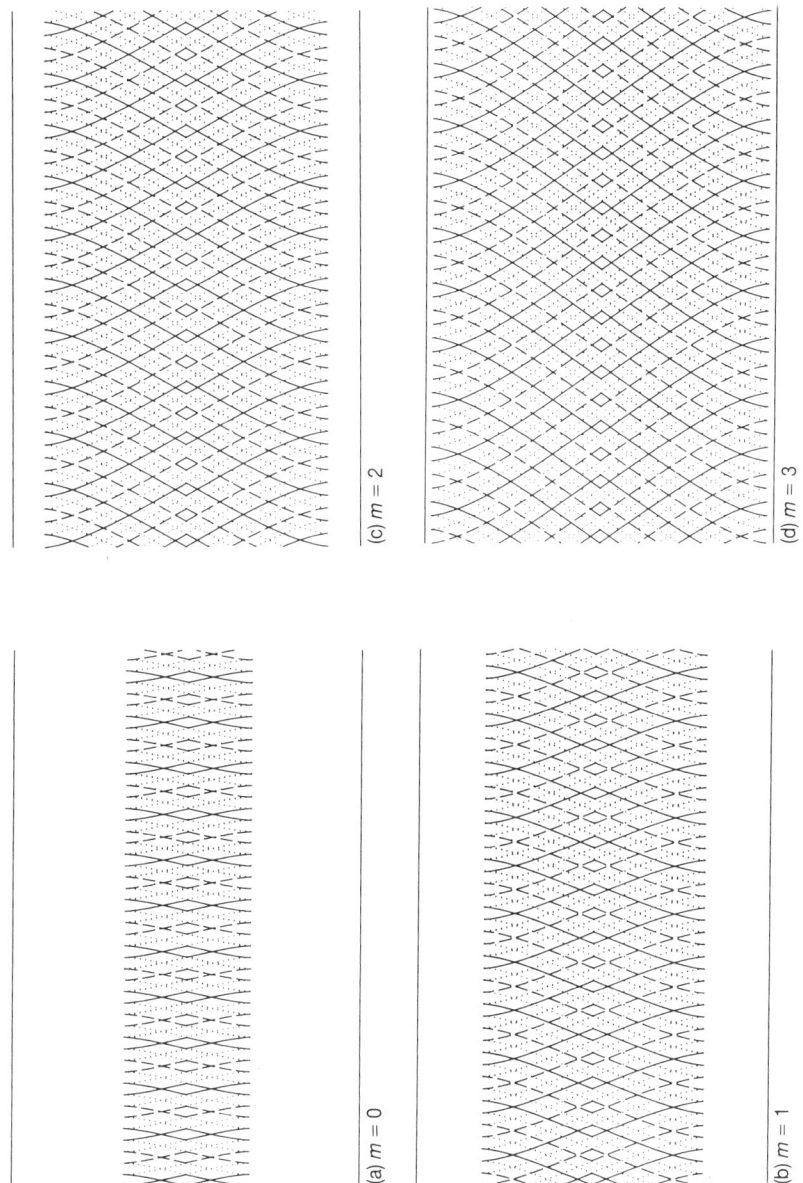

(a) m = 0

(b) m = 1

(c) m = 2

(d) m = 3

Figure 6.7 The wave patterns for the modes of meridional rays in a fiber with a parabolic refractive index profile where $n(0) = 2$ and $\Delta = 0.18$. The wavelength is $\lambda_0 = a/2$ and only the modes with $m = 0$, 1, 2 and 3 exist. The horizontal lines represent interfaces between the core and the cladding. In the radial direction, there are m nodes and $m + 1$ ant-nodes in the wave pattern.

Figure 6.7 a number of wave patterns have been drawn for a fiber with the same exaggerated parameters as in the previous examples; in this case, only four modes with meridional rays exist. If λ_0/a approaches zero, then the mutual differences between the values of g that correspond to a mode also approach zero and the phenomenon loses its discrete character.

6.3.3 The period Z of a meridional ray

The period Z of the path of a meridional ray in a fiber with a parabolic refractive index profile is obtained from the radian frequency of the path as given by equation (6.32). A period corresponds to an angle of $bZ = 2\pi$. Hence

$$Z = \frac{2\pi a g}{n(0)\sqrt{2\Delta}} = \frac{2\pi a}{\sqrt{2\Delta}} \cos\theta(0) \tag{6.46}$$

Since this period depends on the value of g, and thus also on the value of the angle $\theta(0)$ of the ray when it intersects the axis, the fiber is not self-focusing. In Figure 6.8 some rays have been drawn for different values of c and $\theta(0)$ with the same fiber parameters as in the previous examples.

For guided rays the following relationship holds

$$n(0)\sqrt{(1 - 2\Delta)} < g \leq n(0) \tag{6.47}$$

Hence, the maximum value of the period is

$$Z_{max} = \frac{2\pi a}{\sqrt{2\Delta}} \tag{6.48}$$

and the minimum value of the period is

$$Z_{min} = \frac{2\pi a}{\sqrt{2\Delta}} \sqrt{1 - 2\Delta} \tag{6.49}$$

The maximum value of the angle $\theta(0)$ is

$$\theta_{max}(0) = \arcsin\left(\sqrt{2\Delta}\right) \tag{6.50}$$

If realistic fiber parameters are chosen, e.g. $n(0) = 1.5$ and $2\Delta = 0.01$, then we obtain

$$Z_{max} = 20\pi a, \text{ with } \theta(0) = 0 \text{ and } Z_{min} = 19.9\pi a, \text{ with } \theta(0) = 0.1$$

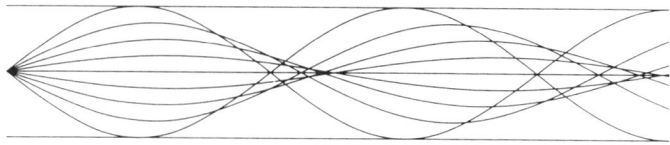

Figure 6.8 Meridional rays in the core of a fiber with a parabolic refractive index profile. $n(0) = 2$, $\Delta = 0.18$.

6.3.4 Meridional rays in a self-focusing fiber

Suppose the refractive index profile described by

$$n(r) = n(0) \operatorname{sech} [(r/a)\sqrt{2\Delta}], \qquad 0 \le r \le a \qquad (6.51\text{a})$$

$$n(r) = n(0) \operatorname{sech} [\sqrt{2\Delta}], \qquad r \ge a \qquad (6.51\text{b})$$

We can deduce from equations (6.13), (6.26) and (6.51a) that

$$\frac{dr}{dr} = \pm \frac{1}{g} \sqrt{\frac{n^2(0)}{\cosh^2 [(r/a)\sqrt{2\Delta}]} - g^2} \qquad (6.52)$$

For a light ray that passes through the point $z = 0$, the solution of equation (6.52) is

$$\sinh [(r/a)\sqrt{2\Delta}] = \frac{1}{g} \sqrt{n^2(0) - g^2} \sin [(z/a)\sqrt{2\Delta}], \qquad 0 \le r \le a$$

$$(6.53)$$

A pattern of rays in a self-focusing fiber is depicted in Figure 6.9.
If Δ is small, then equation (6.53) can be approximated by

$$r \approx \frac{a \sqrt{n^2(0) - g^2}}{g\sqrt{2\Delta}} \sin [(z/a)\sqrt{2\Delta}] \qquad (6.54)$$

The period of the oscillating ray path is deduced to be

$$Z = \frac{2\pi a}{\sqrt{2\Delta}} \qquad (6.55)$$

This period is independent of g, and thus also independent of $\theta(0)$. The fiber is therefore called a self-focusing fiber. A light source of the axis is imaged on the axis every half period (see Figure 6.9). If the light source is not situated on the axis, then only the meridional rays, and not the skew rays, focus periodically.

6.4 Numerical solutions

In general, equations (6.26) and (6.27) cannot be solved analytically; however, numerical methods can be used. Examples of numerical solutions are presented in Figure 6.11.

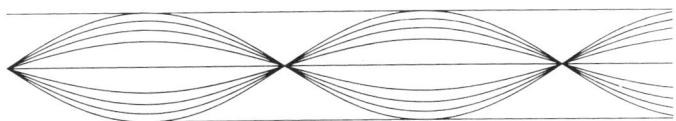

Figure 6.9 Meridional rays in the core of a self-focusing fiber. $n(0) = 2$, $\Delta = 0.18$.

6.5 Ray congruencies, the *h,g*-coordinate system

A ray congruence is determined by the values of the constants h and g and, in general, comprises two kinds of rays – real rays and complex rays.

Real rays
Real rays occur in the oscillating fields that are found in wave optics. For real rays, the constraint is: $p^2(r) > 0$ or

$$n^2(r) > g^2 + h^2(a/r)^2 \qquad (6.56)$$

Complex rays
Complex rays occur in the evanescent fields that are found in wave optics. For complex rays, the constraint is: $p^2(r) < 0$ or

$$n^2(r) < g^2 + h^2(a/r)^2 \qquad (6.57)$$

The areas where real rays can exist are separated from the areas of the complex rays by caustic surfaces. These are cylinders with a radius of r_c that is determined by the condition $p^2(r_c) = 0$. In Figure 6.11, the left-hand and right-hand sides of the inequalities in equations (6.56) and (6.57) are both given as functions of r. The intersections of these curves give the radii that correspond to the radii of the caustic surfaces.

The real rays of a congruence can consist of guided rays, leaky rays and radiated rays. A fiber has a cladding of finite thickness with a coating. Rays that repeatedly hit the interface between this cladding and coating soon die away and do not take part in the transmission. For convenience, we have assumed below that the cladding is infinitely thick and the refractive index profile is given by

$$n(r) = n(0)\sqrt{1 - 2\Delta(r/a)^x}, \qquad 0 \leq r \leq a \qquad (6.58a)$$

$$n(r) = n(0)\sqrt{1 - 2\Delta}, \qquad r \geq a \qquad (6.58b)$$

The results in Sections 6.5.1–6.5.3 apply without further expansion of the theory to a real fiber with a cladding of finite thickness and with an arbitrary refractive index profile. Before the different kinds of congruencies are described, the h,g-coordinate system will be introduced. This is a rectangular coordinate system in which the horizontal axis is the h-axis and the vertical axis is the g-axis. In such a system (see Figure 6.10) we can recognize those areas that correspond to the different kinds of ray congruencies.

There is no basic difference between rays that travel in a positive or a negative z-direction, nor is there a basic difference between rays that rotate to the left or to the right; therefore, we can limit ourselves to those positive values of g and h that correspond to rays in the positive z-direction that rotate to the right.

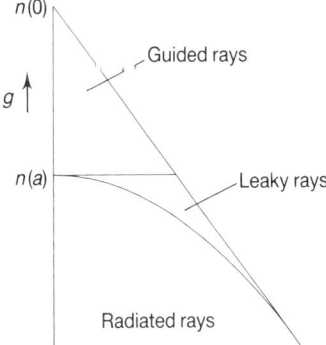

Figure 6.10 The h,g-coordinate system.

6.5.1 Guided rays

The essential characteristic of a congruence of guided rays is the presence of rays in the core and the absence of real rays in the cladding. In the core, there has to be a region where the condition expressed in equation (6.56) is fulfilled; while everywhere outside the core, the condition expressed in equation (6.57) has to be fulfilled. This means that the line $g^2 + h^2(a/r)^2$ must intersect the line $n^2(r)$ at two points in the area where $r < a$; also, no intersection in the area $r > a$ must occur (see Figure 6.11). The latter condition is fulfilled if $g > n(a)$, since no real rays are possible in the cladding. In the g,h-coordinate system, the boundary between areas of guided rays and non-guided rays is given by the line $g = n(a)$ (see Figure 6.10). A congruence of guided rays is characterized by caustic surfaces in the core between which the real rays travel. Some examples of guided rays for the refractive index profile as given in equation (6.58) are depicted in Figure 6.11.

The refractive index profiles in Figures 6.11(a)–(c) respectively correspond to $x = 1.5$, $x = 2$ and $x = 8$. If the condition $g > a$ is not fulfilled, then two kinds of real ray congruencies can exist, namely congruencies of leaky rays or radiated rays. The losses that occur in these two kinds of congruencies are different.

6.5.2 Leaky rays

Typically, a congruence of leaky rays comprises real rays outside a caustic surface in the cladding. As is the case with guided rays, there will be two caustic surfaces in the core between which real rays can travel. Power from the rays in the core tunnels through the region where only complex rays can exist to the region of real rays in the cladding. This effect can be described by the evanescent fields of wave optics. A condition for a caustic surface in the cladding is $p^2(a) < 0$. This condition is fulfilled if the line $g^2 + h^2(a/r)^2$

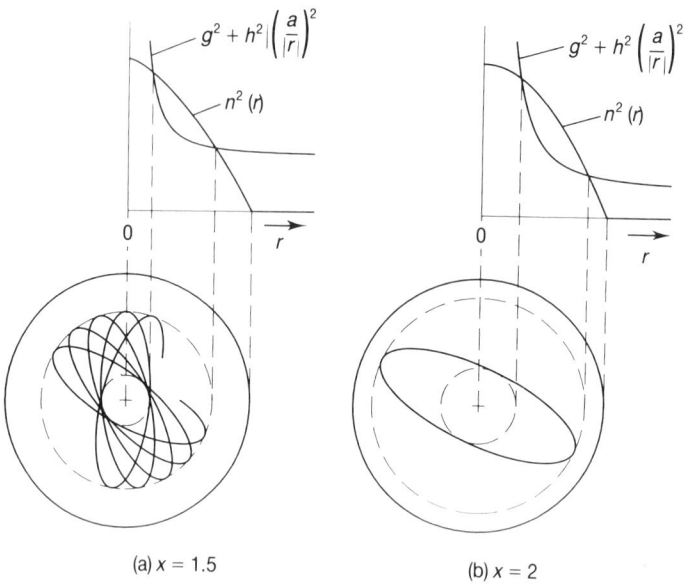

(a) $x = 1.5$ (b) $x = 2$

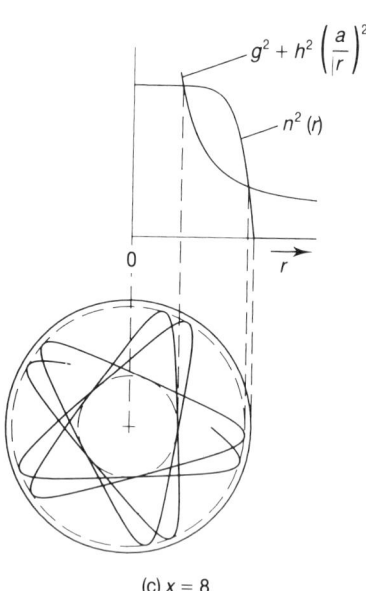

(c) $x = 8$

Figure 6.11 Guided rays in fibers with a refractive index profile as described by equation (6.58).

intersects the line $n^2(r)$ in the area where $r > a$. A boundary of the area of leaky rays in the g,h-coordinate system is obtained from the condition $p^2(a) = 0$. This implies that

$$n^2(a) - g^2 - h^2 = 0 \qquad (6.59)$$

This equation describes a circle with radius $n(a)$ in the g,h-coordinate system and is shown in Figure 6.10. An example of a leaky ray in a fiber with a parabolic refractive index profile has been drawn in Figure 6.12.

Every time a leaky ray touches the outer caustic surface of the core, some of the power tunnels to the region outside the caustic surface in the cladding. This amount of tunnelling power increases with decreasing distance between the caustic surfaces. If the distance is relatively large, then the attenuation can be very modest so that the leaky rays contribute substantially to the transmission. Congruencies of leaky rays in the g,h-coordinate system close to the region of guided rays have little attenuation and congruencies close to the region of radiated rays have great attenuation.

Calculations made with graded index fibers can be done in two different ways: either the leaky rays can be taken into account or they can be neglected – the choice depends on the nature of the problem. The leaky rays can play a role, especially in the case of short fibers.

6.5.3 Radiated rays

Congruencies of radiated rays are characterized by the condition $p^2(a) > 0$, which implies that

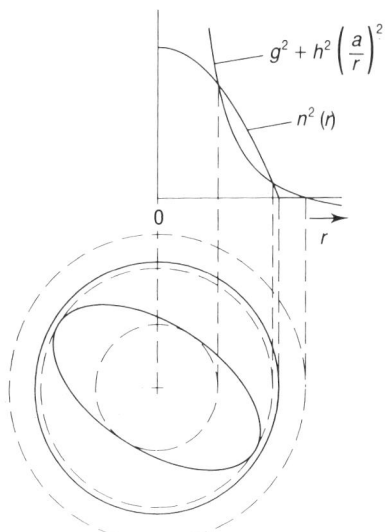

Figure 6.12 A leaky ray in a fiber with a parabolic refractive index profile.

$$g^2 + h^2 < n^2(a) \tag{6.60}$$

Hence, the area of the g,h-coordinate system that corresponds to the congruencies of leaky rays lies inside a circle described by equation (6.59).

6.5.4 Boundaries between real and complex rays

Those regions in the g,h-coordinate system that correspond to guided rays and leaky rays are bounded on the right-hand side of Figure 6.10. This boundary is determined by the point of contact on the line $g^2 + h^2(a/r)^2$ by the line $n^2(r)$ for $0 < r < a$. At that point, the two caustic surfaces in the core coincide. Making the derivatives with respect to r equal in these two functions leads to

$$\frac{dn^2(r)}{dr} = -2h^2 \left(\frac{a}{r}\right)^2 \frac{1}{r} \tag{6.61}$$

We can deduce from equation (6.61) that

$$h = \sqrt{-\frac{1}{2}\left(\frac{r}{a}\right)^3 \frac{dn^2(r)}{d(r/a)}} \tag{6.62}$$

At the point of contact, the following relationship holds:

$$g^2 = n^2(r) - h^2\left(\frac{a}{r}\right)^2 \tag{6.63}$$

Substituting equation (6.62) into equation (6.63) gives

$$g = \sqrt{n^2(r) + \frac{r}{2}\frac{dn^2(r)}{dr}} \tag{6.64}$$

The parametric equations (6.62) and (6.64) determine the boundary of the region in which real solutions can be obtained for the rays in the core.

For a fiber with a parabolic refractive index profile, this boundary is a parabola given by

$$h = -\frac{g^2}{2n(0)\sqrt{2\Delta}} + \frac{n(0)}{2\sqrt{2\Delta}} \tag{6.65}$$

6.6 The local numerical aperture

The local numerical aperture $A(r, \psi)$ is defined as

$$A(r, \psi) \triangleq n(r)\sin\theta_{max}(\psi) \tag{6.66}$$

If the leaky rays are not taken into account, then θ_{max} is the largest angle that guided rays can make with the z-axis; conversely when the leaky rays are taken into account, θ_{max} is the largest angle that they can make with the z-axis.

6.6.1 The numerical aperture with the leaky rays excluded

The largest value of θ corresponds to the minimum value of g. For guided rays, we have

$$g_{min} = n(a) = n(r) \cos \theta_{max} \tag{6.67}$$

In this case, θ_{max} is independent of ψ, so we obtain

$$A(r, \psi) = A(r) = \sqrt{n^2(r) - n^2(a)} \tag{6.68}$$

The term numerical aperture is often used and is denoted by 'NA' when referring to the central numerical aperture. Thus

$$NA = A(0) = \sqrt{n^2(0) - n^2(a)} \tag{6.69}$$

For fibers with a refractive index profile that is described by equation (6.58), the central numerical aperture is

$$NA = n(0)\sqrt{2\Delta} \tag{6.70}$$

6.6.2 The numerical aperture including the leaky rays

The minimal g, in this case, is found on the boundary between the regions of leaky rays and radiated rays, i.e. on the circle described by equation (6.59)

$$g^2 + h^2 = n^2(a) \tag{6.71}$$

Assuming that $\theta = \theta_{max}$ in equations (6.17) and (6.18), and inserting them into equation (6.71), we obtain

$$A(r, \psi) = \sqrt{\frac{n^2(r) - n^2(a)}{1 - (r/a)^2 \sin^2 \psi}} \tag{6.72}$$

In Figure 6.13, the numerical aperture is shown as a function of r, for a fiber with a parabolic refractive index profile, at different values of ψ.

6.7 The relationship between ray congruencies and modes

In Section 6.3.2, we saw that only certain discrete values of g for meridional rays – namely those derived from equation (6.45) – could lead to consistent wave patterns. In that special case, the inner caustic surface in the core degenerated into a single line. In general, both for guided and for leaky rays, there are two caustic surfaces in the core between which the rays travel. The radial wave vector $\mathbf{k}_r$, as given by equation (6.22), determines the phase in the radial direction. A consistent wave pattern requires the phase difference between the inner caustic surface at $r = r_1$ and the outer caustic surface at $r = r_2$ to be a whole number of phase differences π plus $\pi/2$ in connection with the phase change at a caustic surface. This leads to the following relationship:

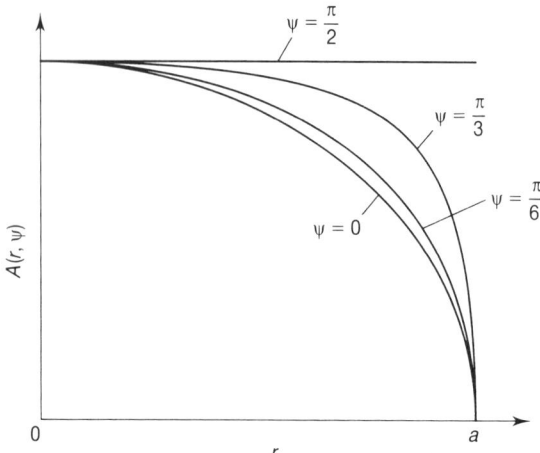

Figure 6.13 The local numerical aperture $A(r, \psi)$ for a fiber with a parabolic refractive index profile and a central numerical aperture NA. If only guided rays are taken into account, then $A(r, \psi)$ is independent of ψ and the curve equals the curve for $\psi = 0$. For both guided and leaky rays $A(r, \psi)$ has been drawn as a function of r with different values of the parameter ψ.

$$\int_{r_1}^{r_2} k_r \, dr = m\pi + \frac{\pi}{2}, \qquad m = 0, 1, 2, 3, \ldots \tag{6.73}$$

A consistent wave pattern in the azimuthal direction requires the following condition to be satisfied:

$$2\pi r k_\varphi = n2\pi, \qquad n = 0, 1, 2, 3, \ldots \tag{6.74}$$

With equation (6.21), this leads to

$$h = \frac{n}{k_0 a} \tag{6.75}$$

Substituting equations (6.13) and (6.75) into equation (6.22), and equation (6.22) into equation (6.73) gives

$$k_0 \int_{r_1}^{r_2} \sqrt{n^2(r) - g^2 - (n/k_0 r)^2} \, dr = m\pi + \frac{\pi}{2} \tag{6.76}$$

A mode is determined by the values of n and m, and the corresponding congruence by the values of g and h that are obtained from equations (6.75) and (6.76). The path of a ray is determined by the values of g and h. The wave pattern is not only dependent on n and m, but also on the wavelength λ_0; this is clear from equation (6.76).

For a fiber with a parabolic refractive index profile, equation (6.76) can be solved analytically in order to give

$$g^2 = n^2(0) - (2m + n + 1) \frac{\lambda_0}{a} \frac{NA}{\pi}$$ (6.77)

Note that equation (6.45) is a special case of equation (6.77) where $n = 0$.

Example 6.1
Assume that $n(0) = 1.5$, $NA = 0.2$, $x = 2$, $a = 25 \ \mu$m and $\lambda_0 = 840$ nm. In Figure 6.14 all the possible pairs of values (g, h) that follow from equations (6.75) and (6.77) have been plotted in a g,h-coordinate system. Every point (x) corresponds to a pair of g,h-values belonging to a mode.

6.8 The WKB method

The Wentzel–Kramers–Brillouin (WKB) method originates from quantum physics. Again, it is assumed that differences in the refractive index of the core are so small that the wave equation (6.1) can be used. Starting with this equation, we can follow the analysis given in Chapter 3 and find that

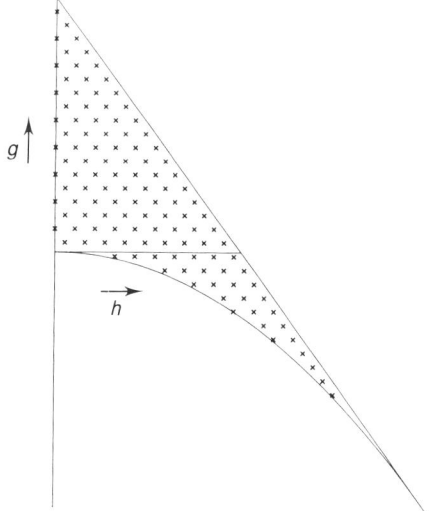

Figure 6.14 The g,h-coordinate system with the coordinates that belong to all the modes of guided and leaky rays. $n(0) = 1.5$, $NA = 0.2$, $x = 2$, $a = 25 \ \mu$m and $\lambda_0 = 840$ nm.

$$E_z = R(r)\,\phi(\varphi)\,Z(z) \tag{6.78}$$

where

$$Z(z) = C_1 \exp(-\gamma z) + C_2 \exp(\gamma z) \tag{6.79}$$

$$\phi(\varphi) = C_3 \cos n\varphi + C_4 \sin n\varphi \tag{6.80}$$

and

$$\frac{d^2 R}{dr^2} + \frac{1}{r}\frac{dR}{dr} + \left(\gamma^2 + k_0^2 \varepsilon_r - \frac{n^2}{r^2}\right) R = 0 \tag{6.81}$$

The only difference from the solutions found for the step index fiber is that ε_r in equation (6.81) depends on the distance r to the axis of the fiber.

Assume (compare with equation 6.6)

$$R(r) = R_0(r) \exp[-jk_0 S(r)] \tag{6.82}$$

Substituting equation (6.82) into equation (6.81) gives

$$\frac{dR_0(r)}{dr^2} - 2jk_0\frac{dR_0(r)}{dr}\frac{dS(r)}{dr} - jk_0 R_0(r)\frac{d^2 S(r)}{dr^2} - k_0^2 R_0(r)\left(\frac{dS(r)}{dr}\right)^2$$

$$+ \frac{1}{r}\frac{dR_0(r)}{dr} - jk_0 R_0(r)\frac{dS(r)}{r\,dr} + \left(\gamma^2 + k_0^2\varepsilon_r - \frac{n^2}{r^2}\right) R_0(r) = 0 \tag{6.83}$$

The imaginary part of equation (6.83) is

$$2\frac{dR_0(r)}{dr}\frac{dS(r)}{dr} + R_0(r)\frac{d^2 S(r)}{dr^2} - R_0(r)\frac{dS(r)}{r\,dr} = 0 \tag{6.84}$$

The solution of equation (6.84) is

$$\frac{dS(r)}{dr} = \frac{C}{rR_0^2(r)} \tag{6.85}$$

where C is an integration constant.

Substituting equation (6.85) into the real part of equation (6.83) leads to

$$\frac{d^2 R_0(r)}{dr^2} - k_0^2 R_0(r)\frac{C^2}{r^2 R_0^4(r)} + \frac{dR_0(r)}{r\,dr} + \left(\gamma^2 + k_0^2\varepsilon_r - \frac{n^2}{r^2}\right) R_0(r) = 0 \tag{6.86}$$

It is assumed that

$$\frac{d^2 R_0(r)}{dr^2} + \frac{dR_0(r)}{r\,dr} \ll \frac{k_0^2 C^2}{r^2 R_0^3(r)} \tag{6.87}$$

With this result and assuming that $\alpha = 0$, equation (6.83) becomes

$$k_0^2 \frac{C^2}{r^2 R_0^4(r)} = -\beta^2 + k_0^2\varepsilon_r - \frac{n^2}{r^2} \tag{6.88}$$

when combined with equation (6.85), this gives

$$\frac{C}{rR_0^2(r)} = \frac{dS(r)}{dr} = \pm \sqrt{\varepsilon_r - (\beta/k_0)^2 - (n/k_0 r)^2} \tag{6.89}$$

With equations (6.20) and (6.75) and $\varepsilon_r = n^2(r)$, equation (6.89) can be written as

$$\frac{C}{rR_0^2(r)} = \frac{dS(r)}{dr} = \pm \sqrt{n^2(r) - g^2 - h^2(a/r)^2} \tag{6.90}$$

We deduce from equation (6.90) that

$$S(r) = \pm \int \sqrt{n^2(r) - g^2 - h^2(a/r)^2} \, dr \tag{6.91}$$

and

$$R_0^2(r) = \frac{C}{r \sqrt{n^2(r) - g^2 - h^2(a/r)^2}} \tag{6.92}$$

With equations (6.91) and (6.92), the amplitude and phase of the r-dependent part $R(r)$ can be found; this is a first-order approximation that complies with the solution based on the eikonal equation. From equation (6.92), it is clear that $R_0(r)$ becomes infinite at the caustic surfaces or turning points of the rays. Note that the condition stated in equation (6.87) cannot hold in this instance.

References

[1] P. Di Vita, 'Theory of propagation in optical fibers: ray approach', *Annales des Télécommunications*, vol. 32, no. 3–4, 1977, pp. 115–34.
[2] H. G. Unger, *Planar Optical Waveguides and Fibres* (Oxford: Clarendon Press, 1977).
[3] D. Marcuse, *Principles of Optical Fiber Measurements* (New York: Academic Press, 1981).
[4] T. Okoshi, *Optical Fibers* (New York: Academic Press, 1982).
[5] D. Gloge and E. A. J. Marcatili, 'Multimode theory of graded-core fibers', *Bell System Technical Journal*, vol 52, November 1973, pp. 1563–78.

Problems

6.1 Substitute equation (6.6) into equation (6.4) and derive equation (6.8). Show that under the proper conditions equation (6.8) can be replaced by the eikonal equation. Give examples of situations where the eikonal equation will certainly not hold.
6.2 Solve the integral given in equation (6.29) and derive the two solutions given by equations (6.30a) and (6.30b).
6.3 Find the intensity of the light in an elementary bundle as depicted in Figure 6.5 as a function of the distance to the fiber axis.
6.4 Complete Figure 6.10 if rays that travel in the negative z-direction and that rotate to the left as well as to the right are also included.

6.5 The refractive index of a graded index glassfiber with a central numerical aperture $NA = 0.2$ is described by

$$n(r) = n(0)\left[1 - 2\Delta\left(\frac{r}{a}\right)^3\right]^{\frac{1}{2}} \qquad\qquad 0 \le r \le a$$

$$n(r) = n(a) \qquad\qquad r \ge a$$

The front end of the fiber is a plane perpendicular to the axis of the fiber. The axis cuts this plane at the point O. A light ray in the surrounding air hits this plane interface at the point P at a distance $a/2$ from the point O. This light ray is perpendicular to the line OP and makes an angle θ_0 with the normal in P to the end-face of the fiber. Find the conditions that θ_0 has to fulfil for the light ray to be: (a) a guided ray; (b) a leaky ray; (c) a radiated ray.

7

Dispersion in graded index fibers

7.1 Mode model
7.2 Ray model
     References
     Problems

In Section 6 of Chapter 1, a system model was presented that can be used to analyze the behaviour of a fiber transmission link. In such a model, each mode is associated with a transmission channel. The transmission channels are mutually power coupled, in general, as a result of mode conversion in which the power from one mode is transferred to another mode, because of fiber irregularities, bends or joints. For the sake of simplicity, it can be assumed that the transmission channels are not mutually coupled; therefore, the phase characteristic of an individual channel equals the phase characteristic of its corresponding mode. In the first part of this chapter, the model will be used to develop expressions for dispersion in graded index fibers; in particular, for fibers with a parabolic refractive index profile under different conditions. In the second part, the transmission channels of the model will be associated with ray congruencies; it will therefore be relatively easy to find expressions for the dispersion of a larger class of fiber refractive index profiles.

7.1 Mode model

7.1.1 Monochromatic sources and absence of material dispersion

For the purposes of this model, it is assumed that the refractive indices of the materials are independent of the frequency of the single harmonic signal. In Chapter 6 it was shown that a ray congruence is determined by the two constants g and h. In this context, constant means that g and h do not depend on where they are situated along the ray in the fiber. The specific phase shift in the z-direction of a wave in a graded index fiber is given by equations (6.17) and (6.20)

$$\beta = k_0 g = \frac{\omega}{c_0} g = k_0\, n(r) \cos \theta(r) \tag{7.1}$$

If it is assumed that the light propagates in a certain mode, with mode numbers n and m, then the constant g of the corresponding ray congruence is given implicitly by equation (6.76), which is repeated below for convenience

$$k_0 \int_{r_1}^{r_2} \sqrt{n^2(r) - g^2 - (n/k_0 r)^2}\ \, dr = m\pi + \frac{\pi}{2} \tag{7.2}$$

In particular, for the fiber with a parabolic refractive index profile, g is given explicitly by equation (6.77), which is repeated below in a slightly different form

$$g = \left[n^2(0) - \frac{2\mathrm{NA}c_0}{a\omega}(2m + n + 1)\right]^{1/2} \tag{7.3}$$

From this equation, it is clear that the constant g is a function of ω; this will henceforth be denoted as $g(\omega)$. Consequently, if the frequency changes and the light continues to propagate in the same mode, then the ray congruence in which the light propagates also changes. As an example, curves for $\beta/k_0 = g(\omega)$ have been plotted in Figure 7.1 for a fiber with a parabolic refractive index profile. For the independent variables on the horizontal axis, the normalized frequency v is used, which for graded index fibers is defined as

$$v = \frac{2\pi a}{\lambda_0} \sqrt{n^2(0) - n^2(a)} = k_0 \mathrm{NA} a \tag{7.4}$$

With this normalized frequency, equation (7.3) reads

$$g = \left[n^2(0) - \frac{2\mathrm{NA}^2}{v}(2m + n + 1)\right]^{1/2} \tag{7.5}$$

Note that the phase shifts of all the modes in a group with a common value of $(2m + n + 1)$ are represented by the same curve. For guided modes, $n(a)$ is the lowest limit of g; therefore, the cut-off frequency is given by the condition $g = n(a)$. When this condition is inserted into equation (7.5), the cut-off frequency v_c becomes

$$v_c = 2(2m + n + 1) \tag{7.6}$$

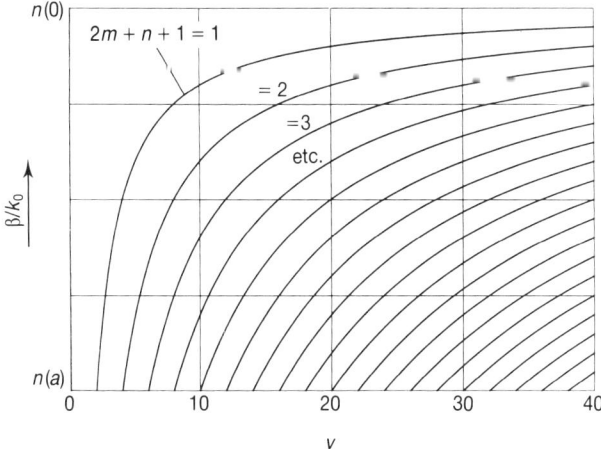

Figure 7.1 Curves of the normalized specific phase shift $\beta/k_0 = g$ for a fiber with a parabolic refractive index profile. $a = 25\ \mu m$, $n(0) = 1.5$, $n(a) = 1.486$, $NA \approx 0.2$.

From equation (7.3), it is clear that β/k_0 tends to $n(0)$ with increasing v. The specific group delay $\tau_g = d\beta/d\omega$ is more important than $\beta(\omega)$ itself.

Differentiating $\beta(\omega)$, as expressed in equation (7.1), with respect to ω gives

$$\tau_g = \frac{d\beta}{d\omega} = \frac{g(\omega)}{c_0} + \frac{\omega}{c_0}\frac{dg(\omega)}{d\omega} \tag{7.7}$$

From equations (7.4) and (7.5) it follows that

$$\omega\frac{dg(\omega)}{d\omega} = \frac{NA^2(2m + n + 1)}{g(\omega)v} \tag{7.8}$$

Equations (7.7) and (7.8) combine to give

$$c_0\tau_g = c_0\frac{d\beta}{d\omega} = g(\omega) + \frac{NA^2(2m + n + 1)}{g(\omega)v} \tag{7.9}$$

In Figure 7.2 the curves for the difference between the normalized specific group delay $c_0\tau_g$ and the value of the refractive index on the axis of the fiber $n(0)$ have been plotted as functions of the normalized frequency v, for a fiber with the same parameters as those used in Figure 7.1. At the cut-off frequency, the value of the difference becomes $[n(0) - n(a)]^2/2n(a)$. This is also the maximum value. Increasing v causes this difference to approach zero. The derivative of τ_g with respect to ω, or with respect to λ_0, is a measure of the specific dispersion D for the mode concerned; differentiating τ_g with respect to λ_0 gives the dispersion

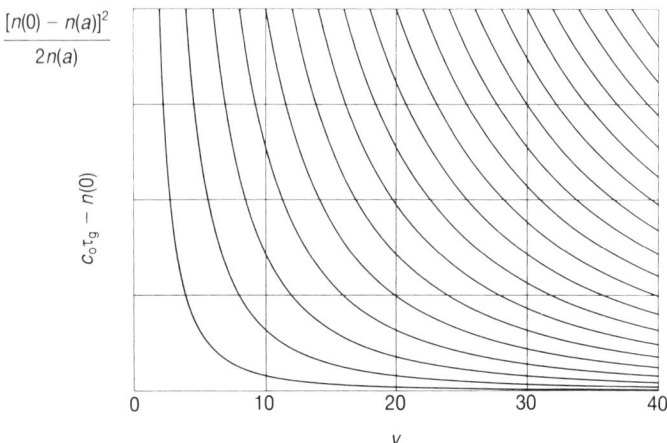

Figure 7.2 Curves of the difference between the normalized specific group delay and $n(0)$, for a fiber with a parabolic refractive index profile. $a = 25\ \mu\text{m}$, $n(0) = 1.5$, $n(a) = 1.486$, $\text{NA} \approx 0.2$. The maximum value of this difference, $[n(0) - n(a)]^2/2n(a)$, then becomes 66×10^{-6}.

$$D = \frac{\mathrm{d}\tau_g}{\mathrm{d}\lambda_0} = \frac{\mathrm{d}^2\beta}{\mathrm{d}\omega\,\mathrm{d}\lambda_0} = \frac{\text{NA}^3(2m + n + 1)^2}{2\pi c_0 a v[n^2(0) - (2\text{NA}^2/v)\,(2m = n + 1)]^{3/2}}$$

$$(7.10)$$

At the cut-off frequency, this specific dispersion D equals

$$D_c = \left(\frac{\mathrm{d}^2\beta}{\mathrm{d}\omega\,\mathrm{d}\lambda_0}\right)_{\text{cut-off}} = \frac{\text{NA}^3(2m + n + 1)}{4\pi c_0 a n^3(a)} \qquad (7.11)$$

As v increases to infinity, D approaches zero. The dispersion described by equation (7.10) is depicted in Figure 7.3 as a function of v for a fiber with the same parameters as in the previous examples. For all the guided modes in this example, the dispersion is less than $0.6\ \text{ps}\,\text{km}^{-1}\,\text{nm}^{-1}$, which is under $1.3 \times 10^{-3}\ \text{ps}\,\text{km}^{-1}\,\text{GHz}^{-1}$. The change of the group delay over the frequency interval that corresponds to the spectrum width of a short, technically feasible pulse is so small that this kind of dispersion is negligible when compared to the group delay differences of different modes. For example, take a short input pulse described by

$$f_i(t) = At^2 \exp[-(t/\sigma)^2] \qquad (7.12)$$

as shown in Figure 7.4, with $\sigma = 100\ \text{ps}$. The corresponding width of the Fourier transform of such a pulse is $\Delta\omega \approx 1/2\sigma = 5 \times 10^9\ \text{s}^{-1}$. For the fiber used in the examples in Figures 7.1 and 7.2 and with a wavelength of $\lambda_0 = 840\ \text{nm}$, it appears that the pulse width increases by less than $5 \times 10^{-4}\ \text{ps}\,\text{km}^{-1}$, for all the guided modes.

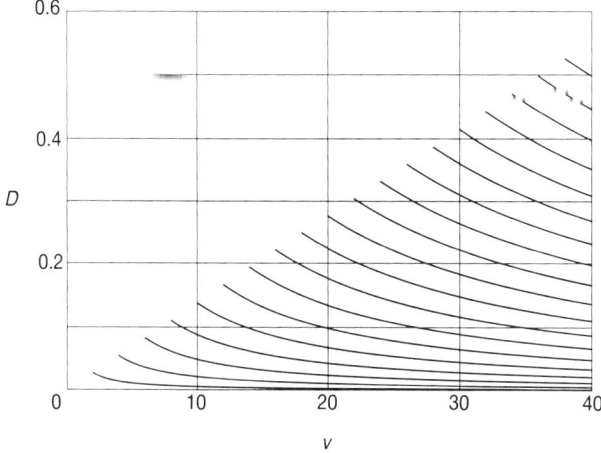

Figure 7.3 Curves for the specific dispersion for a fiber with a parabolic refractive index profile. $a = 25$ μm, $n(0) = 1.5$, $n(a) = 1.486$, NA ≈ 0.2.

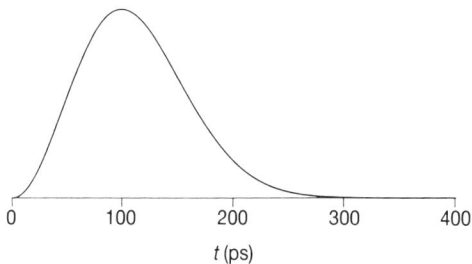

Figure 7.4 A short input pulse, with $\sigma = 100$ ps.

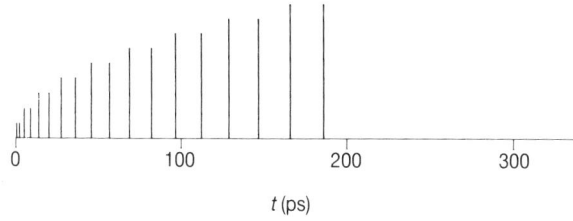

Figure 7.5 The impulse response of a fiber with a parabolic refractive index profile. $a = 25$ μm, $n(0) = 1.5$, $n(a) = 1.486$, NA ≈ 0.2.

The variation of the group delay with the mode numbers n and m can be demonstrated by the impulse response of a fiber. In Figure 7.5 an impulse response of a fiber with a parabolic refractive index profile is illustrated. Only guided modes have been taken into account and it is assumed that the total input power is equally distributed over the guided modes. Again, the same fiber parameters are assumed as in the previous examples and a wavelength of $\lambda_0 = 840$ nm. In that case, 18 groups of guided modes with different group delays are possible and the response of each group of modes is shown as a vertical line. Since each mode transmits the same power, the heights of the vertical lines that represent the responses of a group of modes has been taken as proportional to the number of modes in that group. The horizontal time-scale represents the difference in group delay between the modes in the fiber and the delay of a wave in bulk material with a refractive index of $n(0)$.

If all the guided modes are excited by the same input pulse, e.g. by the pulse described by equation (7.12) and illustrated in Figure 7.4, then all the vertical lines of Figure 7.5 must be replaced by output pulses of the same form as the input pulse and with a height that is proportional to the number of modes in the group concerned. The sum of those pulses gives the final output pulse, and this is illustrated in Figure 7.6.

7.1.2 A non-monochromatic source and no material dispersion

A certain width of the power spectrum of the source also contributes to pulse dispersion. From Figure 7.3 it is clear that the higher-order modes, especially, contribute to dispersion. For instance, if we have a multimode laser with a mutual distance between two spectral lines of 0.3 nm then even for the higher-order modes, the difference in travel time for the two wavelengths is less than 0.18 ps km^{-1}. When the source has a relatively wide power spectrum, e.g. an LED with $\Delta\lambda_0 \approx 50$ nm, then the maximum dispersion will be 30 ps km^{-1} for the higher-order modes. Thus, the greatest dispersion in a mode caused by the width of the source is much smaller than that caused by the different delays of the different modes.

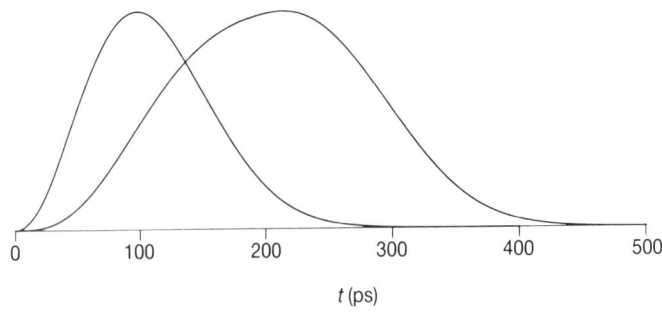

t (ps)

Figure 7.6 The response of 1 km of a fiber with a parabolic refractive index profile to an input pulse as given by equation (7.12) with $\sigma = 100$ ps. $a = 25\ \mu$m, $n(0) = 1.5$, $n(a = 1.486$, NA ≈ 0.2 and a wavelength $\lambda = 840$ nm.

7.1.3 Monochromatic sources and material dispersion

In reality, material dispersion is always involved. The refractive index is not only a function of r, but also a function of the frequency/or wavelength. In general, the shape of a refractive index profile also changes with the frequency. For the sake of simplicity, here it is assumed that the profile is again described by equation (1.2), with the exponent $x = 2$, which is independent of the frequency, i.e. a restriction is again made for fibers with a parabolic refractive index profile. Moreover, the central numerical aperture NA is assumed to be independent of the frequency, and thus $n(0)$, $n(a)$ and Δ depend on the frequency. If the cladding is made of pure silica, for instance, then for a fiber with $a = 25\ \mu$m and NA $= 0.2$, the curves for β/k_0 are like those shown in Figure 7.7. Now instead of equation (7.9) we find for the normalized specific group delay

$$c_0 \tau_g = c_0 \frac{d\beta}{d\omega} = g(\omega) + \frac{NA^2(2m + n + 1)}{g(\omega)v} + \frac{n(0)v}{g(\omega)} \frac{dn(0)}{dv} \tag{7.13}$$

Compared with equation (7.9), this equation contains an extra term that accounts for variations in the refractive index with frequency. The normalized specific group delay is plotted as a function of the normalized frequency in Figure 7.8; again, for a fiber with NA $= 0.2$ and $a = 25\ \mu$m and the same parameters as those in Figure 7.7.

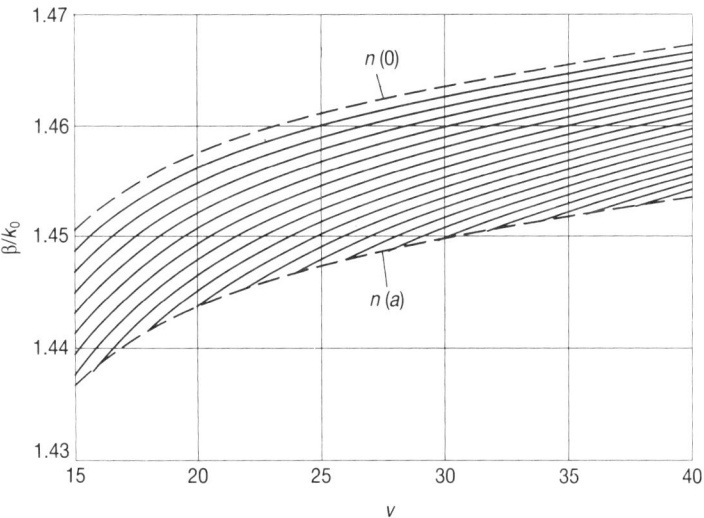

Figure 7.7 Curves for β/k_0 as a function of the normalized frequency v and with $(2m + n + 1)$ as the parameter for a graded index fiber with a parabolic index profile and a cladding of pure silica. NA $= 0.2$ and $a = 25\ \mu$m.

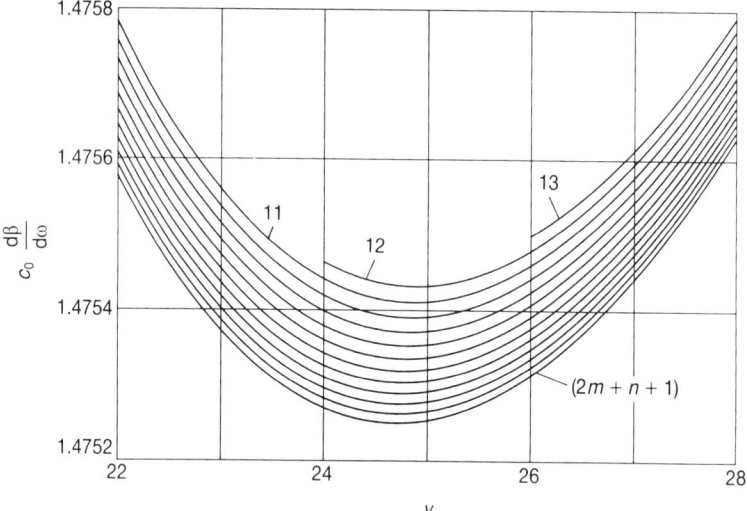

Figure 7.8 Curves for $c_0\tau_g$ as a function of the normalized frequency v and with $(2m + n + 1)$ as the parameter for a graded index fiber with a parabolic index profile and a cladding of pure silica. NA = 0.2 and $a = 25\ \mu$m. The range of v is such that the minima of the curves fall within this range. The corresponding wavelength is about 1.27 μm.

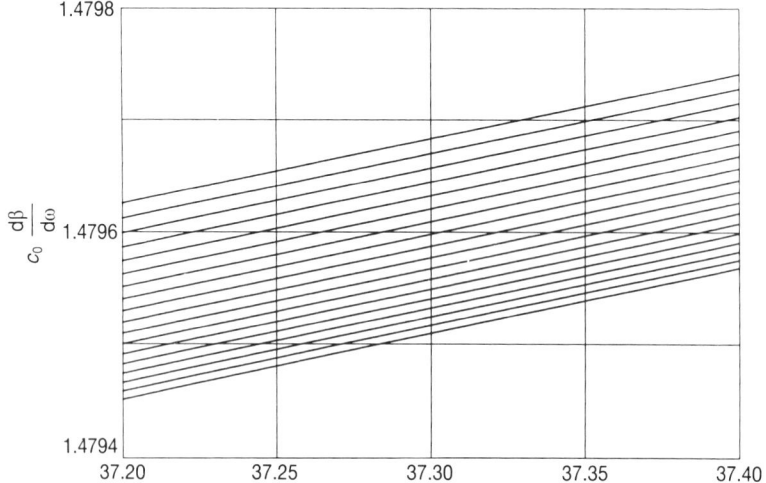

Figure 7.9 Curves for $c_0\tau_g$ as a function of the normalized frequency v and with $(2m + n + 1)$ as the parameter for a graded index fiber with a parabolic index profile and a cladding of pure silica. NA = 0.2 and $a = 25\ \mu$m. The range for v has been taken so that the corresponding wavelength is about 844 nm.

The last term in equation (7.13) has a great influence on the curves of Figure 7.8. In the latter figure, the range of v is such that the minima of the curves appear in this range. The values of n in the minima correspond to the wavelengths of $\lambda_0 \approx 1.27 \mu$m. In Figure 7.9 the normalized specific group delay $c_0 \tau_g$ for the same fiber has been plotted for the values of v that correspond to the wavelengths in the range 840–844 nm.

From Figures 7.8 and 7.9 it is clear that the mode dispersion, for a fiber with a parabolic refractive index profile, generally dominates the material dispersion. Only for a source with a relatively broad frequency spectrum (e.g. an LED) does the material dispersion dominate in the 840 nm range.

7.2 Ray model

An alternative way of describing the dispersion in graded index fibers is with the help of ray congruences. To do this, the system model has to be changed so that the individual channels of the fiber become associated with ray congruences and not with modes as was done in the previous sections. Again we will restrict ourselves to refractive index profiles that are described by equations (6.58a) and (6.58b).

For monochromatic light, the number of radians b_s that occur in the distance s travelled by a guided or leaky ray when it goes from the inner caustic surface (point s_1) to the outer caustic surface (point s_2) in the core, or vice versa, is given by

$$b_s = 2\pi \int_{s_1}^{s_2} \frac{ds}{\lambda} = \frac{\omega}{c_0} \int_{s_1}^{s_2} n(r) \, ds = \frac{\omega}{c_0} \int_{r_1}^{r_2} \frac{n^2(r)}{p(r)} \, dr \qquad (7.14)$$

The distance travelled in the z-direction by the ray when it goes from s_1 to s_2 is

$$l_0(h, g) = g \int_{r_1}^{r_2} \frac{dr}{p(r)} \qquad (7.15)$$

Assuming a phase shift of $-\pi/2$ at every radial directional change, the mean specific phase shift over a distance of $l_0(h, g)$ is given by

$$\bar{\beta} = \frac{b_s - (\pi/2)}{l_0(h, g)} \qquad (7.16)$$

In general , the distance $l_0(h, g)$ is very small compared to the distances that are involved in glassfiber communications; therefore, the phase change b in a fiber of length l can be written as

$$b = \bar{\beta} l \qquad (7.17)$$

Taking equations (7.14)–(7.17) and solving the integral gives the phase shift for a fiber with an n profile according to equation (6.58a)

$$b = \frac{\omega l}{c_0} \frac{xn^2(0) + 2g^2}{(x+2)g} - N\frac{\pi}{2} \tag{7.18}$$

where N is the number of radial direction changes, which is given by

$$N = \frac{l}{l_0(h,g)} \tag{7.19}$$

The variable τ has been introduced in order to simplify the notation as follows:

$$\tau = \frac{1}{c_0} \frac{xn^2(0) + 2g^2}{(x+2)g} \tag{7.20}$$

Using this notation with equation (7.18), the specific phase delay τ_p, as introduced in Chapter 4, for a light ray that belongs to the congruence (h, g) and travelling through a fiber of length l is denoted by

$$\tau_p = \frac{b}{\omega l} = \tau - \frac{N\pi}{2\omega l} \tag{7.21}$$

Note that the phase delay for a given refractive index profile does not depend on the parameter h. The specific group delay τ_g for the ray congruence concerned can be found from the following relationship:

$$\tau_g = \frac{d\beta}{d\omega} = \tau_p + \omega \frac{d\tau_p}{d\omega} \tag{7.22}$$

As in the previous case, we first pay attention to the case where material dispersion is neglected, and afterwards examine the general case with material dispersion.

7.2.1 No material dispersion

$n(0)$, Δ and x are assumed to be independent of frequency, so that $n(r)$ is also independent of frequency. Moreover, it is assumed that the trajectory of a light ray does not depend on frequency. In that case, the angle $\theta(r)$, as well as the distance $l_0(h,g)$ and N, are independent of frequency. Combining equation (7.22) with equation (7.21) gives

$$\tau_g = \tau + \omega \frac{d\tau}{d\omega} = \tau \tag{7.23}$$

Next, we pay attention to the graphs of τ_g as functions of g for three specific values of the exponent x, starting by introducing the factor $\varepsilon = n(a)/n(0)$. The graphs are plotted in Figure 7.10.

First case: x = 2
For a parabolic refractive index profile, the minimum value for τ_g is found at $g = n(0)$ and has the value

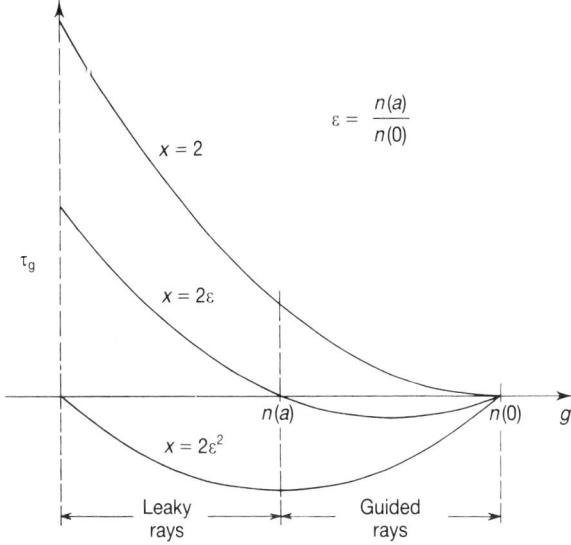

$$\varepsilon = \frac{n(a)}{n(0)}$$

Figure 7.10 The specific group delay τ_g as a function of g for three values of the parameter x.

$$\tau_{g,min} = \frac{n(0)}{c_0} \tag{7.24}$$

If we take only the guided rays into account, then the maximum value of τ_g can be found for $g = n(a)$ as follows:

$$\tau_{g,max} = \frac{1 - \Delta}{\sqrt{(1 - 2\Delta)}} \tau_{g,min} \cong \left(1 + \frac{\Delta^2}{2}\right) \tau_{g,min} \tag{7.25}$$

The difference between $\tau_{g,max}$ and $\tau_{g,min}$ is a measure of the amount of multimode dispersion per unit of fiber length

$$\Delta\tau_g \triangleq \tau_{g,max} - \tau_{g,min} \approx \frac{n(0)}{c_0} \frac{\Delta^2}{2} \tag{7.26}$$

When we also take leaky rays into account, the minimum value for g, which follows from equation (6.64) with $r = a$, is given by

$$g_{min} = n(0) \sqrt{1 - 2\Delta [1 + (x/2)]} \tag{7.27}$$

For $x = 2$ this leads to

$$\tau_{g,max} = \frac{n(0)}{c_0} \frac{1 - 2\Delta}{\sqrt{1 - 4\Delta}} \cong \frac{n(0)}{c_0} (1 + 2\Delta^2) \tag{7.28}$$

Example 7.1
Assume $n(0) = 1.5$ and $2\Delta = 0.01$. The difference between the maximum and minimum values of the specific group delay for guided rays only is then $\Delta\tau_g = 63.1$ ps km^{-1}, and for guided rays plus leaky rays $\Delta\tau_g = 255$ ps km^{-1}.

Second case: $x = 2\varepsilon$
In this case the maximum difference in group delay between two guided congruencies is minimized. The maximum value of τ_g is found for $g = n(0)$ as well as for $g = n(a)$; its value is

$$\tau_{g,max} = \frac{n(0)}{c_0} \tag{7.29}$$

The minimum value of τ_g is found for $g = \sqrt{n(0)n(a)}$; its value is

$$\tau_{g,min} = \frac{n(0)}{c_0} \frac{2\sqrt{n(0)n(a)}}{n(0) + n(a)} \cong \frac{n(0)}{c_0}\left(1 - \frac{\Delta^2}{8}\right) \tag{7.30}$$

It then follows that for guided rays

$$\Delta\tau_g \approx \frac{n(0)}{c_0}\frac{\Delta^2}{8} \tag{7.31}$$

which is four times smaller than for the previous case.

Example 7.2
Once again, if $n(0) = 1.5$ and $2\Delta = 0.01$, then the difference between the maximum and minimum values of the specific group delay for guided rays only is $\Delta\tau_g = 15.8$ ps km^{-1}, and for guided rays plus leaky rays $\Delta\tau_g = 126$ ps km^{-1}.

Third case: $x = 2\varepsilon^2$
In this case, the maximum group delay difference between two congruencies that belong to the union of guided and leaky congruencies is minimized. The maximum value of τ_g is found for $g = n(0)$ as well as for g as given by equation (7.27)

$$\tau_{g,max} = \frac{n(0)}{c_0} \tag{7.32}$$

The minimum value of τ_g is found for $g = n(a)$ and its value is

$$\tau_{g,min} = \frac{n(0)}{c_0} \frac{2n(0)n(a)}{n^2(0) + n^2(a)} \cong \frac{n(0)}{c_0}\left(1 - \frac{\Delta^2}{2}\right) \tag{7.33}$$

It follows that

$$\Delta \tau_g \approx \frac{n(0)}{c_0} \frac{\Delta^2}{2} \tag{7.34}$$

which is the same as the equation for guided rays in the first case.

Example 7.3
Once again, assume $n(0) = 1.5$ and $2\Delta = 0.01$, then the difference between the maximum and minimum value of the specific group delay for guided rays plus leaky rays is $\Delta \tau_g = 63 \text{ ps km}^{-1}$.

The different values found in the examples for $\Delta \tau_g$ can be interpreted as the multimode pulse dispersion. This is the broadening of a pulse per unit of length as it travels along a fiber when a monochromatic source is used, the material dispersion is neglected and when all the modes regarded are excited with equal power.

It is interesting to compare the multimode dispersion values with those for a step index fiber. From equation (7.20) it can be seen that for a step index fiber $(x \rightarrow \infty)$, when only guided rays are taken into account,

$$\Delta \tau_g \approx \frac{n(0)}{c_0} \Delta \tag{7.35}$$

Thus, the multimode dispersion per unit of fiber length in a graded index fiber $(x = 2)$ is reduced by a factor of 2Δ or 4Δ, depending on the case at hand.

7.2.2 With material dispersion

In practice, material dispersion always occurs. For pure silica, the derivative of the refractive index with respect to frequency at a wavelength of $\lambda_0 = 850$ nm is

$$\frac{dn}{d\omega} = 6 \times 10^{-18} \text{ s} \tag{7.36}$$

or

$$\frac{dn}{d\lambda_0} = -\frac{2\pi c_0}{\lambda_0^2} \frac{dn}{d\omega} = -1.57 \times 10^4 \text{ m}^{-1} \tag{7.37}$$

For the class of refractive index profiles under consideration here, the specific group delay τ_g can be found by differentiating equation (7.20) with respect to ω

$$\tau_g = \tau + \omega \frac{d\tau}{d\omega} \tag{7.38}$$

In general, the n-profile also changes with the frequency. For the sake of simplicity, we assume that Δ and x do not depend on the frequency. We assume

$$n^2(r, \omega) = n^2(0, \omega) [1 - 2\Delta(r/a)^x] \tag{7.39}$$

from which it follows that

$$\frac{dn(r, \omega)}{d\omega} = \frac{dn(0, \omega)}{d\omega} [1 - 2\Delta(r/a)^x]^{1/2} \tag{7.40}$$

Assuming that $\theta(r)$ does not depend on ω, i.e. the curve of the light ray is independent of ω, then

$$g = n(r, \omega) \cos \theta(r) \tag{7.41}$$

and

$$\frac{dg}{d\omega} = \frac{g}{n(r)} \frac{dn(r)}{d\omega} \tag{7.42}$$

Under these conditions, it follows that

$$\frac{d\tau}{d\omega} = \tau \frac{1}{n(0)} \frac{dn(0)}{d\omega} \tag{7.43}$$

and

$$\tau_g = \tau \left[1 + \frac{\omega}{n(0)} \frac{dn(0)}{d\omega}\right] \tag{7.44}$$

or

$$\tau_g = \tau \left[1 - \frac{\lambda}{n(0)} \frac{dn(0)}{d\lambda}\right] \tag{7.45}$$

Monochromatic source
If the light comes from a monochromatic source with a frequency of ω, then

$$\tau_g = \tau \left[1 + \frac{\omega_0}{n(0)} \left(\frac{dn(0)}{d\lambda}\right)_{\omega_0}\right] \tag{7.46}$$

Example 7.4
If $\lambda_0 = 850$ nm, then for the pure silica, we see that

$$\frac{\omega_0}{n(0)} \left(\frac{dn(0)}{d\lambda}\right)_{\omega_0} = 0.009$$

When compared with the case without material dispersion, all the group delays and group delay differences will be multiplied by a factor of 1.009.

Non-monochromatic sources
The influence of a light source with a certain spectral width can be studied if we differentiate τ_g with respect to the frequency ω, with the result that

$$\frac{d\tau_g}{d\omega} = 2\frac{d\tau}{d\omega} + \omega\frac{d^2\tau}{d\omega^2} \tag{7.47}$$

When associated with equation (7.43), this leads to

$$\frac{d\tau_g}{d\omega} = \frac{\tau}{n(0)}\left[2\frac{dn(0)}{d\omega} + \omega\frac{d^2n(0)}{d\omega^2}\right] \tag{7.48}$$

If the spectral width of a source with a central wavelength of λ_0 is $\Delta\lambda_0$, then the maximum difference between the group delays for rays that belong to the same congruence is equal to

$$\Delta\tau = \frac{d\tau_g}{d\lambda_0}\Delta\lambda_0 = \frac{d\tau_g}{d\omega}\frac{d\omega}{d\lambda_0}\Delta\lambda_0 = -\frac{2\pi c_0}{\lambda_0^2}\frac{d\tau_g}{d\omega}\Delta\lambda_0 \tag{7.49}$$

where it is assumed that $d\tau_g/d\lambda_0$ is constant over a wavelength interval of $\Delta\lambda_0$.

References

[1] S. Geckeler, 'Dispersion in optical fibers: new aspects', *Applied Optics*, vol. 17. no. 7, April 1978, pp. 1025–9.
[2] J. A. Arnaud, *Beam and Fiber Optics* (New York: Academic Press, 1976).
[3] H.-G. Unger, *Planar Optical Waveguides and Fibers* (Oxford: Clarendon Press, 1977).
[4] D. Marcuse, *Principles of Optical Fiber Measurements* (New York: Academic Press, 1981).
[5] T. Okoshi, *Optical Fibers* (New York: Academic Press, 1982).
[6] A. W. Snyder and J. D. Love, *Optical Waveguide Theory* (London: Chapman & Hall, 1983).
[7] C. K. Kao, *Optical Fibre* (London: Peter Peregrinus, 1988).
[8] M. K. Barnoski, *Fundamentals of Optical Fiber Communications*, 2nd edn (New York: Academic Press, 1981).

Problems

7.1 Derive the expression for the dispersion D as given by equation (7.10) and prove equation (7.11).

7.2 Derive equation (7.8)

7.3 Assume a graded index fiber with a parabolic refractive index profile with $a = 25\ \mu m$, $n(0) = 1.5$, $NA = 0.2$. The material dispersion and the attenuation can be neglected. The guided modes are equally excited by a monomode laser with $\lambda_0 = 1000$ nm. Estimate the maximum bit rate for binary transmission as a function of the length of the fiber.

7.4 Given the refractive index profile as described by equation (6.58a), derive equation (7.18) by solving the integral of equation (7.14).

7.5 Assume that pulsed light from a single mode laser is launched into a graded index fiber and that all the guided congruencies are equally excited. The input pulses are assumed to be short compared with the duration of the output pulses. If the refractive index profile of the fiber is given by equation (6.58a), find the response of the fiber to a single input pulse for the three values of x as in Section 7.2.1. Repeat the calculations for the case that the leaky congruencies are also taken into account.

8

Light sources and detectors

8.1 Choosing the wavelength region
8.2 The light-emitting diode (LED)
8.3 The semiconductor laser diode (LD)
8.4 Semiconductor laser versus LED
8.5 Photodiodes
References
Problems

8.1 Choosing the wavelength region

At the transmitting end of an optical fiber transmission system is a transducer that converts the electrical input signal of the system into an optical signal carrying information. This means that the electro-optical converter shifts the spectrum of the information signal to a wavelength region where the fiber has better transmission properties; this change of wavelength is achieved by means of modulation. Several modulation methods are suitable, but intensity modulation is currently the most widely used; this method is akin to amplitude modulation (see Appendix 2). The receiving end of the optical fiber requires a detector that has its greatest sensitivity in the wavelength region of the optical source at the transmitting end.

Figure 8.1 is used as a starting point for dealing with an optical fiber communication system; this system comprises transmitter, fiber and receiver.

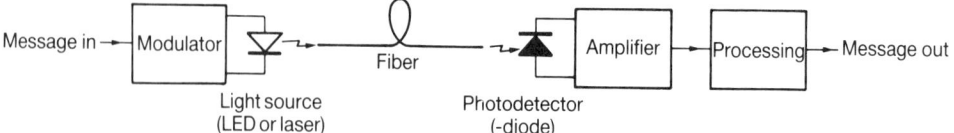

Light source   Photodetector
(LED or laser)   (-diode)

Figure 8.1 Block diagram of an optical fiber communication system.

The operating wavelength of the system is determined by what is desirable on the one hand, and by what is practical on the other. The material properties of both the fiber and the electro-optical devices play a decisive role, as will be shown below.

Two parameters are important – the transmission parameters: attenuation and dispersion. Some typical attenuation characteristics of fiber are shown in Figure 1.9, where the attenuation peaks due to hydroxyl ions are significant (see Figure 1.11). Over the years production technology has improved so that those peaks have almost disappeared (see Figure 8.2). Today, a minimum attenuation of around 0.3 dB km^{-1} can be observed for silica fibers at 1.55 μm, whilst in the wavelength region of 0.8–0.9 μm the attenuation varies between 2 and 3 dB km^{-1}. For wavelengths below 1.55 μm the attenuation behaves in accordance with Rayleigh scattering (see Section 1.3.2), whereas for wavelengths beyond 1.6 μm, the attenuation increases considerably due to infra-red absorption. The material dispersion vanishes at 1.27 μm for silica (see Figures 4.10 and 8.3). Figures 8.2 and 8.3 suggest that an attractive operating wavelength is in the vicinity of 1.3 μm; furthermore, in the region of 1.3–1.5 μm, silica fibers can be designed so that the material and waveguide dispersions cancel each other out

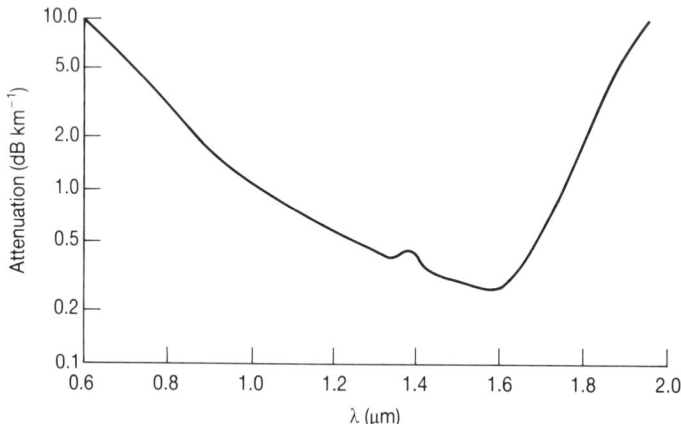

Figure 8.2 Attenuation of a silica fiber as a function of the wavelength. (Source: reproduced with permission from *Optical Communication Systems*, by John Gowar, Prentice Hall, 1984.)

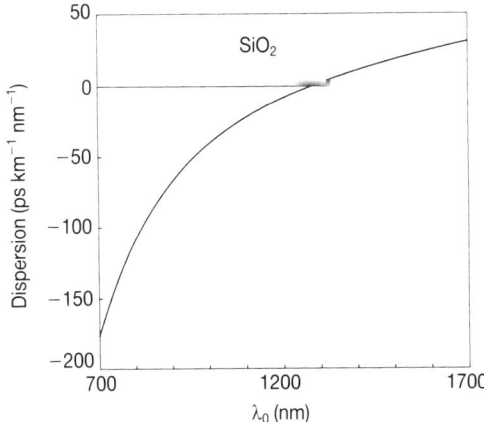

Figure 8.3 The material dispersion of bulk SiO_2 versus wavelength.

(see Section 5.5). In the early days of optical fiber communication these wavelengths were less attractive, because of a high attenuation peak at $1.38 \, \mu m$, due to the hydroxyl ions.

In addition to attenuation, dispersion is an important parameter for transmitting signals, because it determines the maximum bandwidth of the signals. In Chapter 5, it has been shown that the material dispersion together with the waveguide dispersion provides the ultimate dispersion of a single-mode fiber. Figure 8.3 shows the material dispersion for silica glass as a function of wavelength.

It can be concluded that the wavelength region of $1.3–1.5 \, \mu m$ is attractive from a transmission point of view; nevertheless, in practice we often encounter optical fiber transmission systems that operate in the wavelength region of $0.8–0.9 \, \mu m$, despite the larger attenuation and dispersion in that region. The reason for this choice is as follows: an optical fiber communication system, apart from the fiber, consists of a light source and a detector, which are semiconductor devices. These electro-optical and opto-electronic transducers for long wavelengths do not have the same favourable properties as their counterparts for short wavelengths. These disadvantages concern optical, electrical or economical aspects that can sometimes prevent an optimal technical use of a fiber as transmission medium.

8.2 The light-emitting diode (LED)

8.2.1 Principles of operation

Under certain conditions that are quite easily satisfied the p–n junctions of semiconductors emit radiation spontaneously in the visible or near infra-red region of the spectrum, when a forward current flows through the junction [1]. Such a component is called a light-emitting diode (LED). The conduction band of a semiconductor is populated with electrons, which are injected by the forward current; similarly, the valence band is populated with holes. After some time the electrons recombine with the holes spontaneously, and the difference of the energy level of the conduction and the valence band is emitted as a photon. If this energy difference is called E_g, then the wavelength of the emitted radiation can be written as

$$\lambda = \frac{hc}{E_g} \tag{8.1}$$

where h is Planck's constant.

When E_g is expressed in eV, this equation becomes

$$\lambda_0 = \frac{1.24}{E_g} \, \mu m \tag{8.2}$$

This spontaneous emission is called recombination radiation and it occurs mostly in the p-layer close to the p–n junction. Due to irregularities and impurities in the crystal, some non-radiative processes occur via so-called trapped levels [1]; these are energy levels somewhere between the valence band and the conduction band. The ratio of the number of photons to the number of injected carriers is called the internal quantum efficiency. Two causes of non-radiative recombination have been noted, while a third has to do with the type of semiconductor material. Well-known semiconductors such as silicon and germanium have a so-called indirect bandgap. This means that an electron with the lowest energy in the conduction band has a rather different momentum compared to an electron with the highest energy in the valence band. This is illustrated in Figure 8.4(a), which shows that a recombination in indirect bandgap semiconductors can only take place if the difference in momentum is absorbed, e.g. by setting up a lattice vibration in the crystal (a phonon). Apart from absorbing this momentum difference, such a process also carries away some of the total transition energy. However, initiating the two processes at the same time, which requires conservation of total energy and total momentum, makes such a recombination less likely. Consequently, in an indirect bandgap semiconductor, non-radiative recombination tends to be more significant, resulting in a poor quantum efficiency.

The band structure of direct bandgap semiconductor material is depicted in Figure 8.4(b). The electrons at the lowest energy level of the conduction band can be seen to have the same momentum as the electrons at this highest energy level of the valence band, and thus it is probable that direct band-to-band transitions occur; therefore, the internal quantum efficiency is relatively high. Well-designed LEDs can have an internal quantum efficiency as high as 50–80 percent. Due to losses inside the LED material and reflections at the crystal's

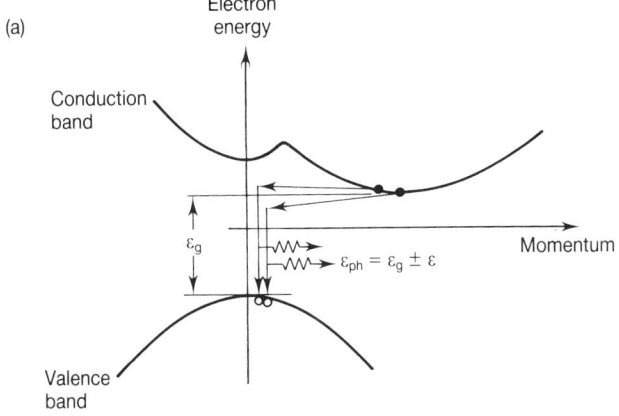

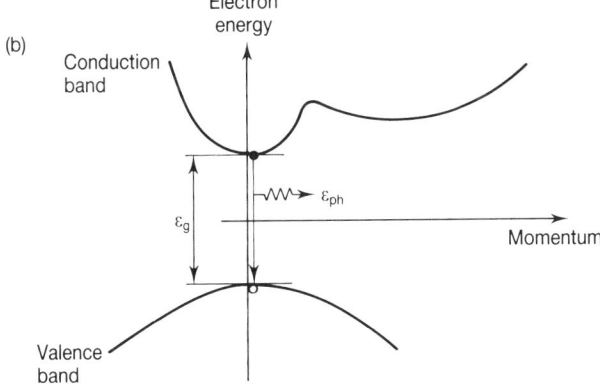

Figure 8.4 Electron energy versus momentum diagrams for: (a) an indirect bandgap semiconductor; (b) a direct bandgap semiconductor. (Source: reproduced with permission from *Optical Communication Systems*, by John Goward, Prentice Hall, 1984.)

surface, the external efficiency will be much lower; this can be found from geometrical optics in the phase space (see Chapter 11).

One objective of designing an LED must be to achieve a high efficiency. Moreover, the radiation has to have a geometry that allows the optical fibers to collect as much of the optical power as possible, in order to maximize the coupling efficiency. A third objective is to design an LED in which the light output can be directly modulated at high rates with an information-carrying signal by the forward current. Finally, its construction must ensure a rapid discharge of heat from the p–n junction, since the light output drops as the junction temperature rises.

8.2.2 LED constructions

Two basic LED constructions have been developed, namely surface emitters and edge emitters. In surface emitters the light is emitted perpendicular to the junction through a thin transparent layer. In edge emitters the light is emitted in the plane of the p–n junction at the edge of the semiconductor crystal.

A surface emitter with a homojunction (one n-type layer and one p-type layer) is shown in Figure 8.5. In an n-type substrate a small area of p-type diffusion appears, and rapid heat transfer away from the junction is assured, because the substrate has been fitted upside down on the heat sink. The emitting area has a diameter of 15–100 μm. A hole has been etched away in the substrate on the light output side, so that the light only has to travel through a thin layer 10–15 μm thick. This allows the fiber to come very close to the emitting surface, which is important for the coupling efficiency, since this efficiency cannot be improved by means of optical appliances if the emitting area is larger than the area of the multimode fiber core [2]. When the emitting area is smaller than the area of a multimode fiber core, then the coupling can be improved considerably with optical devices such as lenses (see Chapter 11). For a single-mode fiber coupling, a butt joint (putting the fiber as close to the emitting surface as possible) always gives the greatest efficiency (see [3] and Chapter 11).

Double heterostructures (DHs) have several different layers of n and p materials, which give them several advantages over homostructures. These advantages are as follows [1, 4]:

• Suitable composition and doping of the material in the various layers produces a bandgap structure like that in Figure 8.6. In this figure, the electrons are injected from the left into the active region (layer 2), where they lose energy, while at the interface between the active region and layer 3 the electron energy increases. The holes, which are injected from the right into the active region, also undergo a drop in energy when coming into this region and they need an increase of energy when they try to leave through layer 1. In this way electrons and holes are confined to the active layer, thus increasing the density of both types of carriers in this region, which results in a greater rate of recombination and a higher efficiency.

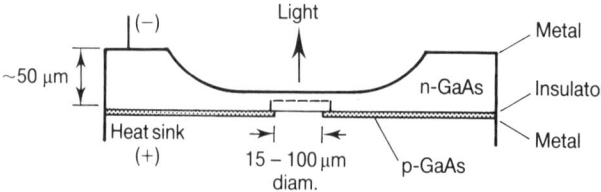

Figure 8.5 Construction of a surface emitter LED with a homojunction.

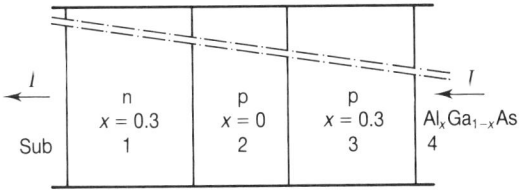

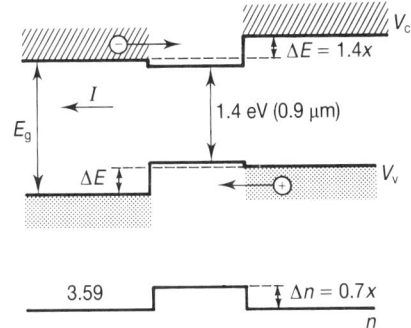

Figure 8.6 Electron energy level diagram for the various layers. (Source: reproduced from *Philips Technisch Tijdschrift*, vol. 36, no. 7, 1976, pp. 198–208, 'Halfgeleider-lasers voor optische communicatie', by G. Acket, J. Daniele, W. Nijman, R. Tijburg and P. de Waard.)

- Less absorption of radiation that is produced in the active layer, because the layers on both sides of the active layer get a larger bandgap than the active layer, thus making the adjacent layers more transparent.
- Wider range of wavelengths of the emitted radiation, by varying the composition of the semiconductor material of the active layer. Of course, this feature is not limited to the DH structure, but is achieved more readily with it.

Figure 8.7 shows an example of a DH surface-emitting LED. Surface emitters behave like a Lambertian source, radiating equal amounts of power per unit of projected area and per unit of solid angle in each direction of a hemisphere (see also Section 11.1). The radiation pattern is spherical.

In Figure 8.8 the light output of an edge emitter LED is shown; it is well directed in the plane of the junction to give an improved coupling into the fiber. A low emission angle is achieved by the optical guidance of the DH structure, which is another advantage of this structure. By giving the layers adjacent to the active layer a somewhat lower refractive index, much of the radiation is reflected back into the active region. The active layer is very thin (approximately 0.05 μm), whereas the optical guiding layers are 2–3 μm thick. If one end-face of the device is provided with a reflector and the other end-face with

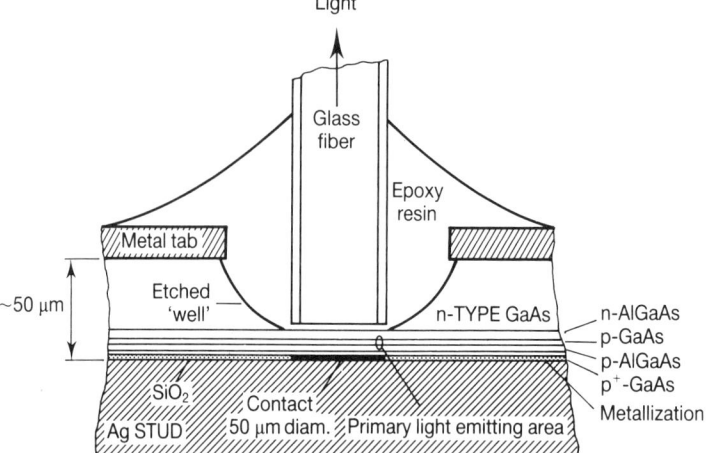

Figure 8.7 A DH surface-emitting LED with fiber pigtail attached. (Source: reproduced with permission from *Optics Communications*, vol. 4, no. 4, December 1971, pp. 307–9, 'Small-area, double heterostructure aluminium-gallium arsenide electroluminescent diode source for optical-fibre transmission lines', by C. Burrus and B. Miller.)

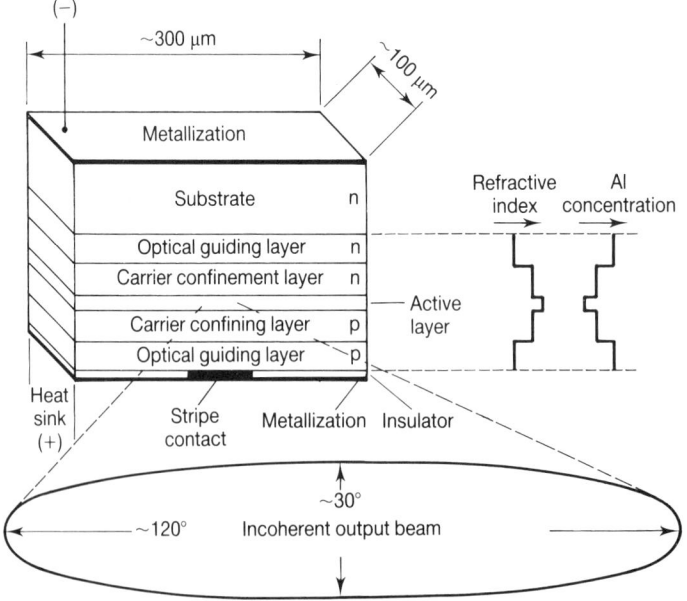

Figure 8.8 Edge emitter DH LED with carrier confinement and optical guiding layers.

an anti-reflection coating, the light leaves the device at the anti-reflecting end with a fairly high intensity. As shown in Figure 8.8, the light beam is Lambertian in the plane of the junction, where there is no optical guiding. In the plane perpendicular to the junction the beam is restricted to one half power angle of 30° by guiding the wave in this direction.

The semiconductor materials used for an LED in the short wavelength region (0.8–0.9 μm) are GaAs and $Al_xGa_{1-x}As$. The latter material is a compound consisting of the mole fraction x of AlAs and a mole fraction $(1 - x)$ of GaAs. The elements Al and Ga are in group III of the periodic table of elements, whereas As belongs to group V; therefore, the compound is called a III–V compound. For longer wavelengths (1.3–1.6 μm) the III–V compound $Ga_xIn_{1-x}P_yAs_{1-y}$ is used. Both compounds are direct bandgap semiconductors over a substantial part of the mole fraction ranges of x and y. In Figure 8.9(a) the influence on the emitted wavelength of the fraction of Al in $Al_xGa_{1-x}As$ is shown; in addition, Figure 8.9(b) shows how the refractive index depends on this fraction. From these two figures it follows that both the carrier confinement and optical guiding are achieved simultaneously, if the confining layers contain less Al than the active layer. Different layers can be used to control the two effects independently, so that each layer performs one of the tasks (see also Figure 8.8).

From Figure 8.9(a) it can be seen that $Al_xGa_{1-x}As$ becomes an indirect bandgap semiconductor for Al mole fractions larger than 0.38 [5].

As dopants for III–V compounds the following materials are used: S, Se, Te (group VI), Zn, Cd, Mg (group II) and Si, Ge, Sn (group IV).

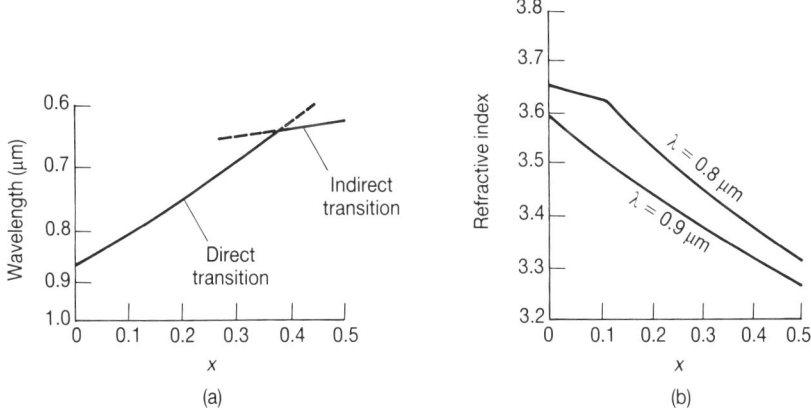

Figure 8.9 (a) The wavelength emitted by $Al_xGa_{1-x}As$ as a function of the mole fraction x. (b) The refractive index of $Al_xGa_{1-x}As$ as a function of the mole fraction x. (Source: reproduced with permission from *Introduction to Optical Fiber Communications*, by Y. Suematsu and K. Iga, Wiley 1982.)

8.2.3 Optical spectrum and coherence function

The energy of the electrons in the conduction band and the holes in the valence band display a Fermi statistic [1]. When electrons recombine with holes the energy distribution of the emitted photons shows a gamma statistic [6], with a standard deviation of $\sqrt{3}(k\Theta)$ (k is Boltzmann's constant; Θ is the absolute temperature). Impurities increase the standard deviation to a practical value of about $2k\Theta$. Differentiating equation (8.1) yields

$$|\Delta\lambda| = \frac{\lambda^2}{hc}\,\Delta E_g \tag{8.3}$$

When $\Delta E_g = 2k\Theta$ is inserted, the spectral width is found to be between 25 and 40 nm at room temperature for the short wavelength values (0.8–$0.9\ \mu$m). In the wavelength region 1–$1.3\ \mu$m the spectral width becomes 50–100 nm. It is obvious that the optical spectrum becomes broader with increasing temperature. This means that the spectral widths can show considerably higher values than those given above when the device is not properly cooled, because the current densities in the junction can be as high as a few kiloamps per cm^2, thereby giving rise to high junction temperatures. The spectra resemble closely the Gaussian bell shape. In Figure 8.10 several typical LED spectra are depicted at various wavelengths.

Let $\omega_c = 2\pi v$, then the electromagnetic wave emitted by the LED can be described as a stochastic band-pass process [6]

$$w(t) = x(t)\cos\omega_c t - y(t)\sin\omega_c t \tag{8.4}$$

where $x(t)$ and $y(t)$ are real baseband processes. Define the complex process

$$z(t) \stackrel{\triangle}{=} x(t) + \mathrm{j}y(t) \tag{8.5}$$

then $z(t)$ can be considered to be the complex envelope of the stochastic process $w(t)$. Since spontaneous emission occurs in an LED, the phase of $w(t)$ is

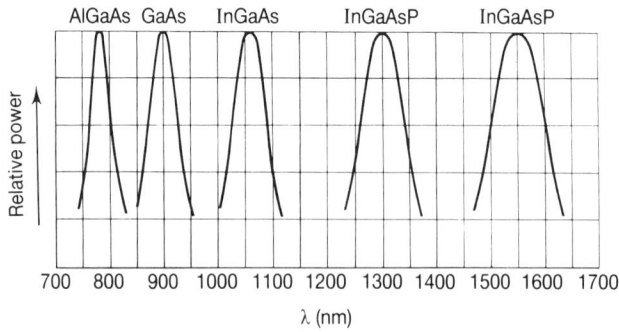

Figure 8.10 Typical LED spectra.

random and the processes $x(t)$ and $y(t)$ are independent. The following assumptions are therefore reasonable:

$$E[x(t)] - L[y(t)] = 0 \tag{8.6}$$

$$R_{xx}(\tau) = R_{yy}(\tau) \tag{8.7}$$

and

$$R_{xy}(\tau) = R_{yx}(\tau) \tag{8.8}$$

with $R_{xx}(\tau)$ and $R_{yy}(\tau)$ the autocorrelation functions of the processes $x(t)$ and $y(t)$, respectively. The cross-correlation functions of these processes are denoted by $R_{xy}(\tau)$ and $R_{yx}(\tau)$. Equations (8.6)–(8.8) guarantee the wide sense stationarity of $w(t)$ [6]. From equations (8.5), (8.7) and (8.8) it follows that

$$R_{zz}(\tau) = 2R_{xx}(\tau) \tag{8.9}$$

For relationships between the various correlation functions and the power spectra, it follows that [6]

$$R_{ww}(\tau) = R_{xx}(\tau) \cos \omega_c \tau \tag{8.10}$$

$$S_{ww}(\omega) = \frac{S_{xx}(\omega - \omega_c) + S_{xx}(\omega + \omega_c)}{2} \tag{8.11}$$

and

$$S_{zz}(\omega) = 2S_{xx}(\omega) \tag{8.12}$$

In optics, equation (8.10) is called the coherence function of the source, rather than the autocorrelation function.

The spectrum $S_{ww}(\lambda)$ can be measured easily with the aid of a monochromator and a photodiode.

8.3 The semiconductor laser diode (LD)

8.3.1 Laser constructions

By putting the LED in an optically resonant cavity the device can act as a laser. Its laser action only occurs if the forward current is so large that population inversion takes place, i.e. the product of the number of electrons in the conduction band and the number of holes in the valence band is larger than the product of the number of electrons in the valence band and the number of holes (empty states) in the conduction band. The current at which the laser action starts is called the threshold current. For currents below the threshold the device behaves like an LED. A typical DH laser is shown in Figure 8.11. The parallel reflecting end-faces are produced by cleaving the crystal and setting up a Fabry–Perot resonator. Since the refractive index of GaAs is around 3.6, the

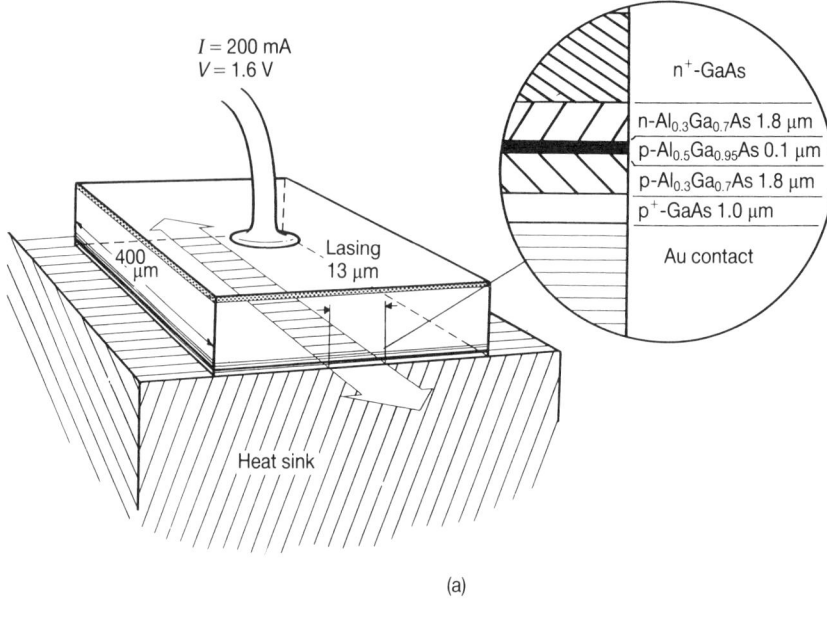

$I = 200$ mA
$V = 1.6$ V

Lasing
13 μm

400 μm

Heat sink

n⁺-GaAs

n-Al₀.₃Ga₀.₇As 1.8 μm
p-Al₀.₅Ga₀.₉₅As 0.1 μm
p-Al₀.₃Ga₀.₇As 1.8 μm
p⁺-GaAs 1.0 μm

Au contact

(a)

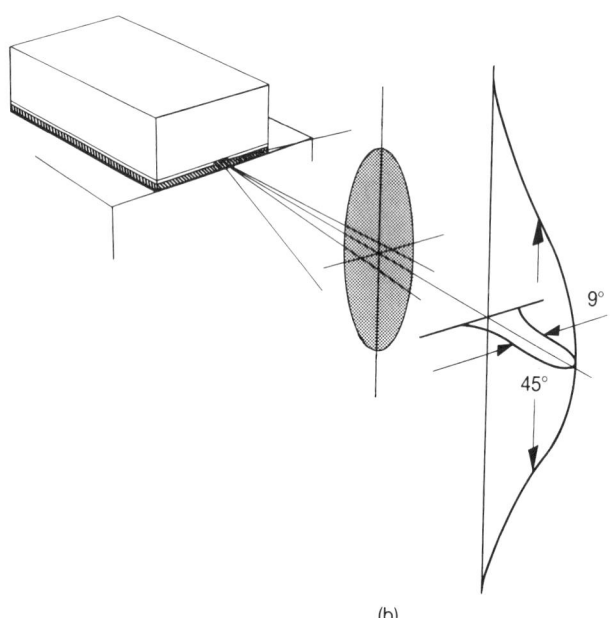

9°

45°

(b)

Figure 8.11 Stripe-geometry DH laser. (a) Construction. (b) Far-field emission pattern.

Fresnel reflection from an GaAs–air interface is about 32 percent, which is sufficient for laser action. The dimensions shown in Figure 8.11 are typical for the semiconductor lasers used for optical fiber communication. Proton bombardment causes certain regions of the semiconductor material to be made highly resistant, so that the current is confined to the 13 μm wide stripe. Layers of SiO$_2$ can be used as an alternative means for confining the current. The stripe-geometry technique not only reduces the active area – by keeping the operating current low – but also separates the active area from the surface, in such a way that the lifetime of the device is extended. The construction of the LED in Figure 8.8 includes an active region surrounded by optical and carrier confinement layers. The emitted light is transverse electrically (TE) polarized in the plane of the junction.

The stripe-geometry LD is one of the simplest and oldest types of laser. Many other designs have been developed, the most important of which will be discussed below.

In the buried-heterostructure laser (BH laser), which is depicted in Figure 8.12, the original design is a planar structure. Much of this planar structure is then etched away, so that a small mesa, containing the active region, remains on the substrate. Afterwards, the etched part is filled with a semiconductor material that has a lower refractive index than the active region. In this way optical guiding in the plane of the junction is also achieved. Lasers with such a property are called index-guided lasers, whereas the previous type is called a gain-guided laser. In the BH laser the active region is only 1 μm square. This produces only a small threshold current (10–15 mA) and a symmetrical far-field radiation pattern.

The next type to be considered is also an index-guided laser, but this time the etching is restricted to two small channels, one on each side of the active region, which are filled with a lower refractive index material. This so-called double-channel, planar, buried-heterostructure laser diode (DC-PBHLD) is shown in Figure 8.13.

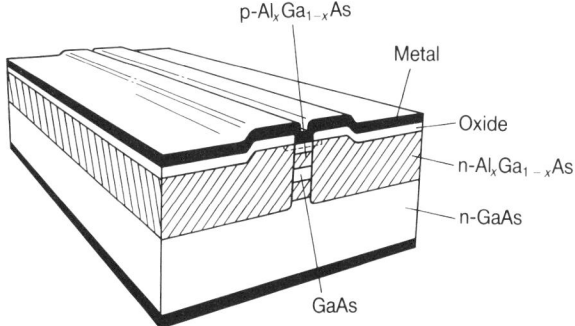

Figure 8.12 The buried-heterostructure (BH) laser.

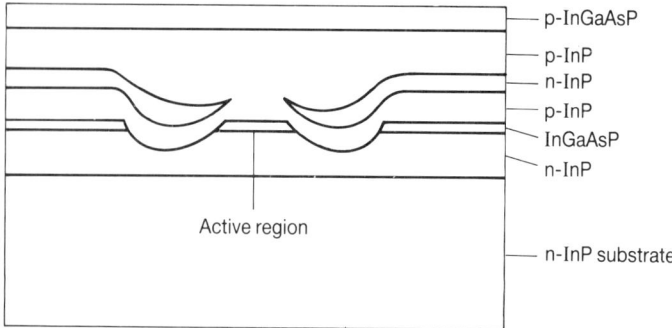

Figure 8.13 Cross-section of a DC PBH laser diode. (Source: reproduced with permission from *Electronics Letters*, vol. 22, no. 1, January 1986, pp. 16–18, 'Improved high-frequency response of InGaAsP double-channel buried-heterostructure lasers', by A. Valster, L. Meuleman, P. Kuindersma and T. van Dongen.)

It will be explained in the next section that lasers can have an optical spectrum that consists of several spectral lines, which is reduced to a single-line spectrum by introducing extra selectivity. In the distributed feedback (DFB) laser this is done by a periodically corrugated index in a layer which is near the active region. This periodic corrugation acts as an optical grating. Figure 8.14 shows the DBF laser as a DC-PBH laser diode. The single-mode character is also maintained when the laser is modulated.

In the distributed Bragg reflection (DBR) laser the corrugated section is situated outside the active region and is displaced in the longitudinal direction (see Figure 8.15).

The same semiconductor materials are used for an LD as for an LED. The

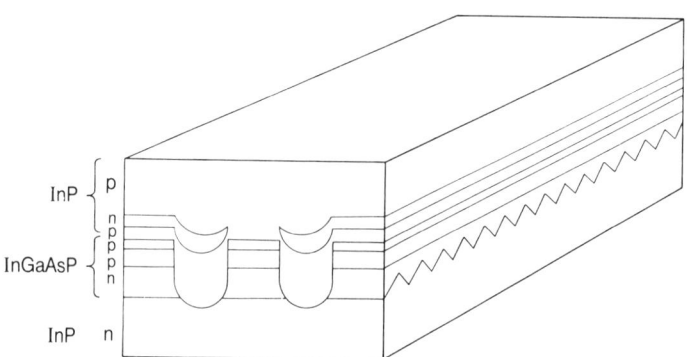

Figure 8.14 Structure of a DFB laser diode.

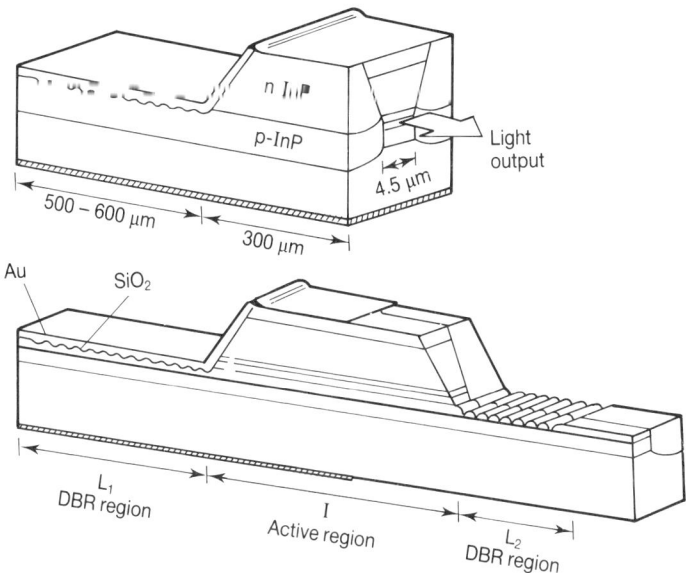

Figure 8.15 The DBR laser diode. (Source: reproduced with permission from *Electronics Letters*, vol. 17, no. 11, May 1981, pp. 366–8, 'CW operation of 1.5–1.6 μm wavelength GaInAsP/InP buried-heterostructure integrated twin-guide laser with distributed Bragg reflector', by K. Kobayashi, K. Utaka, Y. Abe and Y. Suematsu.)

quantum efficiency of semiconductor lasers ranges between 5 and 50 percent depending on the definition of this efficiency (see Chapter 9).

8.3.2 Coherence function and optical spectrum

The wavelength of a semiconductor laser is governed by the same laws as those for an LED, namely equations (8.1) and (8.2). One field component of the light emitted by an unmodulated laser is written as

$$E(t) = E_0 \cos[\omega_c t + \phi(t)] \tag{8.13}$$

where the amplitude E_0 is supposed to be constant. The phase $\phi(t)$ changes slowly with respect to $\omega_c t$ and is a stochastic process that performs a random walk [7]. Although the stochastic process (8.13) is non-stationary, it is ergodic in the autocorrelation function of its envelope [8]. The most important conse- quence of this is that, as far as the autocorrelation of this envelope is concerned, the ensemble mean equals the time average. Therefore, in equation (8.13) the autocorrelation of the envelope $R_{zz}(t_1, t_2)$ can be written as $R_{zz}(\tau) = R_{zz}(t_2 - t_1)$ and the process may be considered ergodic as far as the optical spectrum is concerned. If the laser oscillates in a single longitudinal mode the autocorrelation assumes the form [7]

$$R_{zz}(\tau) = A \exp\left(-|\tau|/\tau_c\right) \tag{8.14}$$

The time constant τ_c is called the coherence time, and multiplying τ_c by the speed of light c produces the coherence length L_c. The following spectrum is associated with equations (8.13) and (8.14):

$$S(\omega) = \frac{2A\,\tau_c}{1 + (\omega - \omega_c)^2 \tau_c^2} \tag{8.15}$$

Spectra of this shape are said to have a Lorentz profile centered around $f_c = \omega_c/2\pi$ and can be as small as 50 MHz. In the wavelength domain this corresponds to a linewidth less than 1 pm, whereas the coherence length is more than 10 m.

For stimulated emission the electromagnetic wave associated with the stimulated photon has to be in phase with the electromagnetic wave that caused the emission. Moreover, the oscillation conditions must be satisfied, i.e. the loop gain in the cavity has to have an absolute value of 1 and the loop phase shift must be an integer multiple of 2π. This phase condition can only be met for discrete values of the wavelength. If the free space wavelength is λ_0, then the wavelength in the cavity becomes $\lambda = \lambda_0/n$, where n is the refractive index of the cavity material ($n \approx 3.6$ for AlGaAs). A standing wave occurs in the cavity, if twice the length L of the cavity corresponds to an integer multiple m of the wavelength λ. Mathematically, this condition looks like

$$m\frac{\lambda_0}{n} = 2L \tag{8.16}$$

or

$$\lambda_0 = \frac{2nL}{m} \tag{8.17}$$

Equation (8.17) shows that a laser spectrum can consist of several lines, each corresponding to one of the various values of m. Those spectral lines are called longitudinal modes and the mutual distance between them is found by differentiating λ_0 with respect to m. However, it must be kept in mind that the refractive index also depends on λ_0, so it is easier to start with

$$\frac{\mathrm{d}m}{\mathrm{d}\lambda_0} = \frac{\mathrm{d}}{\mathrm{d}\lambda_0}\left(\frac{2nL}{\lambda_0}\right) = \frac{-2NL}{\lambda_0^2} \tag{8.18}$$

with

$$N \triangleq n - \lambda\frac{\mathrm{d}n}{\mathrm{d}\lambda} \tag{8.19}$$

The quantity N is called the group index (see equation 4.58). For large values of m, the mutual distance between the longitudinal modes becomes

$$|\Delta\lambda| \approx \frac{\lambda_0^{\,2}}{2NL} \approx \frac{\lambda_0^{\,2}}{2nL} \qquad\qquad (8.20)$$

The last approximation in this equation follows from Figure 4.9. Practical values of the various parameters in equation (8.20) lead to mode spacing values in the range 0.1–1 nm, depending on the wavelength.

Increasing the temperature has two effects: the bandgap decreases and the refractive index increases. The net effect is a displacement of the spectrum to longer wavelengths at a rate of 0.2–0.7 nm K^{-1}, depending on the operating wavelength. When the laser oscillates in the different longitudinal modes, this influences the coherence function, and from Fourier analysis we know that a periodic function results in the time domain too. A typical example is shown in Figure 8.16, where the equivalent baseband spectrum $S_{zz}(\omega)$ is presented, together with its inverse Fourier transform, the autocorrelation function of the envelope $R_{zz}(\tau)$.

The individual lines of the spectrum in Figure 8.16 show a Lorentz profile, while in that figure both the envelope in the frequency domain and the lines in the time domain have a Gaussian profile. In the time domain the envelope is an exponential. At the bottom of Figure 8.16 the relationships between the functions in the time and frequency domain are presented. A typical value for the distance between the lines in the wavelength domain is about 0.5 nm, while the width of the envelope can be 5 nm. The shape of the envelope of this spectrum can be explained as follows. We used the phase condition (equations 8.16–8.20) to show the existence of various spectral lines; and there is also the

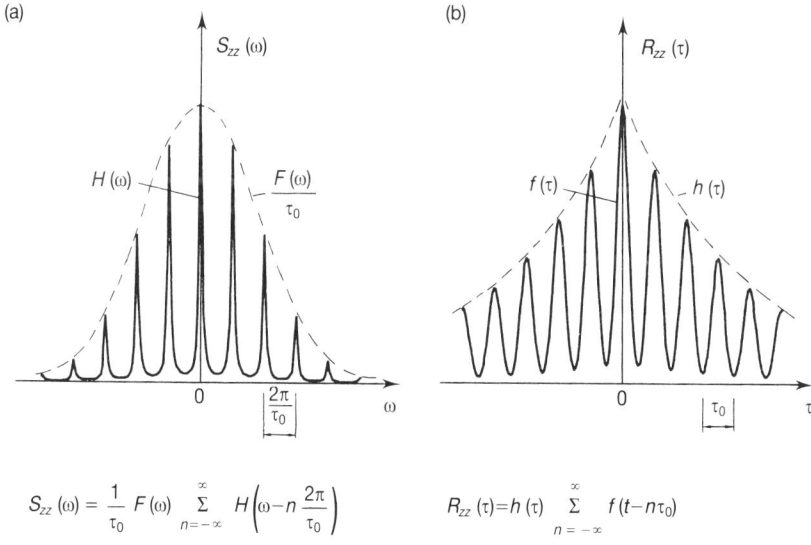

$$S_{zz}(\omega) = \frac{1}{\tau_0} F(\omega) \sum_{n=-\infty}^{\infty} H\!\left(\omega - n\,\frac{2\pi}{\tau_0}\right) \qquad\qquad R_{zz}(\tau) = h(\tau) \sum_{n=-\infty}^{\infty} f(t - n\tau_0)$$

Figure 8.16 A typical equivalent baseband line spectrum (a) and the corresponding coherence function (b) of a multimode laser.

amplitude condition (loop gain ≥1). This gain depends on the wavelength, due to the spread in the energy levels of the carriers on both sides of the bandgap, as mentioned in Section 8.2.3. Consequently, the envelope of the spectral lines has a similar shape to the LED spectrum, for those wavelengths where the loop gain exceeds unity. As a result the complete laser spectrum width is narrower than the LED spectrum. Further spectral narrowing is achieved by means of extra selectivity, as for the DFB and DBR lasers.

In addition to its longitudinal modes, and the related multiple line spectrum, a laser can also oscillate in various lateral and transverse modes [9]. These modes will be related to the continuity conditions in, respectively, the plane of the p–n junction and the direction perpendicular to this plane. In the transverse direction the active area is usually substantially smaller than the wavelength, and therefore a semiconductor laser nearly always operates in the fundamental transverse mode. For the BH laser also the lateral dimension of the active area is substantially smaller than the wavelength, but this certainly does not apply to the stripe-geometry laser. High-order lateral modes lead to satellite spectral lines near the longitudinal mode lines; in fact, so near that they are difficult to measure. From a spectral point of view these modes are not so important, although they show a property that is important in other respects. A laser operating in the fundamental transverse and later mode has a radiation pattern that has a Gaussian shape [9]. High-order lateral or transverse modes disturb this pattern; in the radiation pattern, maxima and minima occur and, moreover, the pattern becomes wider [1] (see Figure 8.17). A wide radiation pattern with a deviation from the Gaussian shape leads to a poor laser–fiber coupling efficiency (see Chapter 11).

8.4 Semiconductor laser versus LED

When deciding whether to choose an LED or an LD as the light source in a particular optical communication system, the main features to be considered are the following:

• The optical power versus current characteristics of the two devices differ considerably (see Figure 8.18). Near the origin the LED characteristic is linear, although it becomes non-linear for larger power values. However, the laser characteristic is linear above the threshold. Sometimes a multimode laser is slightly non-linear above the threshold, this is caused by mode-hopping – at certain current values the maximum emission suddenly hops to an adjacent spectral line. Single-mode lasers show an excellent linear characteristic above the threshold. Linearity of the source is important for analog systems, but is less important for digital systems. The power-to-current characteristic of an LD depends greatly on temperature (see Figure 8.18), but this dependence is not so great for an LED. The power supplied by both devices is similar (about 10–20 mW). However, the maximum coupling efficiency of a fiber is

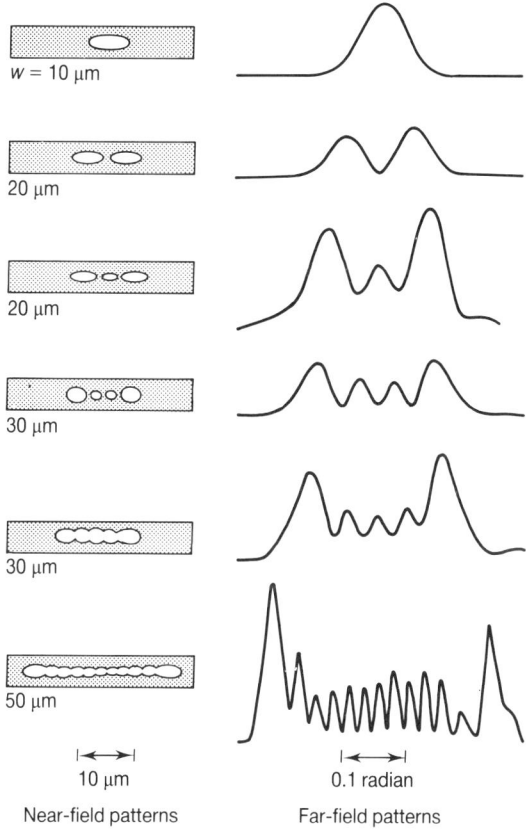

$w = 10\ \mu m$

$20\ \mu m$

$20\ \mu m$

$30\ \mu m$

$30\ \mu m$

$50\ \mu m$

|←→|
$10\ \mu m$

|←→|
0.1 radian

Near-field patterns   Far-field patterns

Figure 8.17 Near-field and far-field patterns of lateral modes in a laser diode. (Source: reproduced with permission from *Japanese Journal of Applied Physics*, vol. 12, 1973, pp. 1585–92, 'A GaAs–Al$_x$Ga$_{1-x}$As double heterostructure planar stripe laser' by H. Yonezu *et. al.*)

much smaller for an LED than for an LD: for an LED it is 5–10 percent, but for an LD it can be up to 90 percent. This difference in coupling efficiency has to do with the difference in radiation geometry of the two devices (see also Chapter 11).

• As an LED emits spontaneous radiation, the speed of modulation is limited by the spontaneous recombination time of the carriers. LEDs have a large capacitance and modulation bandwidths are not very large (a few hundred megahertz). The capacitance can be reduced by biasing the diode with a forward current, which increases the modulation speed. For a laser above the threshold the electrons remain in the conduction band for a very short time, due to the stimulated recombination; therefore, very fast modulation is possible (up to 10 GHz).

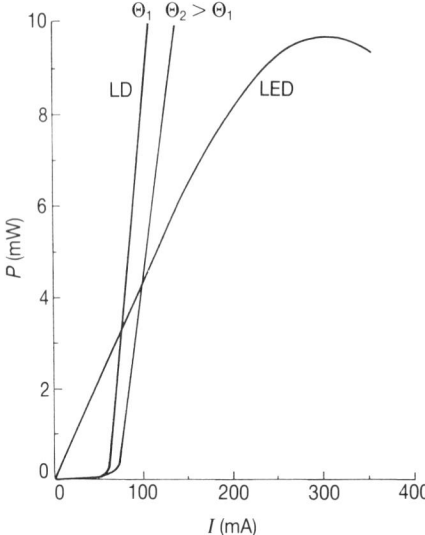

Figure 8.18 Power versus current characteristic of an LD and an LED.

- LDs have narrower spectra than LEDs, and the single-mode lasers, in particular have a very narrow spectrum. This explains why the pulse broadening at transmission through an optical fiber is very small. Therefore, with an LD as a light source, wideband transmission systems can be designed. The spectrum of an LD remains more stable with temperature than that of an LED.
- Changes of power output for an LD with temperature can be prevented by stabilizing the heat sink temperature with a Peltier element and a control circuit. This generally requires more complicated electronic circuits than for an LED.
- The expected lifetime of both an LD and an LED is around 10^7 hours, which is sufficient for practical purposes. LEDs can withstand power overloading for short duration better than LDs.
- At current prices, LEDs are less expensive than LDs.

8.5 Photodiodes

8.5.1 Introduction

The photodetector is an important component of optical fiber communication systems; it converts the optical signal into an electrical information signal, which is amplified and then processed in order to restore the transmitted signal (see

Figure 8.1). The most important requirements of the detector are as follows:

1. Great sensitivity in the wavelength region where the light source operates. Originally, wavelengths of 0.8–0.9 μm – the region where AlGaAs LEDs and LDs emit – were important. Nowadays, interest is moving towards the region 1.3–1.6 μm (see Section 8.1).
2. Sufficient bandwidth or speed of response, in order to process the information carrying signal. The bandwidth of today's commercial systems is a few hundred megahertz and this will extend to a few gigahertz in the coming years.
3. Little excess noise from the detector. Moreover, the dark current, the leakage current, and the parasitic induction and capacitance should be kept as small as possible.
4. Low sensitivity to temperature changes.

Semiconductor photodiodes meet all these requirements. They are small and compact, relatively cheap and can be coupled to a fiber easily. Si diodes are highly sensitive in the wavelength region 0.8–0.9 μm; they also have a large bandwidth (several gigahertz), have small dark and leakage currents, low capacitance, little excess noise, and long-term stability. Since their response decreases rapidly above 1 μm, they are unsuitable for the long wavelengths in the range 1.3–1.6 μm. Ge or InGaAsP diodes are satisfactory for these wavelengths, as will be shown in the next section.

8.5.2 Principles of the photo-detection process

In semiconductor photodiodes the basic detection process consists of generating hole–electron pairs by photons; this process has to be initiated in the vicinity of a p–n junction. Two different detection modes are available. In the photovoltaic mode the electrons are gathered by diffusion to one side of the junction, and the holes go to the other side. This mode corresponds to an unbiased diode. Each type of carrier is concentrated on a different side of the junction; this produces a voltage across the junction and a current flows if the device is loaded. In the second mode, the photoconductive mode, the holes are separated from the electrons by a high electric field, such as in the depletion layer of a reverse-biased p–n junction. Drift of the carriers in the depletion layer induces a current in the outer circuit.

The relationship between the current and the voltage of a photodiode is given by the expression

$$I = I_0 \left[\exp(eV/\gamma k\Theta) - 1 \right] - I_p \tag{8.21}$$

where I_0 is the dark (leakage) current and γ is a factor representing the quality of the junction ($\gamma = 1$ for an ideal junction and $\gamma = 2$ for a junction where the dark current consists mainly of surface leakage). In equation (8.21) I_p represents the photocurrent, which will be evaluated later on. Two conditions of

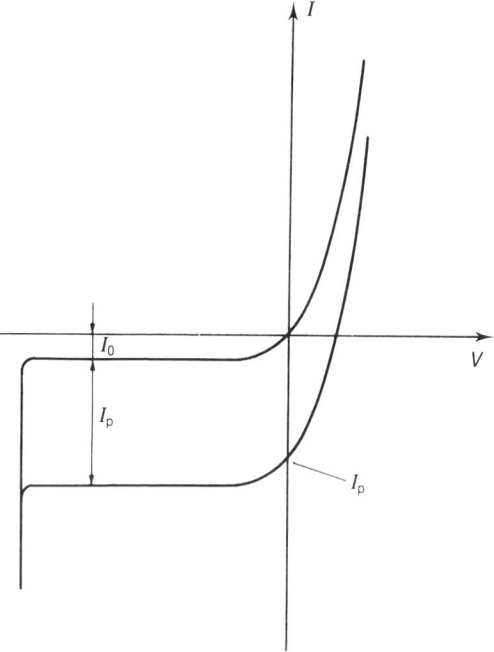

Figure 8.19 The $I-V$ characteristic of a photodiode under illuminated ($I_p \neq 0$) and dark ($I_p = 0$) conditions.

equation (8.21) are depicted in Figure 8.19: $I_p = 0$ and $I_p \neq 0$. In the photovoltaic mode the diode can be used in two different ways. For further amplification and processing we can use the open voltage or the short-cut current (see also Section 12.1). In the first case there will be a small voltage across the diode, while in the second case the voltage will be zero. In both cases the holes and electrons are separated by diffusion, and thus the diode reacts slowly to variations in optical power.

When a reverse bias is applied to the photodiode the photoconductive mode results and the carriers drift faster due to the electric field. The photodiode is therefore faster acting in this mode than in the photovoltaic mode and we shall restrict discussion here to the photoconductive mode. A large quantum efficiency (the mean number of hole–electron pairs per incident photon) requires the absorption of a large amount of the incident light, and this means having a thick depletion layer. On the other hand, the drift time has to be reduced in order to give the diode a faster response, and this requires a thinner depletion layer. Thus, there must be a compromise between the quantum efficiency and the speed of response.

If the photon energy is equal to or larger than the bandgap energy of the semiconductor, then a hole–electron pair can be generated. In a well-designed photodiode this process occurs in the depletion region of the p–n junction, where the incident light is absorbed. The high field in that region causes the holes and electrons to be separated, as shown in Figure 8.20. Carriers generated outside this depletion layer, but within the diffusion length of each side of this layer, can drift into this region and, in doing so, also contribute to the photocurrent. By increasing the reverse bias voltage, the depletion region widens and the capacitance of the junction decreases. Moreover, more photons are absorbed and the drift velocity of the carriers increases, so that the photodiode becomes faster and more sensitive. If the field strength in the depletion region is large enough (for Si this field strength is greater than 10^5 V cm^{-1}), a hole or electron can gain sufficient energy from the field to cause an ionizing collision, thereby creating a new electron–hole pair. These new carriers, in their turn, can again cause ionizing collisions until an avalanche of carriers occurs. As long as the diode voltage is below its breakdown voltage, the

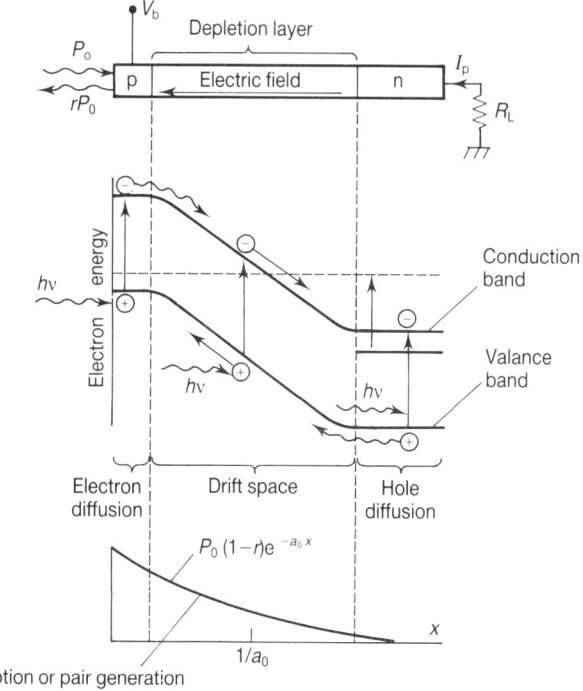

Figure 8.20 The operating principle of a semiconductor photodiode. (Source: reproduced with permission from *Journal of Luminescence*, vol. 7, 1973, pp. 390–414, 'Sensitive high speed photodetectors for the demodulation of visible and near infrared light', by H. Melchior.)

total number of carriers is limited and proportional to the number of primary carriers (these are the carriers generated directly by a photon). The average gain G depends on the frequency of the information signal and can be written as [10]

$$G(\omega) = \frac{G_0}{\sqrt{1 + (\omega G_0 \tau)^2}}$$ (8.22)

where τ is a time constant that depends on the structure of the diode.

For a photodiode two parameters are of importance: the quantum efficiency η and the responsivity R. The quantum efficiency is defined as the average number of primary carrier pairs per incident photon impinging on the diode, and the responsivity as the average photocurrent per unit of incident optical power.

In the following sections these parameters will be expressed in terms of physical constants, material properties and dimensions of the photodiode.

The quantum efficiency relates to the actions of the photons as they penetrate into a photodiode. First of all, assuming that they come from air, they will strike the thin p-layer (see Figure 8.20). At this interface, Fresnel reflection $r = (n_1 - n_2)^2/(n_1 + n_2)^2$ occurs, where n_1 and n_2 are the refractive indices of air and the p-layer, respectively. Applying an anti-reflection coating to the p-layer reduces the reflection. The fraction $(1 - r)$ of the incident photons enters the p-layer, where some are absorbed. Let us assume that the number absorbed is proportional to the local photon density and call the absorption coefficient α (unit m^{-1}); then, the photon density decreases exponentially with the distance traversed through the material. When the thickness of the p-layer is w_p the fraction $(1 - r)\exp(-\alpha w_p)$ is left on arrival in the depletion region. Absorption in this region generates an electron–hole pair. If the width of the depletion region is w_d, then the fraction $1 - \exp(-\alpha w_d)$ of the photons penetrating this region generates an electron–hole pair. Hence, the quantum efficiency is

$$\eta = (1 - r)\exp(-\alpha w_p)\,[1 - \exp(-\alpha w_d)]$$ (8.23)

It can be concluded that the quantum efficiency will be large if the Fresnel reflection is small, the p-layer is thin and the depletion layer is thick. It has been shown earlier that, as far as the depletion layer is concerned, there is a trade-off with the speed of response.

In order to arrive at the responsivity we assume that an optical power P_0 impinges on the photodiode. During each unit of time, $P_0/h\nu$ photons arrive, with an optical frequency of ν. According to the definition of quantum efficiency, this means that $\eta P_0/h\nu$ primary carrier pairs are created per unit of time. The number of secondary carrier pairs is $\eta P_0 G/h\nu$, resulting in a current of $e\eta P_0/h\nu$ (e = electron charge = 1.6×10^{-19} coulomb). Therefore, the responsivity becomes

$$R = \frac{e\eta G}{h\nu} = \frac{e\eta G}{hc}\lambda$$ (8.24)

For an ideal photodiode ($\eta = 1$) without internal gain ($G = 1$), $R = \lambda_0/1.24$ A W^{-1}, when λ_0 is expressed as μm. In Figure 8.21 the responsivities of a typical Si photodiode, a typical Ge photodiode and a typical InGaAsP photodiode, all without internal gain, are depicted as functions of the wavelength. Figure 8.21 also shows the greatest responsivity that can be calculated from equation (8.24) with $\eta = 1$ (dashed line). With regards to these characteristics, three main regions can be distinguished.

A middle region can be seen where the responsivity depends almost linearly on λ_0 and thus obeys equation (8.24), assuming η to be constant. However, this quantum efficiency depends on λ_0, since the absorption coefficient α depends on the wavelength (see Figure 8.22). In the short-wavelength region ($\lambda_0 < 0.4 \ \mu$m) the value of R decreases more rapidly than λ_0; this is caused by an increased photon absorption in the p-layer (see the first exponential in equation 8.23 and Figure 8.22). A third region is found beyond the maximum responsivity value, where the photon energy has become smaller than the bandgap energy, so that the photons fail to create electron–hole pairs. The responsivity in this region rapidly declines as the absorption coefficient decreases (see Figure 8.22).

When designing an optical communication system, the best decision is to choose a photodiode material with a bandgap energy that is a little smaller than the photon energy of the longest-wavelength component of the source. This guarantees maximum responsivity, while keeping the dark current as low as possible. From Figure 8.22 it follows in principle that all the materials illustrated

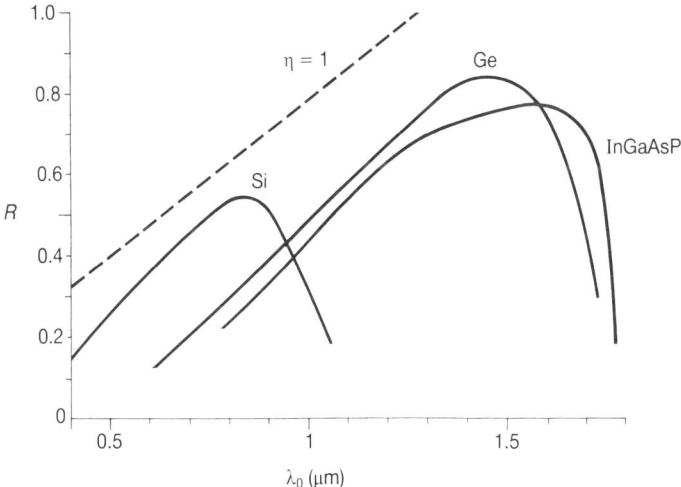

Figure 8.21 The responsivities of a Si photodiode, a Ge photodiode and a InGaAsP photodiode, all without internal gain, as functions of the wavelength.

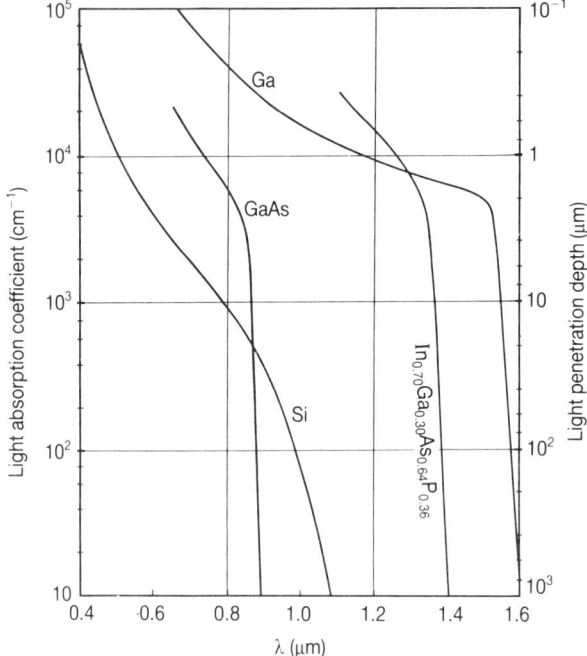

Figure 8.22 The light absorption coefficient and penetration depth of various photodiode materials.

are useful in the wavelength region of $0.8-0.9\,\mu m$, whereas for the longer wavelengths of $1.3-1.6\,\mu m$ only Ge and InGaAsP can be considered. The latter material has the advantage of a bandgap that can have different values depending on the composition of the compound. Due to its smaller bandgap, Ge gives a greater dark current than Si. Below $1\,\mu m$, Si is preferable on account of its smaller dark current, and also because its technology is more widely developed.

8.5.3 Construction of photodiodes without internal gain

Photodiodes without internal gain have a simple p–n or p–i–n structure. In the p–n diode a thin p$^+$ diffusion is applied to an n-layer; while in the PIN diode the p$^+$ diffusion is applied to a ν-layer. A ν-layer is a very lightly doped, almost intrinsic n-layer (see Figure 8.23). When light power enters the device in the direction perpendicular to the junction it is called a front-illuminated photodiode. The distribution of the electric field in a PIN diode is shown in Figure 8.24. In a p–n diode the depletion region is about $1-3\,\mu m$ wide, which is optimal for detecting both visible light with Si and near infra-red light with Ge. For shorter wavelengths, light is absorbed in the p$^+$ layer, which leads to a poor

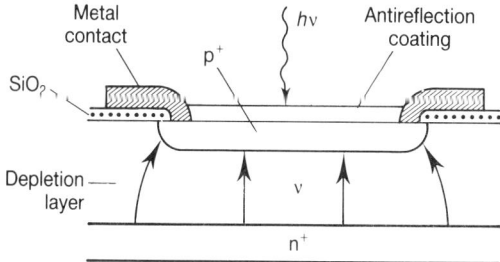

Figure 8.23 Construction of a front-illuminated PIN photodiode. (Source: reproduced with permission from *Journal of Luminescence*, vol. 7, 1973, pp. 390–414, 'Sensitive high speed photodetectors for the demodulation of visible and near infrared light', by H. Melchior.)

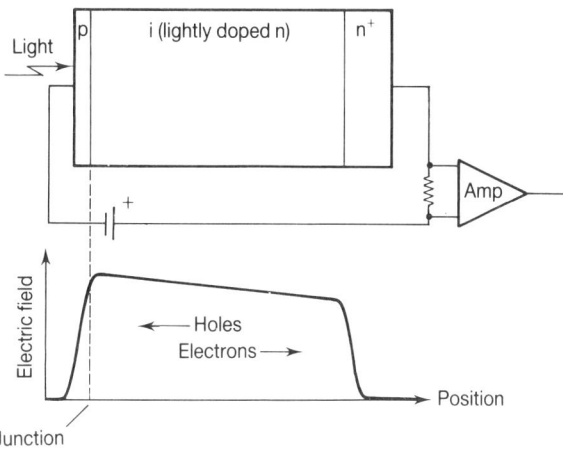

Figure 8.24 Field distribution in a PIN photodiode. (Source: reproduced with permission from *Proceedings of the IEEE*, vol. 65, no. 12, December 1977, pp. 1670–8, 'Receiver design for optical fiber systems', by S. Personick.)

quantum efficiency. In the wavelength region of 0.8–0.9 μm, light penetrates deeper into the semiconductor material and, since PIN diodes have a wider depletion region, they are preferred. For Si diodes a depletion region width of 20–50 μm is required to achieve a high quantum efficiency ($> 70\%$) for those wavelengths. Nevertheless, response times of around 1 ns are possible. At $\lambda_0 = 1\ \mu$m the absorption length required in Si is 500 μm, thus giving a low bandwidth for Si diodes.

A large quantum efficiency with a thin depletion layer is possible for a side-illuminated photodiode, such as that shown in Figure 8.25. The light enters

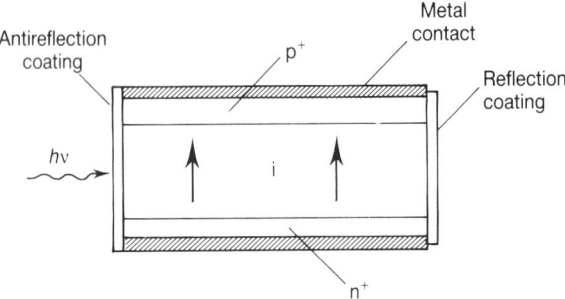

Figure 8.25 Side-illuminated PIN photodiode. (Source: reproduced with permission from *Journal of Luminescence*, vol. 7, 1973, pp. 390–414, 'Sensitive high speed photodetectors for the demodulation of visible and near infrared light', by H. Melchior.)

the diode along the plane of the junction. Such a Si photodiode can have a quantum efficiency of 90 percent and a bandwidth of 1 GHz at $\lambda_0 = 1\ \mu$m, but it has the disadvantage of being difficult to couple light into the narrow depletion region.

Ge diodes are useful over all the wavelengths in the region now under consideration for optical fiber communication (0.8–1.6 μm). The relatively large dark current and the excess noise of this material are serious drawbacks.

8.5.4 Construction of photodiodes with internal gain

Photodiodes with internal gain are called avalanche photodiodes (abbreviated to APDs) and they are more complicated to construct than the PIN diode. Except for a drift region in the depletion layer, where most of the primary carriers are generated, these devices are provided with a high-field region, in which multiplication of the carriers occurs. A diagram of the field distribution is shown in Figure 8.26 [11]. It is important to produce little excess noise in these devices and this means restricting the current entering the multiplication region to carriers of only the higher ionization rate (see [12] and Chapter 12). In Si, electrons show the higher ionization rate, but in Ge and InGaAsP the holes have a higher ionization rate. The reach-through structure (RAPD) has excellent noise properties and consists of $p^+-\pi-p-n^+$ layers as shown in Figure 8.27(a). For low reverse-bias voltages most of the voltage drops across the $p-n^+$ junction. When the voltage increases the depletion region extends mainly into the p-layer and, at a certain voltage, reaches through the almost intrinsic π-layer. Semiconductor layers that are completely depleted have a constant space charge. From Maxwell's law $\nabla.\mathbf{D} = \rho$, where ρ is the space charge, it follows that the derivative of the field in such a layer becomes constant. Further voltage increases do not change the shape of the field distribution, but only raise

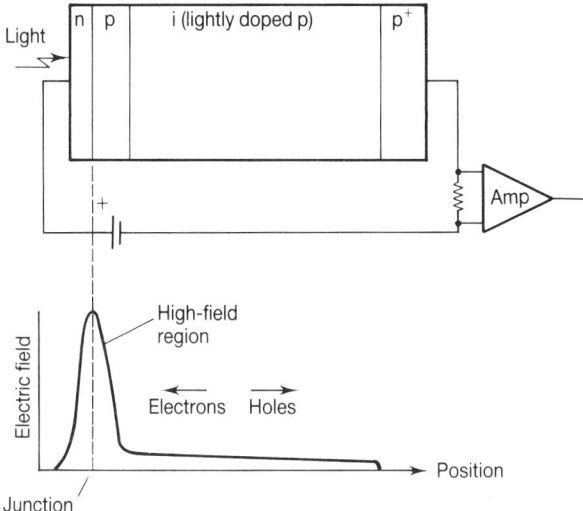

Figure 8.26 Field distribution in an APD. (Source: reproduced with permission from *Proceedings of the IEEE*, vol. 65, no. 12, December 1977, pp. 1670–8, 'Receiver design for optical fiber systems', by S. Personick.)

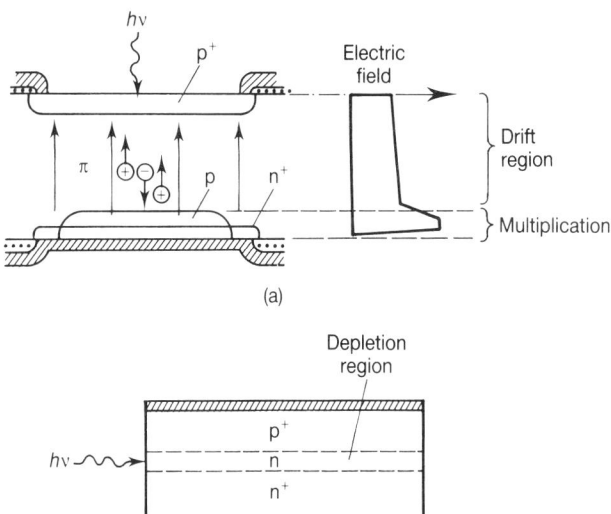

Figure 8.27 (a) Front-illuminated, reach-through APD (RAPD). (b) Side-illuminated p$^+$–n–n$^+$ APD. (Source: reproduced with permission from *Journal of Luminescence*, vol. 7, 1973, pp. 390–414, 'Sensitive high speed photodetectors for the demodulation of visible and near infrared light', by H. Melchior.)

the amplitude, which means that this voltage increase occurs mainly in the π-region. Since the π-region is much wider than the p-layer, the field strength in the multiplication region increases relatively slowly with increasing voltage. The field in the π-region is lower than in the vicinity of the p–n junction, but it is high enough to cause the carriers to drift quickly enough to keep the response time short. As the light is almost completely absorbed in the π-layer, a very pure electron current is injected into the high-field region near the p–n$^+$ junction.

APDs are also made with a side-illuminated p$^+$–n–n$^+$ structure (see Figure 8.27b).

Si APDs can have a quantum efficiency as high as 85 percent with a bandwidth that approaches 1 GHz. They are rather susceptible to temperature changes, as can be seen from Figure 8.28, where the average gain is shown as a function of the reverse voltage for an APD and an RAPD. The latter type is less sensitive in this respect than a standard APD. Its mean gain decreases with rising temperature and this effect is explained as follows. An increasing temperature causes the lattice vibrations to have large amplitudes. Consequently, the collision-free distance travelled by the carriers decreases, so that between two collisions the carriers cannot obtain sufficient energy from the field to cause an ionizing collision.

Materials for the long wavelengths of 1.3–1.6 μm, such as Ge and InGaAsP, must have a smaller bandgap, which produces a large dark current; it is then easier for the thermal energy $k\Theta$ to create an electron–hole pair. Moreover, for those materials the ionization rate ratio of electrons to holes is larger than the ionization rate ratio of holes to electrons for Si, whereas this situation is somewhat more favourable for InGaAsP than for Ge (see [4] and Chapter 12). Due to this fact the excess noise from long-wavelength APDs is larger than from Si APDs. Thus the advantage of the APD, namely an improved signal-to-noise ratio, vanishes.

References

[1] J. Gowar, *Optical Communication Systems* (London: Prentice Hall, 1984).

[2] M. C. Hudson, 'Calculation of the maximum optical coupling efficiency into multi-mode optical waveguides', *Applied Optics*, vol. 13, no. 5, May 1974, pp. 1029–33.

[3] Kuo-Liang Chen and D. Kerps, 'Coupling efficiency of surface emitting LED's to single-mode fibers', *Journal of Lightwave Technology*, vol. LT-5, no. 11, November 1987, pp. 1600–4.

[4] S. E. Miller and A. G. Chynoweth, *Optical Fiber Telecommunications* (New York: Academic Press, 1979).

[5] Y. Suematsu and K. Iga, *Introduction to Optical Fiber Communications* (New York: Wiley, 1982).

[6] A. Papoulis, *Probability, Random Variables and Stochastic Processes* (New York: McGraw-Hill, 1984).

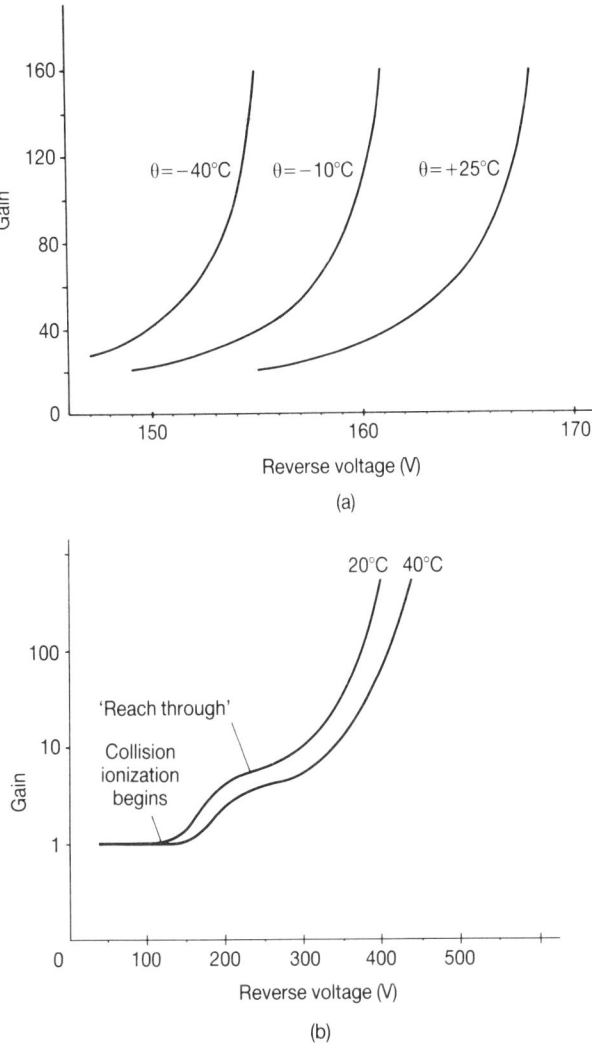

Figure 8.28 The average gain of an APD as a function of the reverse voltage for several temperatures: (a) a standard APD; (b) a RAPD.

[7] A. Maitland and M. Dunn, *Laser Physics* (Amsterdam: North-Holland, 1969).

[8] W. van Etten, 'The ergodicity of laser light in connection with optical fibre transmission', *Optical and Quantum Electronics*, vol. 13, no. 6, November 1981, pp. 519–21.

[9] T. H. Zachos and J. E. Ripper, 'Resonant modes of GaAs junction lasers', *IEEE Journal of Quantum Electronics*, vol. QE-5, January 1969, pp. 29–37.

[10] R. B. Emmons, 'Avalanche-photodiode frequency response', *Journal of Applied Physics*, vol. 38, 1967, pp. 3705–14.

[11] S. D. Personick, 'Receiver design for optical fibre systems', *Proceedings of the IEEE*, vol. 65, no. 12, December 1977, pp. 1670–8.
[12] R. J. McIntyre, 'Multiplication noise in uniform avalanche diodes', *IEEE Transactions on Electronic Devices*, vol. ED-13, January 1966, pp. 164–8.

Problems

8.1 Consider a step index silica fiber. The relative refractive index difference of core and cladding amounts to 1 percent. Only guided rays are considered. The fiber is excited by an LED, which operates at a wavelength of 800 nm and is at room temperature.

   Calculate the material dispersion and the mode dispersion per unit length. Which is the larger of the two?

8.2 A laser oscillates in a single longitudinal mode. Its coherence length appears to be 0.3 m at a wavelength of 800 nm.

   (a) Calculate the spectral width of this light source, defined by the width of the spectrum at half maximum.
   (b) The laser is used as a light source in an optical fiber communication system; the length of the fiber is 100 km. Calculate the maximum bandwidth of the system, assuming that the dispersion of the link is determined by the material dispersion of silica. NB: Assume that the bandwidth is the inverse of the dispersion and use the spectral width found in (a).

8.3 A multimode laser has a 300 μm long cavity. The group index of the laser material is 3.6. What is the distance between adjacent spectral lines at $\lambda_0 = 800$ nm and $\lambda_0 = 1300$ nm?

8.4 An optical fiber communication system operates at a nominal wavelength of 800 nm. The light source consists of a GaAs laser with a cavity length of 400 μm. Its optical spectrum comprises six spectral lines. The link consists of SiO_2 fiber with a length of 5 km, while a Si diode is used as the detector.

   (a) Calculate the distance between adjacent spectral lines of the laser (take the group index of the laser material equal to its refractive index).
   (b) Calculate the material dispersion of the link.
   (c) What is the maximum bit rate for digital communication over the link, if it is assumed that material dispersion is the limiting factor?

8.5 A front-illuminated Si PIN photodiode ($\varepsilon_r = 12$) is used at a wavelength of 850 nm. The quantum efficiency amounts to 0.8.

   Calculate the thickness of the depletion layer if:

   (a) the diode is provided with an ideal anti-reflection coating (no Fresnel reflection);
   (b) no anti-reflection coating is applied.

9

Modulation of semiconductor light sources

9.1   The rate equations
9.2   The laser condition
9.3   The efficiency of lasers
9.4   The turn-on delay of a laser and the behaviour of an LED
9.5   Transient behaviour of a laser
9.6   Modulation of a laser by small signals
9.7   Amplitude noise of lasers
      References
      Problems

In this chapter we will consider the relationship between the electrical input and the optical output of semiconductor light sources, both for an LED and a laser diode. In other words, we describe the interaction between electrons and photons in such devices. This is necessary for understanding the transient response of these light sources, as well as the modulation of the light output by analog or digital information signals. During this treatment we shall describe light properties in terms of photons, by means of electromagnetic waves. We shall start by deriving some fundamental equations for a laser and showing how these equations can also be used to arrive at some interesting properties of LEDs.

9.1 The rate equations

As a starting point for our analysis we take the so-called rate equations [1]. These are a pair of coupled differential equations which describe the density of injected carriers N_e and the density of stimulated photons N_p in the cavity of a laser. The equations are valid for laser diodes with the following assumptions:

1. The laser operates in a single mode above the threshold.
2. The cavity is ideal and has a homogeneous population inversion and the density of carriers and photons is homogeneous.
3. The stimulated photon gain per unit of time is a linear function of the carrier density, as far as it exceeds a minimal value N_0.
4. Noise is excluded from consideration.

In their simplest form the rate equations read:

$$\frac{dN_e}{dt} = \frac{I}{eV} - A(N_e - N_0)N_p - \frac{N_e}{\tau_s} \tag{9.1}$$

$$\frac{dN_p}{dt} = A(N_e - N_0)N_p - \frac{N_p}{\tau_p} + \beta\frac{N_e}{\tau_s} \tag{9.2}$$

After modification these equations can be changed in order to make them valid for multimode operation [2]. The left-hand side of equation (9.1) represents a change of the carriers in the cavity per unit of time. Assuming a uniform distribution of the injected carriers over the volume V of the active region, the first term on the right-hand side of this equation gives the injected carrier density per unit of time, where I is the total forward current through the device. The second term on the right-hand side of this equation shows the number of carriers that recombine to emit a coherent photon, when stimulated by the local wave. This term agrees with the third assumption, where $A(N_e - N_0)$ is the stimulated photon gain; note that the units of A are $m^3\,s^{-1}$. Finally, the third term is the spontaneous recombination, and is the driving force for the whole process. The spontaneous recombination rate is proportional to the carrier density N_e; the proportionality factor is $1/\tau_s$, where τ_s is the mean lifetime of the carriers that recombine spontaneously.

Let us now look at the photons, in other words equation (9.2). The left-hand side represents the change of the photon density in the cavity per unit of time. On the right-hand side the first term is the same as the second term on the right-hand side of equation (9.1). The reversed sign of these two terms shows that the stimulated recombination of a carrier represents the generation of a photon. Photons can be described in another way as electromagnetic waves, travelling at the speed of light. In such a model some photons are considered to be lost from a limited space, such as in the cavity of a laser. The number of photons disappearing per unit of time is proportional to the photon density. The process depends on the time constant τ_p, which is the mean lifetime of the stimulated photons, and is expressed by the second term. In the next section a further explanation will be given for the photon lifetime. The last term of equation (9.2) relates to the fraction β of the spontaneous recombinations that contributes to the coherent light. This term is very small; however, it will be shown in Section 9.5 that it is very important as far as the transient behaviour of a laser is concerned and can be neglected when considering other effects.

9.1.1 The photon lifetime

When a number of photons are present in an optical cavity and they are described by an electromagnetic wave, then this wave propagates in the cavity with the speed of light c. As far as a laser is concerned, we are interested in a coherent wave, i.e. when we start with a wave propagating in a certain direction perpendicular to the end-faces of the cavity, then the amplitude and the phase of that wave after an even number of reflections from the mirrors formed by the end-faces of the cavity are important. Based on considerations of phase, it is possible to determine the longitudinal modes of a laser (see Chaper 8). Now we shall analyze the amplitude condition of the oscillator, which is equivalent to the power reflection condition.

In Figure 9.1 a laser cavity is shown diagrammatically. The power is partly reflected from the end-faces according to the ratio R; the length of the cavity is L. Let us assume that I_0 photons are injected, at a certain point x_1 in the cavity. The power loss per unit of cavity length is α. When the wave returns to its starting point x_1 after a round trip in the cavity the power of the wave is $I_0 R^2 \exp(-2\alpha L)$. From the point of injection the elapsed time for the round trip reads

$$t_1 = \frac{2L}{c_1} = \frac{2Ln}{c_0} \tag{9.3}$$

After two round trips the power of the wave will have changed by a factor of $R^4 \exp(-4\alpha L)$; this means that the power drops off exponentially as a function of time. The time constant τ_p of this exponential is called the lifetime of the

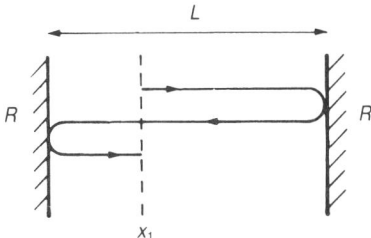

Figure 9.1 Diagram of an electromagnetic wave in a laser cavity with partially reflecting end-faces.

photons in the cavity. This time constant can be calculated, because the power reduction factor after each round trip and the round trip delay are known. Combining these factors yields

$$\exp\left(-t_1/\tau_p\right) = R^2 \exp\left(-2\alpha L\right) \tag{9.4}$$

Substituting equation (9.3) in equation (9.4) and solving for τ_p produces

$$\tau_p = \frac{n}{c_0} \frac{1}{\alpha + (1/L)\ln(1/R)} \tag{9.5}$$

Example 9.1

An impression of the value of the photon lifetime can be obtained by considering some practical parameters. As the cavity material we take GaAs which has $n = 3.6$. For a GaAs–air interface it has been found that $R \approx 0.32$. In the literature $\alpha \approx 50$ cm^{-1} is quoted for the given material and the appropriate wavelengths, while $L = 300$ μm is a common value for semiconductor lasers. Altogether, this leads to: $\tau_p \approx 1.3 \times 10^{-12}$ s.

9.1.2 The steady state of a laser diode

In a steady state, when the LD is unmodulated so that it emits constant power, changes in the photon density N_p and the carrier density N_e vanish, i.e. $dN_e/dt = dN_p/dt = 0$, giving $N_e = \bar{N}_e$ and $N_p = \bar{N}_p$. It is interesting to solve the rate equations in this situation in order to get some idea of the basic laser principles. Moreover, the steady-state solution is needed to interpret the transient behaviour; see Sections 9.5 and 9.6 for an analysis of a modulated laser.

Starting from the LED region, just where the laser action starts, i.e. at the threshold current I_d, equations (9.1) and (9.2) are valid for $N_p \approx 0$, so that

$$I_d = \frac{eV\bar{N}_e}{\tau_s} \tag{9.6}$$

From equation (9.2) it follows that for $I > I_d$ the carrier density is constant and independent of the photon density

$$\bar{N}_e \approx \frac{1}{A\tau_p} + N_0 \tag{9.7}$$

In this expression the contribution of the spontaneous emission $\beta N_e/\tau_s$ has been neglected with respect to the stimulated emission N_p/τ_p. Then, from equation (9.1) it can be concluded that above the threshold

$$I = [eVA(\bar{N}_e - N_0)]\bar{N}_p + \frac{eV}{\tau_s}\bar{N}_e \tag{9.8a}$$

Using equation (9.6) this can also be written as

$$I - I_{\mathrm{d}} = eVA(\bar{N}_{\mathrm{e}} - N_0)\bar{N}_{\mathrm{p}} = \frac{eV}{\tau_{\mathrm{p}}}\bar{N}_{\mathrm{p}} \qquad (9.8\mathrm{b})$$

Therefore, the photon density is proportional to the current increment

$$\frac{\mathrm{d}\bar{N}_{\mathrm{p}}}{\mathrm{d}I} = \frac{\tau_{\mathrm{p}}}{eV} \qquad (9.9)$$

This linear relationship agrees with the fact that each additional injected carrier produces a stimulated photon. As a consequence, the light output power characteristic is a linear function of the current, as far as the current exceeds the threshold current; thus, it is a linear function of $I - I_{\mathrm{d}}$. The slope of this characteristic is large when τ_{p} is large, while τ_{p} is large when α is small and L large.

9.2 The laser condition

Let us now have a look at the photon density gain $A(N_{\mathrm{e}} - N_0)$. For the sake of convenience it is easier to use the gain per unit of length in the direction perpendicular to the mirrors (x-direction). Consider a slice of the cavity with a thickness of $\mathrm{d}x$ and area S (see Figure 9.2). The passage of a wave changes the number of photons in this slice by $\mathrm{d}N_{\mathrm{p}}S\,\mathrm{d}x$. The attenuation due to absorption is proportional to the number of photons in the volume $S\,\mathrm{d}x$ and to the absorption coefficient α, so that the attenuation is $-N_{\mathrm{p}}S\,\mathrm{d}x\,\alpha\,\mathrm{d}x$. The gain in the slice is proportional to the gain coefficient $A(N_{\mathrm{e}} - N_0)$, to the volume $S\,\mathrm{d}x$ and to the time that the wave needs to traverse the slice of thickness $\mathrm{d}x$. This

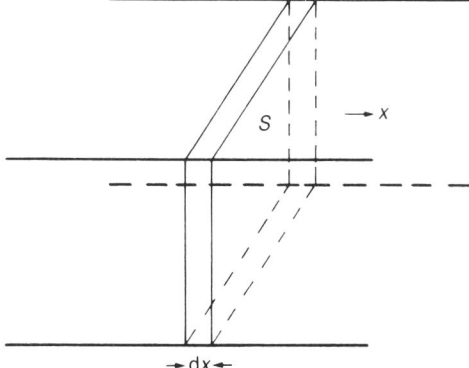

Figure 9.2 A slice through the laser cavity parallel to the mirrors, with a thickness of dx and area of S.

time equals dx/c and the gain in the slice becomes $A(N_e - N_0)N_pS\,dx\,dx/c$. For a full description of the change of the number of photons in the slice $S\,dx$ the following differential equation is obtained:

$$dN_p\,S\,dx = -N_pS\,dx\,\alpha\,dx + A(N_e - N_0)N_pS\,dx\frac{dx}{c} \qquad (9.10)$$

which is equivalent to

$$\frac{dN_p}{dx} = N_p\left[-\alpha + \frac{A(N_e - N_0)}{c}\right] \qquad (9.11)$$

As has been mentioned already, the first term in the brackets in equation (9.11) represents the loss of cavity material per unit of length. The second term in the brackets has to be interpreted as the gain per unit of length and is defined as

$$g \triangleq \frac{A(N_e - N_0)}{c}, \qquad \text{for } N_e > N_0 \qquad (9.12)$$

Going back to Figure 9.1, specifically to the power of the optical wave after two successive passages at x_1. The total attenuation between two passages has already been shown to be $R^2\exp(-2\alpha L)$. However, over the distance $2L$ there is also a gain of $\exp(2gL)$. For a steady-state oscillation the product of the gain and the attenuation for one round trip equals unity: $\exp(2gl)$ $R^2\exp(-2\alpha L) = 1$, leading to the oscillation condition

$$g_d = \alpha + \frac{1}{L}\ln\frac{1}{R} \qquad (9.13)$$

From equation (9.7) it is concluded that $\bar{N}_e$ is constant; then, according to equation (9.12), g is also a constant and equation (9.13) is interpreted as the value of the specific gain at the threshold g_d. Below the threshold N_e is proportional to the current, so that g as a function of the current behaves as shown in Figure 9.3. It must be emphasized that this figure only represents the mathematical relationship between g and I, and does not apply to certain values of g. It must be kept in mind that with the introduction of the rate equations, the gain factor for the number of stimulated photons is a linear function of the carrier density N_e, as long as this number exceeds the minimal value N_0.

The threshold current is determined from $\alpha + \ln(1/R)/L$ and gets smaller as L gets larger. Beyond a certain value of the length, however, no further reduction in the threshold current occurs. This value appears to be 300–400 μm [3], which is why the length of the cavity in most semiconductor lasers is in that region.

For a current I_p the gain g is greater than zero. In the first instance laser action does not occur, because the attenuation of the cavity material and the mirror losses prevent the round trip gain from reaching unity. If g is larger than α, then only the material absorption is compensated, but not the mirror losses. Only when $I > I_d$ are both losses overcome and laser action takes place at a

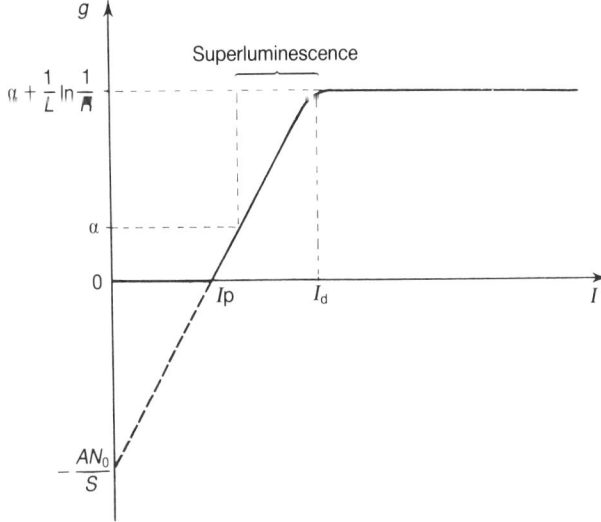

Figure 9.3 The gain coefficient of a laser as a function of the forward current.

saturation of g [3]. The part of the characteristic between $g = \alpha$ and $g = g_d$ is called the region of superluminescence and is where superluminescent diodes (SLDs) operate. Such diodes have an optical gain, but the losses and reflections prevent any laser action from taking place ($g_d > g > \alpha$). SLDs have a narrow optical spectrum (5–10 nm) and a rather well-directed light bundle (10°–60°).

9.3 The efficiency of lasers

Various definitions of the efficiency of semiconductor lasers are in use, and it is therefore important to understand the definition used by a particular supplier.

The internal quantum efficiency η_i is the ratio of the number of stimulated photons to the number of injected carriers. It is usually between 60 and 80 percent.

Another version is the differential quantum efficiency, which is defined as the ratio of the change in number of the external photons per unit of time to the change in number of the injected carriers per unit of time. Expressed mathematically this becomes

$$\eta_d \triangleq \frac{d(P/h\nu)}{d(I/e)} = \frac{1}{V_g} \frac{dP}{dI} \qquad (9.14)$$

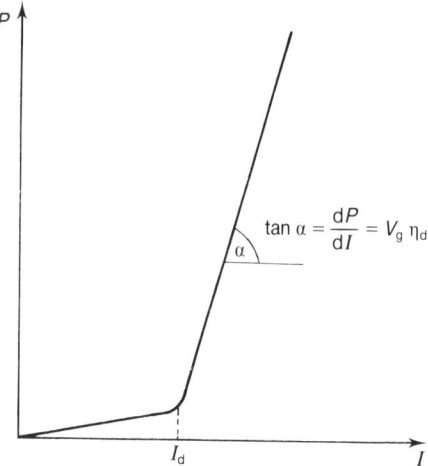

Figure 9.4 The *P–I* characteristic of a semiconductor laser with the slope expressed as the differential quantum efficiency.

where V_g is the potential difference that corresponds to the bandgap energy. The differential quotient dP/dI equals the slope of the curve for light power versus current as shown in Figure 9.4. On one hand the photons are amplified by a factor of g_d per unit length, and on the other hand they are absorbed by a factor of α per unit length. The efficiency of this process is $(g_d - \alpha)/g_d$, so that the external differential quantum efficiency per mirror becomes

$$\eta_d = \frac{\eta_i}{2} \frac{\ln(1/R)}{\alpha L + \ln(1/R)} \tag{9.15}$$

If α, L and R are known quantities, then the internal quantum efficiency is found by measuring the slope of the *P–I* characteristic and using equation (9.15).

The next quantity to discuss is the total device efficiency, i.e. the ratio of the number of external photons to the number of injected carriers

$$\eta_t \triangleq \frac{P/h\nu}{I/e} = \frac{P}{IV_g} = \eta_d\left(1 - \frac{I_d}{I}\right) \tag{9.16}$$

Finally, there is the laser efficiency, i.e. the ratio of the optical output power to the electrical input power of the device

$$\eta_l = \frac{P}{IV} = \eta_t \frac{V_g}{V} \tag{9.17}$$

where V is the voltage across the laser; in practical situations this is 2–2.5 V.

Example 9.2

Taking the parameters given at the end of Section 9.1.1 and using them in the equation (9.15), it can be seen that $\eta_d = 0.43\eta_i/2$. With $\eta_i = 0.6$ it gives $\eta_d = 0.13$. Furthermore, if $I = 1.25 I_d$, then $\eta_t = 0.026$ and with a bandgap equal to 1.45 eV the laser efficiency becomes $\eta_1 \approx 1.9-1.5$ percent. Moreover, it is interesting to calculate the slope of the $P-I$ characteristic. For the values given above it is $dP/dI = 0.18$ W A^{-1}.

As a starting point we took the least favourable value of the internal quantum efficiency η_i. Choosing $\eta_i = 0.9$, all the efficiencies improve by a factor of 1.5, just like the slope of the $P-I$ characteristic.

9.4 The turn-on delay of a laser and the behaviour of an LED

When a laser below the threshold is given a current pulse, the light output pulse starts a few nanoseconds later. This phenomenon is called the turn-on delay and it depends on the behaviour of the semiconductor in the LED region (below the threshold). Therefore, these two subjects are treated together in this section.

In the LED region the relationship between photons and carriers is written as

$$\frac{dN_e}{dt} = \frac{I}{eV} - \frac{N_e}{\tau_s} \tag{9.18}$$

Compared with equation (9.1), the second term on the right-hand side has disappeared, because there are no stimulated photons. The second term on the right-hand side of equation (9.18) with a positive sign represents the change in the spontaneous photon density per unit of time and this rate is proportional to the carrier density N_e. The solution of the differential equation reads

$$N_e(t) = \frac{I}{eV}\,\tau_s[1 - \exp(-t/\tau_s)] + N_e(0)\exp(-t/\tau_s) \tag{9.19}$$

or similarly

$$\frac{N_e(t)eV}{\tau_s} = I - I\exp(-t/\tau_s) + \frac{N_e(0)eV}{\tau_s}\exp(-t/\tau_s) \tag{9.20}$$

Suppose that the bias current is $I_v < I_d$ for $t < 0$ and a transient is applied at $t = 0$, such that the current is $I_m > I_d$ for $t > 0$. Let us assume that the carrier density becomes $\bar{N}_e$ at the point of time t_d; the left-hand side of equation (9.20) then equals I_d, according to equation (9.6). Consequently, equation (9.20) can be written as

$$I_d = I_m - I_m\exp(-t_d/\tau_s) + I_v\exp(-t_d/\tau_s) \tag{9.21}$$

so that

$$t_d = \tau_s \ln \frac{I_m - I_v}{I_m - I_d} \qquad (9.22)$$

From the above reasoning it follows that t_d is the delay between the current transient and the start of the laser action. It can be seen from equation (9.22) that this turn-on delay becomes smaller as the bias I_v gets closer to the threshold current I_d, and it completely disappears at $I_v = I_d$.

Let us now consider the LED region and suppose that a harmonic excitation of the current is superimposed on a bias. Equation (9.18) then shows that the optical power also has a harmonic progress superimposed on a constant bias. The following equation relates the harmonic component of the optical output power to that of the driving current:

$$p \propto \frac{i}{1 + j\omega\tau_s} \qquad (9.23)$$

The value of the spontaneous recombination time τ_s depends on the concentration of the dopant and for AlGaAs is about 10^{-9} s. This corresponds to a modulation bandwidth of about 150 MHz. There appears to be a trade-off between the radiance of an LED and the modulation bandwidth, so that their product will be a constant [4, p. 233]. The measured bandwidth is usually smaller than $1/(2\pi\tau_s)$, because the diode also has parasitic and diffusion capacitances, which, together with the dynamic resistance, augment the effective time constant.

9.5 Transient behaviour of a laser

The rate equations given by equations (9.1) and (9.2) are non-linear differential equations, which are not as a rule easy to solve. In Figure 9.5 some numerical solutions are presented [5]. They involve transient phenomena, i.e. N_e and N_p are given as functions of time at a step excitation at $t = 0$. This means that the current is zero for $t < 0$ and the current is constant and larger than the threshold

Figure 9.5 shows several phenomena that have been dealt with in earlier sections. The number of carriers increases exponentially for $t > 0$, until the laser action starts. From that time the carrier density remains steady, even when the photon density changes considerably. Moreover, the turn-on delay is demonstrated by the fact that the photon density remains very small for $t < t_d$.

The strong fluctuation of photon density at small values of β (10^{-4}) is noteworthy. At smaller values of β, e.g. 10^{-2}, the oscillating character of N_p is completely absent. This shows that the part of the spontaneous emission that contributes to the stimulated emission plays a prominent role as far as transient phenomena are concerned.

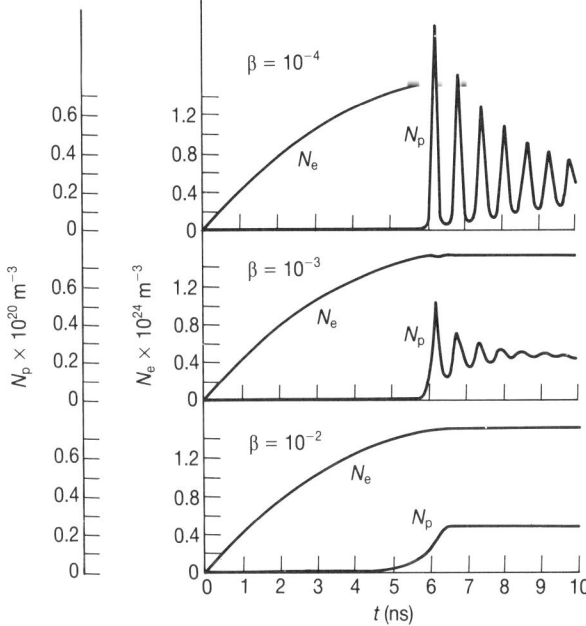

Figure 9.5 The photon and carrier densities in a laser as a function of time at a step excitation at $t = 0$, for various values of β, which is the fraction of the spontaneous photons that contribute to the coherent light. $I = 0$ is taken for $t < 0$, whereas $I = 1.3I_d$ is taken for $t > 0$. (Source: reproduced with permission from *Electronics Letters*, vol. 11, no. 10, May 1975, pp. 206–8, 'Dynamic behaviour of semiconductor lasers', by P. Boers, M. Vlaardingerbroek and M. Danielsen.)

9.6 Modulation of a laser by small signals

In this section we shall analyze the relationship between the light output power and the current input signal, when a small signal current ΔI is superimposed on the current $\bar{I}$. For this purpose we denote

$$
\left.
\begin{aligned}
N_e &= \bar{N}_e + \Delta N_e \\
N_p &= \bar{N}_p + \Delta N_p \\
I &= \bar{I} + \Delta I
\end{aligned}
\right\}
\tag{9.24}
$$

Inserting these expressions in equations (9.1) and (9.2) yields

$$\frac{d\Delta N_e}{dt} \approx \frac{\bar{I}}{eV} + \frac{\Delta I}{eV} - A(\bar{N}_e - N_0)\Delta N_p - A(\bar{N}_e - N_0)\bar{N}_p$$

$$- A\bar{N}_p \, \Delta N_e - \frac{\bar{N}_e}{\tau_s} - \frac{\Delta N_e}{\tau_s} \tag{9.25}$$

$$\frac{d\Delta N_p}{dt} \approx A(\bar{N}_e - N_0)\Delta N_p + A(\bar{N}_e - N_0)\bar{N}_p + A\bar{N}_p \, \Delta N_e$$

$$- \frac{\bar{N}_p}{\tau_p} - \frac{\Delta N_p}{\tau_p} \tag{9.26}$$

In these equations second-order terms, such as the contribution of spontaneous emission, have been neglected.

In equation (9.25), the first, fourth and sixth terms on the right-hand side cancel each other out when taking into account the steady-state solution given in equation (9.8) of Section 9.1.2. The same holds for the second and the fourth terms on the right-hand side of equation (9.26), when equation (9.7) is included; moreover, this causes the first and fifth terms to vanish. After these steps have been performed the following equations remain

$$\frac{d\Delta N_e}{dt} = \frac{\Delta I}{eV} - \frac{\Delta N_p}{\tau_p} - A\bar{N}_p \, \Delta N_e - \frac{\Delta N_e}{\tau_s} \tag{9.27}$$

$$\frac{d\Delta N_p}{dt} = A\bar{N}_p \, \Delta N_e \tag{9.28}$$

Using these approximations the non-linear set given by equations (9.1) and (9.2) is transformed into a linear set. Laplace transform techniques are invoked to solve the latter set. We are interested in the relationship between the current ΔI and the light output power ΔP. This power is proportional to the change in the photon density in the cavity; therefore, the ratio of ΔP to ΔI is required and ΔN_e has to be eliminated from equations (9.27) and (9.28). Inserting the Laplace operator $p = d/dt$ gives

$$\Delta P \propto \Delta N_p = \Delta I \, \frac{A\bar{N}_p/eV}{p^2 + p[(1/\tau_s) + A\bar{N}_p] + (A\bar{N}_p/\tau_p)} \tag{9.29}$$

When ΔI is a step function, the optical power changes as

$$\Delta P = \Delta P(\infty) \, [1 + B \exp(-at) \sin(\omega_c t + \phi)] \tag{9.30}$$

with

$$B = \frac{1}{\sqrt{(\tau_p/A\bar{N}_p) - \tfrac{1}{4} \, [(\tau_p/\tau_s A\bar{N}_p) + \tau_p]^2}} \tag{9.31}$$

$$a = \frac{1}{2}\left(A\bar{N}_p + \frac{1}{\tau_s}\right) \tag{9.32}$$

$$\omega_c = \sqrt{\left(\frac{A\bar{N}_p}{\tau_p} - a^2\right)} \tag{9.33}$$

$$\phi = \arctan \frac{\sqrt{(A\bar{N}_p/\tau_p)} \ \frac{1}{2}[(1/\tau_s) + A\bar{N}_p]^2}{\frac{1}{2}[(1/\tau_s) + A\bar{N}_p]} \tag{9.34}$$

The response to a step excitation is a damped oscillation, which tends to give an asymptotic value of $P(0) + \Delta P(\infty)$, where $P(0)$ is the optical power emitted before the step was initiated. As $\bar{N}_p$ cannot be measured directly, these expressions are written in terms of the currents. By means of equations (8.6) and (8.9) and assuming $\bar{N}_e \gg N_0$, it follows that

$$a \approx \frac{1}{2\tau_s} \frac{\bar{I}}{I_d} \tag{9.35}$$

$$\omega_c \approx \sqrt{\frac{1}{\tau_s\tau_p}\left(\frac{\bar{I}}{I_d} - 1\right) - a^2} \tag{9.36}$$

The currents $\bar{I}$ and I_d differ little, but the time constants τ_s and τ_p can differ by a factor of 10^3 in order to give

$$\omega_c \approx \sqrt{\frac{1}{\tau_s\tau_p}\left(\frac{\bar{I}}{I_d} - 1\right)} \tag{9.37}$$

It can be seen from equation (9.35) that the oscillations vanish after a period of time approximately equal to τ_s or a few nanoseconds. The larger the current $\bar{I}$ with respect to I_d, the smaller is the time constant $1/a$. The frequency f_c increases as the lifetime of the photons τ_p or the lifetime of the carriers τ_s decreases. Moreover, f_c increases with the current $\bar{I}$.

Example 9.3
Let us take $\bar{I} = 1.25I_d$, $\tau_s = 10^{-9}$ s, $\tau_p = 1.3 \times 10^{-12}$ s to be the values used in Sections 9.1.1 and 9.4. Then $f_c = 2.2$ GHz.

Substituting $p = j\omega$ in equation (9.29), we find the transfer characteristic of the laser for small, harmonic signals

$$H(\omega) = \frac{\Delta P(\omega)}{\Delta I(\omega)} \propto \frac{1}{1 + j\omega\tau_p[(1/\tau_s A\bar{N}_p) + 1] - \omega^2(\tau_p/A\bar{N}_p)}$$

$$= \frac{1}{1 + j\omega\tau_p(\bar{I}/(\bar{I} - I_d)) - \omega^2\tau_s\tau_p(I_d/(\bar{I} - I_d))} \tag{9.38}$$

This function has been drawn in Figure 9.6. in order to show a maximum value for the frequency of

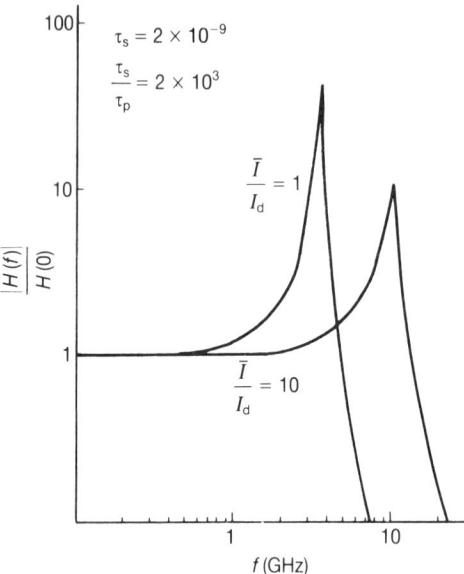

Figure 9.6 Modulation transfer characteristics of a laser for small, harmonic signals.

$$f_m = \frac{1}{2\pi} \sqrt{\frac{1}{\tau_s \tau_p} \left(\frac{\bar{I}}{I_d} - 1\right) - 2a^2} \approx \frac{1}{2\pi} \sqrt{\frac{1}{\tau_s \tau_p} \left(\frac{\bar{I}}{I_d} - 1\right)} = f_c \qquad (9.39)$$

Therefore, the transfer characteristic becomes maximal at a frequency nearly equal to the oscillation frequency of the response to a step function transient.

The modulation characteristic is rather flat below the resonance frequency f_m, increases sharply in the vicinity of the resonance and drops sharply beyond f_m.

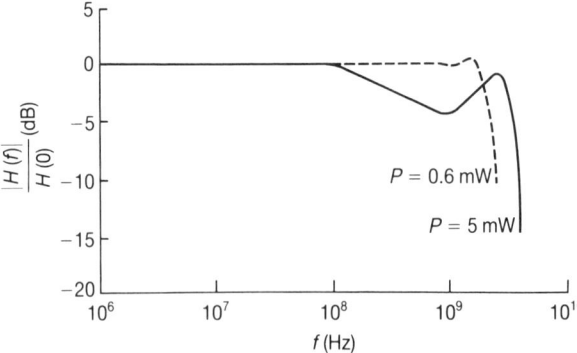

Figure 9.7 The measured modulation transfer characteristic of two laser types. The resonance peak is masked by parasitic effects.

In Figure 9.7 the measurements of the modulation transfer characteristic are given for two laser types. When compared with Figure 9.6, the peak at f_m can be seen to be less pronounced than in the theoretical curve. The resonance effect is partly masked by a first-order, low-pass filtering, caused by parasitic and diffusion capacitances of the device.

9.7 Amplitude noise of lasers

The steps described in Section 9.6 can also be used to obtain an impression of the amplitude noise, or the optical intensity noise, of semiconductor lasers. When such lasers are looked at as selective amplifiers of noise [6, 7], the driving noise in this case is due to spontaneous emission and ultimately the injection of carriers. From the viewpoint of stochastic processes, the driving noise behaves like shot noise, which has a white spectrum. Consequently, the optical amplitude fluctuations show a spectrum that is proportional to the square of the function shown in Figure 9.6. It has been shown in equation (9.7) that above the threshold the carrier density is constant and, as a consequence, the driving noise power is steady, independent of the forward current, and thus independent of the optical output power.

The so-called relative intensity noise (abbreviated to RIN) is defined as the variance of the optical power fluctuations, divided by the mean optical power squared, i.e.

$$\text{RIN} \triangleq \frac{\sigma_p^2}{P^2} \tag{9.40}$$

where P is the mean optical power of the laser. From this latter definition it follows that the power spectrum of the RIN is written as

$$S_{\text{RIN}}(\omega) \propto \frac{|H(\omega)|^2}{P^2} \tag{9.41}$$

This equation implies that in the flat part below the resonance frequency of the characteristic in Figure 9.6, the RIN spectrum is inversely proportional to P^2, as follows from equations (9.38) and (9.41).

References

[1] H. Kressel and J. Butler, *Semiconductor Lasers and Heterojunction LEDs* (New York: Academic Press, 1977).

[2] M. J. Adams and M. Osinski, 'Longitidunal mode competition in semiconductor lasers, rate equations revisited', *IEE Proceedings, Part I*, vol. 129, no. 6, December 1982, pp. 271–4.

[3] J. R. Biard, W. N. Carr and B. S. Reed, 'Analysis of a GaAs laser', *Transactions of the Metallurgical Society of AIME*, vol. 230, March 1964, pp. 286–90.

[4] R. T. Kersten, *Einführung in die Optische Nachrichtentechnik* (Berlin: Springer-Verlag, 1983).
[5] P. M. Boers, M. T. Vlaardingerbroek and M. Danielsen, 'Dynamic behaviour of semiconductor lasers', *Electronics Letters*, vol. 11, no. 10, 15 May 1975, pp. 206–8.
[6] A. Maitland and M. Dunn, *Laser Physics* (Amsterdam: North-Holland, 1969).
[7] E. Wolf (ed), *Progress in Optics*, vol. VI, chap. VI (Amsterdam: North-Holland, 1967).

Problems

9.1 The electron density in a laser is 1.5×10^{24} m^{-3}. The laser has a cavity length of 300 μm and $\tau_s = 10^{-9}$ s, while the cavity width is 1 μm.

Calculate the cavity thickness in order to achieve a threshold current of 10 mA.

9.2 An AlGaAs laser has a cavity length of 400 μm. The absorption of the cavity material amounts to 30 cm^{-1}. What is the optical gain in the laser for:

(a) uncoated facets in air;
(b) coating of the facets so that the reflectivity amounts to 0.8?

9.3 For the laser of Problem 9.2(a) the derivative of the P–I curve is 0.2 W A^{-1}. Calculate the internal quantum efficiency of this laser.

9.4 Consider the laser of Problem 9.1. Calculate the turn on delay for bias zero and modulation current $I_m = 2I_d$.

9.5 A laser shows a peak in its transfer function $H(\omega) = \Delta P(\omega)/\Delta I(\omega)$ at a frequency of 5 GHz and $I/I_d = 2$. The carrier lifetime is 10^{-9} s.

Calculate the photon lifetime.

10

Transfer characteristic and impulse response of fiber communication systems

---

10.1 Transmission via a single-mode fiber
10.2 Transmission via multimode fibers
References
Problems

10.1 Transmission via a single-mode fiber

10.1.1 Systems with a coherent light source

Let us consider an optical communication system consisting of a modulated light source, a single-mode optical fiber and a photodiode as detector. We assume that the light source is modulated such that the optical output power varies proportionally to the input signal $m_i(t)$. Furthermore, it is assumed that $m_i(t) \geq 0$, $\forall t$ and that the source emits a bundle of coherent light. As far as the optical fields are concerned, the optical fiber is considered to be a linear, time-invariant, dispersive transmission medium. Finally, the photodiode is considered to be a component giving a photocurrent that is proportional to the average optical power received. Analyzing this system shows that three non-linear operations are involved. Firstly, the optical power of the source being proportional to $m_i(t)$ implies that the input field to the fiber is proportional to the square root of $m_i(t)$. Second, the photocurrent is proportional to the optical power received, therefore this current is proportional to the square of the output field of the fibre. Third, the photodiode current is proportional to the average power received, which can be considered as envelope detection of the optical wave. The linearity of the fiber is determined by the relationship between the optical field at, respectively, the input and output. In this way, the fiber can be considered as a bandpass system excited by a modulated input signal, as described in Appendix 2. The system as outlined above leads to the diagram in Figure 10.1.

The optical electric field E_i at the input of the fiber is described in polar coordinates, just like the field E_o at the fiber output. The linear relationship between the field at the input and output is given by means of the transfer function.

$$G(\omega) \triangleq \frac{E_0(r, \phi, \omega)}{E_i(r, \phi, \omega)} \tag{10.1}$$

Since this is a bandpass characteristic, $G(\omega)$ can be written as

$$G(\omega) = G_1(\omega - \omega_c) + G_1^*(-\omega - \omega_c) \tag{10.2}$$

with $G_1(\omega)$ being the equivalent baseband characteristic (see Appendix 2). Furthermore, the impulse response belonging to this equivalent baseband transfer function is defined

$$g(t) \triangleq \frac{1}{2\pi} \int_{-\infty}^{\infty} G_1(\omega) \exp(j\omega t) \, d\omega \tag{10.3}$$

When the light source is modulated by an input signal $m_i(t)$ and the carrier frequency of the unmodulated source is ω_c, then the field at the fiber input is written as

$$E_i(r, \phi, t) = a(r, \phi)\sqrt{2m_i(t)} \cos \omega_c t \tag{10.4}$$

where $a(r, \phi)$ is the field on the fiber end-face. This equation also includes the coupling efficiency from light source to fiber. We assume $a(r, \phi)$ to be independent of the modulating signal $m_i(t)$ [1, p. 320]. From Appendix 2 we know that the complex envelope of the output signal of a bandpass system is written as the convolution of the complex envelope of the input signal and the

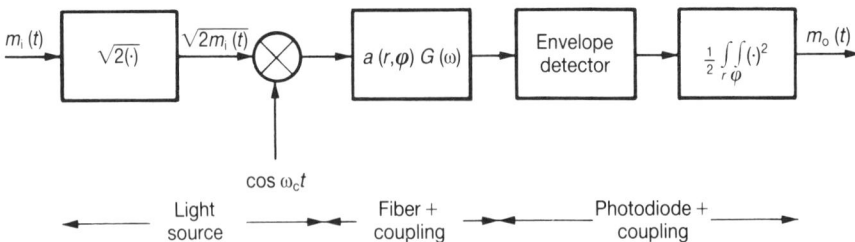

Figure 10.1 Model of an optical single-mode fiber system with a coherent light source. This model is used to calculate the signal transfer.

complex impulse response $g(t)$ of the system as given by equation (10.3). So the envelope of the field at the output of the fiber becomes

$$\left|\hat{E}_o(r, \phi, t)\right| = \sqrt{2}\left|a(r, \phi)\int_{-\infty}^{\infty} g(\tau) \sqrt{m_i(t - \tau)} \, d\tau\right| \tag{10.5}$$

where $\hat{E}_o(r, \phi, t)$ is the complex envelope of the output field.

Finally the output signal of the photodiode reads

$$m_o(t) = c\left|\int_{-\infty}^{\infty} g(\tau) \sqrt{m_i(t - \tau)} \, d\tau\right|^2 \tag{10.6}$$

In this equation c includes the integration of the optical output power of the fiber over r and ϕ, the fiber–photodiode coupling efficiency, the quantum efficiency and the avalanche gain of the photodiode. Equation (10.6) gives the relationship between the input and output signals of the overall system, consisting of light source, fiber and detector. Note that the relationship is a non-linear one. If, however, a certain input signal is multiplied by a positive factor, then the output signal keeps the same shape and is multiplied by the same factor, but superposition is not valid. In [2] a method has been developed to calculate the bandpass characteristic of a single-mode fiber.

The relationship given by equation (10.6) is based on the assumption that the light source is unbiased. Mostly, however, the source is biased in order to enable $m_i(t)$ to have negative values or to shorten or eliminate the turn-on delay of a laser. Biasing implies that the light field, apart from the information-carrying part, also consists of a certain constant part M. Hence equation (10.6) changes to

$$m_o(t) = c\left|\int_{-\infty}^{\infty} g(\tau)\sqrt{M + m_i(t - \tau)} \, d\tau\right|^2, \quad M + m_i(t) \geq 0, \forall t \tag{10.7}$$

The term originating from the biasing can be removed at the receiving end in order to yield the output signal

$$m_o(t) = c\left|\int_{-\infty}^{\infty} g(\tau)\sqrt{M + m_i(t - \tau)} \, d\tau\right|^2 - c\left|\int_{-\infty}^{\infty} g(\tau)\sqrt{M} \, d\tau\right|^2 \tag{10.8}$$

Changing the amplitude of the input signal also changes the shape of the output signal.

As an illustration, we shall consider the transmission of a Gaussian pulse by a single-mode fiber system. From calculations we can obtain an impression of the maximum data rates, or bandwidths, for single-mode fiber systems. The transfer function of the fiber is written as

$$G(\omega) = \exp[-\gamma(\omega)l] = \exp[-\alpha(\omega)l - j\beta(\omega)l]$$

$$= \exp[-a(\omega)]\exp[-jb(\omega)], \quad \omega \geq 0 \tag{10.9}$$

where $\gamma(\omega)$ is the propagation constant and l the length of the fiber. Expanding $\gamma(\omega)$ into a Taylor series about ω_c gives

$$\gamma(\omega) = \gamma(\omega_c) + (\omega - \omega_c)\,\gamma'(\omega_c) + \frac{(\omega - \omega_c)^2}{2!}\,\gamma''(\omega_c) + \cdots \tag{10.10}$$

Then it follows

$$G_1(\omega) = \prod_{m=0}^{\infty} \exp\left[-\frac{\omega^m}{m!}\,\gamma^{(m)}(\omega_c)l\right] \tag{10.11}$$

As input signal we take the Gauss pulse

$$m_i(t) = \exp\left(-\frac{t^2}{2\sigma^2}\right) \tag{10.12}$$

If we define

$$x(t) \triangleq \sqrt{m_i(t)} = \exp\left(-\frac{t^2}{4\sigma^2}\right) \tag{10.13}$$

then the Fourier transform of $x(t)$ reads

$$X(\omega) = 2\sigma\sqrt{\pi}\,\exp(-\sigma^2\omega^2) \tag{10.14}$$

In the relevant frequency region the attenuation of the fiber is assumed to be constant, so that the equivalent baseband characteristic becomes

$$G_1(\omega) \approx A\,\exp\left(-jb_0 - jb_0'\omega - j\frac{b_0''}{2}\,\omega^2\right) \tag{10.15}$$

with

$$b_0 \triangleq \beta(\omega_c)l, \qquad b_0' \triangleq \beta'(\omega_c)l$$
$$b_0'' \triangleq \beta''(\omega_c)l, \qquad A \triangleq \exp(-\alpha l) \tag{10.16}$$

In equation (10.15) we assumed that $\beta^{(m)}(\omega_c) = 0$ for $m \geq 3$. For the case in question, we can use equation (10.6) where the convolution is avoided by multiplying the transformed functions involved in the frequency domain. The calculations proceed as follows:

$$\int_{-\infty}^{\infty} g(\tau)\sqrt{m_i(t - \tau)}\,\mathrm{d}\tau = \frac{2\sigma\sqrt{\pi}\,A}{2\pi}$$

$$\times \int_{-\infty}^{\infty} \exp\left(-jb_0 - jb_0'\omega - j\frac{b_0''}{2}\,\omega^2 - \sigma^2\omega^2 + j\omega t\right)\mathrm{d}\omega$$

$$= \frac{2\sigma\sqrt{\pi}\,A}{2\pi}\,\exp(-jb_0)$$

$$\times \int_{-\infty}^{\infty} \exp\left\{-\frac{1}{2}\,\omega^2(jb_0'' + 2\sigma^2)\right\}\exp\{j\omega(t - b_0')\}\,\mathrm{d}\omega$$

$$= A\,\exp(-jb_0)\,\frac{\sigma\sqrt{2}}{\sqrt{jb_0'' + 2\sigma^2}}$$

$$\times \exp\left[-\frac{(t-b'_0)^2}{2(jb''_0 + 2\sigma^2)}\right] \tag{10.17}$$

The detected signal then reads

$$m_o(t) = \frac{2cA^2\sigma^2}{\sqrt{(b''_0)^2 + 4\sigma^4}} \exp\left[-\frac{(t-b'_0)^2}{4\sigma^4 + (b''_0)^2} \, 2\sigma^2\right] \tag{10.18}$$

Therefore the output pulse also becomes Gaussian; however, it is delayed over b'_0 with respect to the input pulse. The width of the input pulse is determined by σ. Defining the width of the output pulse analogously we find that

$$\sigma_o = \frac{1}{2\sigma}\sqrt{4\sigma^4 + (b''_0)^2} = \sigma\sqrt{1 + (b''_0/2\sigma^2)^2} \tag{10.19}$$

It appears that σ_o as a function of σ becomes minimal and this minimum is found for the following value of the input pulse width:

$$\sigma_{o,min} = \sqrt{b''_0/2} \tag{10.20}$$

The minimum pulse width at the output reads

$$\sigma_{o,min} = \sigma\sqrt{2} = \sqrt{b''_0} \tag{10.21}$$

For the given assumptions and approximations, the pulse broadening when transmitting a Gaussian pulse amounts to a factor of $\sqrt{2}$. A closer look at σ_o as a function of σ, using equation (10.19), shows that this function has two asymptotes. Accordingly, the function is approximated in two different regions as

$$\sigma_o \simeq \sigma, \qquad \text{for } \sigma \gg \sqrt{b''_0/2} \tag{10.22a}$$

$$\sigma_o \simeq \frac{b''_0}{2\sigma}, \qquad \text{for } \sigma \ll \sqrt{b''_0/2} \tag{10.22b}$$

By means of these approximations an insight is gained into the function's behaviour. Figure 10.2 depicts the function and clearly shows its non-linear character, because a certain value of σ_o can originate from two different values of σ. For linear systems this is impossible.

In [2] it is found that for a silica fiber $\beta''(\omega_c) \approx 5 \times 10^{-26}$ s^2/m at a wavelength of 0.8 μm. In Chapter 7 the material dispersion was found to be $d\tau_g/d\lambda \approx 75$ ps km^{-1} nm^{-1} = 75×10^{-6} s m^{-2}. Noting that

$$\frac{d\tau_g}{d\omega} = -\frac{\lambda^2}{2\pi c}\frac{d\tau_g}{d\lambda} = \frac{d^2\beta}{d\omega^2} = \beta'' \tag{10.23}$$

the dispersion of one mode (or congruence) has approximately the same value as in [2].

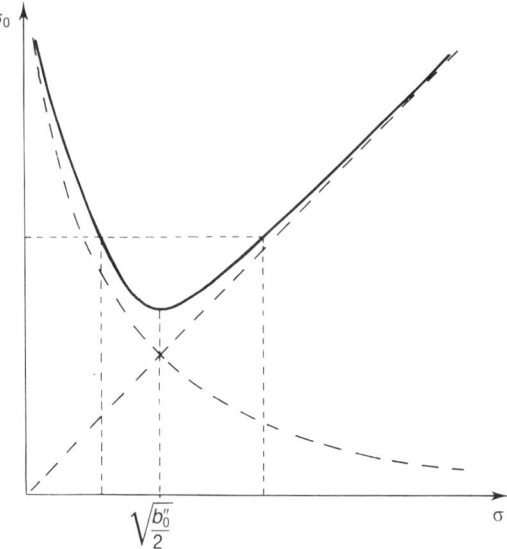

Figure 10.2 The output pulse width as a function of the input pulse width. The excitation consists of a Gaussian pulse and a monochromatic source.

Example 10.1
Let us consider a silica fiber with $l = 10$ km. We then get $b_0'' = 5 \times 10^{-22}$ and the minimum value of σ_0 occurs at $\sqrt{b_0''/2} \approx 1.6 \times 10^{-11} = 16$ ps. As this pulse width is rather small, we mostly operate beyond the minimum of Figure 10.2.

For bit rates that are not too high (<5 Gbit s^{-1}) equation (10.22a) is satisfactory for describing the output pulse width; this implies that there is no pulse broadening. Moreover, on the basis of equation (10.18) it can be concluded that the amplitude of the output pulse is independent of the input pulse width. We can therefore state that a single mode of an optical fiber behaves like a linear system consisting of a pure delay, provided that the fiber is excited with a monochromatic light source.

10.1.2 Systems with a partially coherent light source

In the preceding section we considered systems with a monochromatic source. The coherence function of such a source is a cosine and these sources are called completely coherent. A white light source is a completely incoherent source, with a coherence function equal to a delta function. Both models have more a theoretical than a practical value. In practice the region of the light sources is between these two extremes and is known as the region of partial coherence.

Examples of such sources are given in Sections 8.2.3 and 8.3.2. Introducing a bias for the light source right from the beginning, the complex envelope of the field at the fiber output reads

$$\hat{E}_i(r, \phi, t) = \sqrt{2} a(r, \phi) \sqrt{M + m_i(t)} \, z(t) \qquad (10.24)$$

with $z(t)$ the complex envelope of the source field, being a stochastic process. This yields for the complex envelope of the fiber output field

$$|\hat{E}_o(r, \phi, t)| = \sqrt{2} \left| a(r, \phi) \int_{-\infty}^{\infty} g(\tau) \sqrt{M + m_i(t - \tau)} \, z(t - \tau) \, d\tau \right| \quad (10.25)$$

so that the output signal of the photodiode is given by

$$m_o(t) = c \left| \int_{-\infty}^{\infty} g(\tau) \sqrt{M + m_i(t - \tau)} \, z(t - \tau) \, d\tau \right|^2$$

$$= c \int_{-\infty}^{\infty} \int_{-\infty}^{\infty} g(\tau_1) g^*(\tau_2) \sqrt{M + m_i(t - \tau_1)}$$

$$\times \sqrt{M + m_i(t - \tau_2)} \, z(t - \tau_1) \, z^*(t - \tau_2) \, d\tau_1 \, d\tau_2 \qquad (10.26)$$

From this equation it can be seen that the output signal of the system is also a stochastic process. The expectation of this process reads [3]

$$E[m_o(t)] = c \int_{-\infty}^{\infty} \int_{-\infty}^{\infty} g(\tau_1) g^*(\tau_2) \sqrt{M + m_i(t - \tau_1)}$$

$$\times \sqrt{M + m_i(t - \tau_2)} \, R_{zz}(\tau_1 - \tau_2) \, d\tau_1 \, d\tau_2 \qquad (10.27)$$

Because $m_o(t)$ is a stochastic process, a signal-dependent noise contribution is added to this mean value. In this chapter, however, we restrict ourselves to the expectation of $m_o(t)$ as given by equation (10.27). The analysis starts by remembering that $g(\tau)$ is a delta function; according to the statement at the end of the previous section, a single mode behaves as a pure delay. In other words, the duration of $g(\tau)$ is very short with respect to the duration of $m_i(t)$. For time differences $(\tau_1 - \tau_2) \neq 0$ for which $m_i(t - \tau_1)$ and $m_i(t - \tau_2)$ are almost equal, the product $g(\tau_1) g^*(\tau_2)$ is negligible. The product of the two square roots of equation (10.27) are then approximated either by $M + m_i(t - \tau_1)$ or $M + m_i(t - \tau_2)$ [4]. Both approximations, however, lead to an asymmetrical integrand in τ_1 and τ_2, which produces a complex value for $E[m_o(t)]$. Such a complex solution does not have any actual meaning. The following approximation is similar to the two earlier ones, but it has the advantage of giving a real solution [5]

$$\sqrt{M + m_i(t - \tau_1)} \, \sqrt{M + m_i(t - \tau_2)}$$

$$\simeq \left[M + \frac{1}{2} m_i(t - \tau_1) + \frac{1}{2} m_i(t - \tau_2) \right] \qquad (10.28)$$

Substituting equation (10.28) into equation (10.27) yields

$$E[m_o(t)] \simeq cM \int_{-\infty}^{\infty} \int_{-\infty}^{\infty} g(\tau_1)g^*(\tau_2) \, R_{zz}(\tau_1 - \tau_2) \, d\tau_1 \, d\tau_2$$

$$+ \frac{1}{2} c \int_{-\infty}^{\infty} \left[\int_{-\infty}^{\infty} g^*(\tau_2) \, R_{zz}(\tau_1 - \tau_2) \, d\tau_2 \right] g(\tau_1)m_i(t - \tau_1) \, d\tau_1$$

$$+ \frac{1}{2} c \int_{-\infty}^{\infty} \left[\int_{-\infty}^{\infty} g(\tau_1) \, R_{zz}(\tau_1 - \tau_2) \, d\tau_1 \right] g^*(\tau_2)m_i(t - \tau_2) \, d\tau_2$$

$$(10.29)$$

The third term of this equation is the complex conjugate of the second term. Writing

$$E[m_o(t)] \simeq h_0 + \int_{-\infty}^{\infty} h_1(\tau)m_i(t - \tau) \, d\tau \qquad (10.30)$$

it can then be seen that the system is approximately linear with

$$h_0 = cM \int_{-\infty}^{\infty} \int_{-\infty}^{\infty} g(\tau_1)g^*(\tau_2) \, R_{zz}(\tau_1 - \tau_2) \, d\tau_1 \, d\tau_2 \qquad (10.31a)$$

$$h_1(t) = c \, \mathrm{Re}\left[g(t) \int_{-\infty}^{\infty} g^*(\rho) \, R_{zz}(t - \rho) \, d\rho \right] \qquad (10.31b)$$

We shall assume below that the spectrum $S_{zz}(\omega)$ of the complex envelope of the source reads

$$S_{zz}(\omega) = \frac{1}{B \sqrt{2\pi}} \exp\left(-\frac{\omega^2}{2B^2} \right) \qquad (10.32)$$

Written in this form the total emitted power from the source is unity and is independent of the bandwidth B. The integral in equation (10.31b) is the convolution of $g^*(t)$ and $R_{zz}(t)$, so that

$$g(t) \int_{-\infty}^{\infty} g^*(\rho) \, R_{zz}(t - \rho) \, d\rho = \mathfrak{F}^{-1} \left\{ \frac{1}{2\pi} \, G_1(\omega) \, * \, [G_1^*(-\omega) \, S_{zz}(\omega)] \right\}$$

$$(10.33)$$

where $\mathfrak{F}^{-1} \{ \cdot \}$ represents the inverse Fourier transform operation. For $G_1(\omega)$ we use equation (10.15) with the factor $\exp(-jb_0'\omega)$ omitted because it only causes a delay and has no influence on the shape of the output signal. By means of this $G_1(\omega)$ we get

$$\frac{1}{2\pi} \, G_1(\omega) \, * \, [G_1^*(-\omega) \, S_{zz}(\omega)]$$

$$= \frac{1}{2\pi \sqrt{2\pi} \, B} G_1(\omega) \, * \, \left\{ A \exp(jb_o) \exp\left[-\frac{\omega^2}{2} \left(\frac{1}{B^2} - jb_0'' \right) \right] \right\}$$

$$= \frac{A^2}{2\pi} \exp\left\{ -\frac{\omega^2}{2} \left[(b_0''B)^2 + jb_0'' \right] \right\} \qquad (10.34)$$

As a general property of Fourier transforms we have

$$\text{Re}\{f(t)\} = \frac{1}{2} \mathscr{F}^{-1}\{F(\omega) + F^*(-\omega)\} \tag{10.35}$$

so that the transfer function of the optical fiber communication system under consideration becomes

$$H_1(\omega) = \frac{cA^2}{2\pi} \exp\left[-\frac{\omega^2}{2}(b_0''B)^2\right] \cos\left(\frac{\omega^2}{2}b_0''\right) \tag{10.36}$$

For relatively wideband sources, such as an LED, the impact of the cosine is small and the system bandwidth is determined from the exponential. If this bandwidth is defined by means of the $1/e$ value, we find that

$$\omega_g = \frac{\sqrt{2}}{b_0''B} \tag{10.37}$$

The bandwidth is therefore inversely proportional to the source bandwidth and the fiber length, while the transfer function becomes

$$H_1(\omega) \simeq \frac{cA^2}{2\pi} \exp\left[-\frac{\omega^2}{2}(b_0''B)^2\right] \tag{10.38}$$

with

$$h_1(t) \simeq \frac{cA^2}{2\pi\sqrt{2\pi}\,b_0''B} \exp\left[-\frac{t^2}{2(b_0''B)^2}\right] \tag{10.39}$$

If a Gaussian pulse of width σ is applied to the input of such a system (see equation 10.12), then the output pulse is also Gaussian, with

$$\sigma_o = \sqrt{\sigma^2 + (b_0''B)^2} = \sqrt{\sigma^2 + [\beta''(\omega_c)B]^2 l^2} \tag{10.40}$$

It can be seen that the square of the output pulse width contains the sum of the squares of the input pulse width σ and the width of the system impulse response $b_0''B$.

When, however, the source has a very narrow spectrum, the cosine factor of equation (10.36) dominates the transfer function. In this case, the transfer function becomes.

$$H_1(\omega) \approx \frac{cA^2}{2\pi} \cos\left(\frac{\omega^2}{2}b_0''\right) \tag{10.41}$$

An impression of this system bandwidth can be obtained by looking at the frequency where the argument of the cosine becomes unity, thus yielding

$$\omega_g = \sqrt{\frac{2}{b_0''}} \tag{10.42}$$

This frequency is the inverse of σ for the Gaussian input pulse where σ_o becomes minimal in the case of $B = 0$ (see equation 10.20). Although from the analysis given in the preceding section it follows that the system is no longer linear in this frequency range, equation (10.42) gives a good insight into the frequency limits of such systems. We must emphasize that the bandwidth given by equation (10.42) is inversely proportional to the square root of the fiber length.

Example 10.2

Let us consider a fiber with $\beta''(\omega_c) = 5 \times 10^{-26} \text{ s}^2 \text{ m}^{-1}$ (see Section 10.1.1) and a fiber length of 10 km. The bandwidth of the source, an LED, is assumed to be $B = 10^{14}$ Hz. Equation (10.37) can be used to calculate the system bandwidth, which yields $\omega_g = 28 \times 10^6 \text{ s}^{-1}$. If, on the other hand, a very narrow source is used we get $\omega_g = 63 \times 10^9 \text{ s}^{-1}$.

So far, the analysis has been based on the approximations from equation (10.15), i.e. on the first three terms of the Taylor expansion. For silica fibers b_0'' vanishes somewhere in the vicinity of 1.3 μm, and this parameter determines the output pulse, or the system bandwidth. Of course, the given method will remain valid for these wavelengths, but the fourth term of the Taylor expansion has to be taken into account, which makes the various operations and formulae more complex [6, 7].

10.1.3 Systems with a polychromatic source

In this section we shall consider systems where the light source has a line spectrum, like a laser oscillating in different longitudinal modes. The spectrum of the complex envelope of such a laser can be written as

$$S_{zz}(\omega) = \sum_{i=-N}^{N} a_i S(\omega - i \, \Delta\omega) \tag{10.43}$$

where a_i is the amplitude of the ith line, $\Delta\omega$ the distance between adjacent spectral lines and $S(\omega)$ is given by equation (10.32). Following the procedure of the preceding section (equations 10.32–10.38), we arrive at the result

$$H(\omega) = \exp\left[-\frac{\omega^2}{2} (b_0'' B)^2\right] \cos\left(\frac{b_0''}{2} \omega^2\right) \sum_{i=-N}^{N} a_i \cos(i\Delta\omega \, b_0'' \omega) \tag{10.44}$$

In order to obtain an impression of this rather complicated expression, we consider the following example.

Example 10.3

Consider a fiber with $b_0'' = 5 \times 10^{-22}$ (see also Example 10.1). The bandwidth of the individual laser lines is taken as $B = 100$ MHz, whilst the distance between the adjacent spectral lines is $\Delta\lambda = 0.3$ nm at $\lambda = 1$ μm. The amplitudes of the spectral lines are given a Gaussian decay of $a_i = \exp(-i^2P)$. Figure 10.3 shows $|H(f)|/H(0)$ for various values of the parameters N and P; the optical spectra are included in the figures. The characteristics where $P = 10$ can be considered as the transfer functions belonging to a single-mode laser spectrum, according to equation (10.41). Within the depicted frequency domain (1 MHz–1 GHz) as well as below this region, the characteristics for $P = 10$ are flat. If the domain is extended we are able to see a cut-off frequency of about 10 GHz, which agrees with the calculation based on equation (10.42). The other characteristics show a flat response up to a certain cut-off frequency, beyond this the response starts oscillating. The amplitude of the oscillations increases to unity in the range 1–10 GHz. However, such a characteristic is useless for transmission; only the frequency range before the first minimum is important. Depending on the values of N and P this range shows a bandwidth of 10^8–10^{10} Hz.

From this example it can be concluded that the transfer function of a single-mode fiber system depends on the number of laser lines. A single-mode laser gives a large bandwidth, but two relatively weak side-modes can greatly reduce this bandwidth.

In Example 10.3 a rather large value of b_0'' has been used. When a fiber is used in the vicinity of its zero-dispersion wavelength, or if a dispersion-flattened fiber is used, then b_0'' is smaller and the bandwidth larger.

In deriving the results it was assumed that the responses of the laser and the photodiode are independent of the frequency. From Chapters 9 and 12 it can be seen that this is not exactly true; therefore, the transfer functions found here have to be multiplied by the transfer functions of the laser and photodiode.

10.2 Transmission via multimode fibers

Signal transmission via multimode fibers does not have the clearly defined behaviour of single-mode fibers, as will become clear in this section. One reason for this is mode coupling or mode mixing: random imperfections in the fiber, splices and connectors cause the optical power to be coupled from one mode to another. Depending on the fiber, as well as the number and quality of splices and connectors, the mode coupling may or may not be important, as will be shown below.

10.2.1 Systems without mode coupling

When a multimode fiber is used as the transmission medium and there is no mode coupling, then each mode can be considered to have its own specific transfer characteristic (see Section 10.1). This characteristic is calculated as in [8]

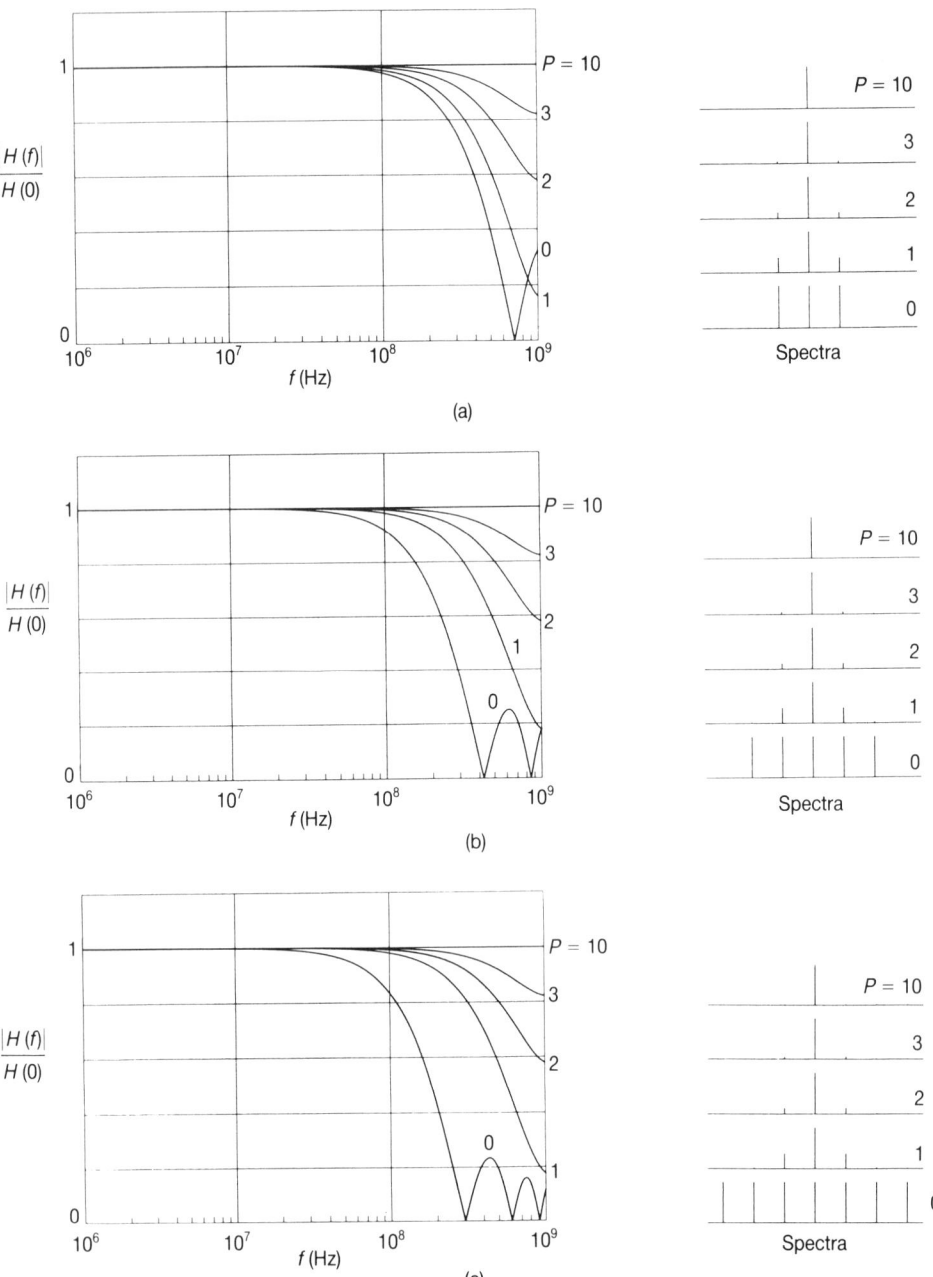

Figure 10.3 The transfer function for connecting a laser, a fiber and a photodiode in series: $b_0'' = 5 \times 10^{-22}$, $B = 100$ MHz, $\lambda = 1$ μm, $\Delta\lambda = 0.3$ nm. (a) Three spectral lines ($N = 1$). (b) Five spectral lines ($N = 2$). (c) Seven spectral lines ($N = 3$).

and [9]. Moreover, we concluded that in many practical situations the impulse response of a single mode may be adequately described by a delta function, provided that the spectral width of the source is small.

As a model for a multimode fiber we now consider a system consisting of a number of orthogonal transmission modes that do not exchange optical power mutually. Furthermore, it is assumed that the various modes are dispersionless, but the propagation delay time of the various modes may differ, so that the system suffers from multimode dispersion. As in equation (10.5), the envelope of the field at the fiber output becomes

$$|\hat{E}_o(s, t)| = \sqrt{2} \left| \sum_k a_k(s) \alpha_k \sqrt{m_i(t - \tau_k)} \right| \tag{10.45}$$

where s is the two-dimensional position at the fiber end-face and τ_k is the delay time of the kth mode. In a similar fashion to equation (10.6), the detector output signal reads

$$m_o(t) = \int_S \left| \sum_k a_k(s) \alpha_k \sqrt{m_i(t - \tau_k)} \right|^2 ds$$

$$= \int_S \sum_k \sum_l a_k(s) a_l^*(s) \alpha_k^2 m_i(t - \tau_k) \, ds \tag{10.46}$$

where S is the area of the fiber end-face.

Since the modes are orthogonal, it follows that

$$\int_S a_k(s) \, a_l^*(s) \, ds = \delta_{kl} \tag{10.47}$$

where δ_{kl} is the Kronecker delta function. In this way the output signal becomes

$$m_o(t) = \sum_k c_k \alpha_k^2 m_i(t - \tau_k) \tag{10.48}$$

Each signal travelling through the fiber in one of the various modes has its own independent transmission path (see Figure 10.4). The distribution of the optical power over the various modes is indicated by the coefficients $\{\alpha_k\}$. This total input power is thus written as

$$P_{in}(t) = \sum_k \alpha_k^2 m_i(t) \tag{10.49}$$

From equation (10.48) it follows that in the model above, a multimode optical fiber communication system behaves like a linear system. Using this equation as a starting point, the model can be refined in different ways [10]. Equation (10.48) is based on the system impulse response

$$h(t) = \sum_k c_k \alpha_k^2 \delta(t - \tau_k) \tag{10.50}$$

This consists of a sequence of delta functions. Since the receiving circuit shows a

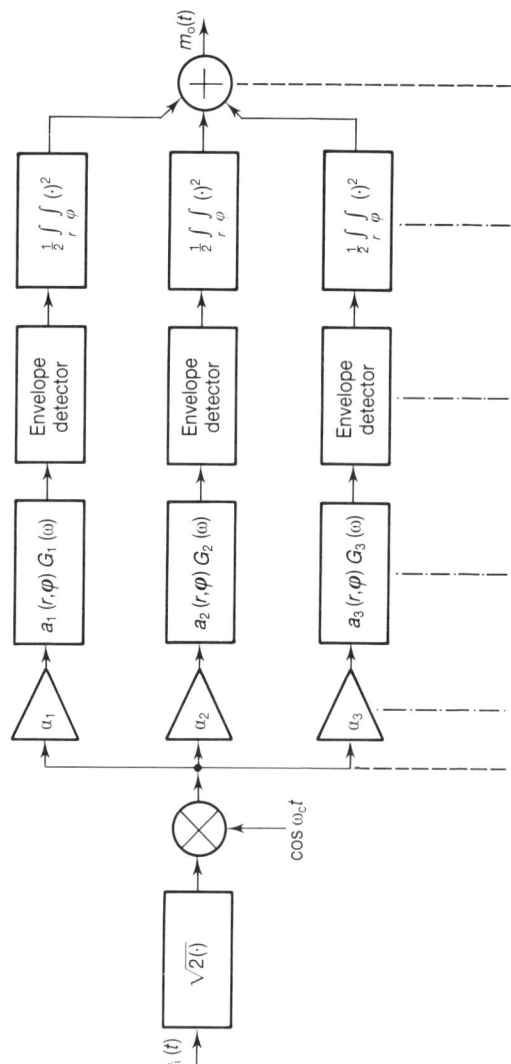

Figure 10.4 Model for the signal transfer in a multimode fiber.

finite bandwidth and the source is never completely coherent, the delta sequence has to be convolved with a function of finite width in order to obtain the actual impulse response. In this way, a continuous response occurs.

As a first refinement we assume that all values of c_k are the same, all modes are equally excited with the same power and all modes show the same attenuation, i.e. all α_k are equal. Moreover, it is assumed that the delay time differences between adjacent modes are constant. The impulse response of this model is rectangular and the transfer characteristic a $(\sin x)/x$ function (see Figure 10.5a). Such a characteristic may be expected for a step index fiber in which there is no mode coupling and all modes have the same attenuation.

When a mode-dependent attenuation that increases exponentially with the delay time is introduced we get the characteristic in Figure 10.5(b). The absolute

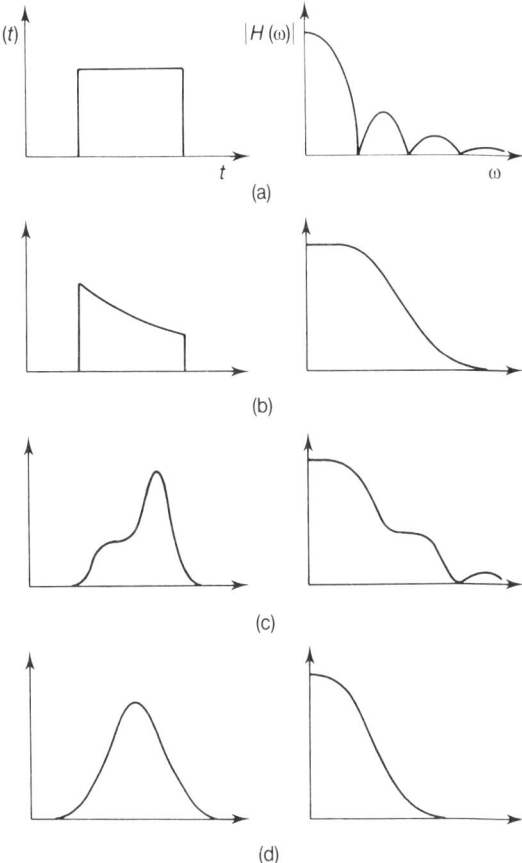

Figure 10.5 Some typical impulse responses and transfer characteristics of multimode fibers.

value of the transfer function looks like $1/(1 + \omega^2\tau^2)$, where τ is the time constant for the exponential attenuation as a function of the delay time.

Figure 10.5(c) shows another possible impulse response and transfer characteristic. Such a characteristic is encountered in fibers where the core has a higher refractive index than the base material (usually SiO_2). When the preform collapses, a part of the dopant material evaporates from the innermost layers of the preform tube; as a consequence, the refractive index of the fiber shows a dip in the center of the core. Low-order modes have a smaller delay time than higher-order modes. During the fabrication of the preform it is possible to compensate for this effect by giving the inner layers a somewhat larger refractive index than is needed for the profile. Overcompensation leads to a refractive index value that is too large for the middle of the core. In that case, the impulse response would be the reverse of Figure 10.5(c), whereas the absolute value of the transfer characteristic would not change.

10.2.2 Systems with mode coupling

In this section some effects are described that result from an exchange of power between the different modes or the ray congruences of optical fiber systems intended for digital transmission. The starting point for the analysis is the system model described in Sections 1.6 and 4.6, neglecting the waveguide dispersion and, for the time being, the material dispersion. We will restrict ourselves to pulse-shaped signals, described by their power, i.e. the mean power over a time interval that is short compared with the pulse duration and long compared with a period of the modulated light. Formulae will be derived below for the energies, the delays and the widths of the pulses travelling from source to destination via the various channels that are associated with the modes. The theory is applied to the case of two coupled modes. Though this number of modes is far from representative for a multimode fiber, the results of this analysis nevertheless give an insight into the effects of mode coupling for digital transmission. Moreover, a model of two coupled modes serves well in the case of two nearly degenerate orthogonal modes in a single-mode fiber. As derived and described in Section 3.5, a hybrid mode EH_{nm} or HE_{nm} of a circular symmetrical fiber has two orthogonal degenerate modes. A real fiber can never be perfectly circular and symmetrical; therefore, the two orthogonal modes always have slightly differing velocities. The mode dispersion that results from the velocity difference is especially important in cases where the waveguide dispersion and the material dispersion compensate each other (see Section 5.5).

The signal that is carried by the ith channel is denoted as $s_i(l, t)$. It has a power dimension and is a function of the length l and time t. Some of the properties of the signal – namely its energy, delay and width of the pulse – can be defined with the aid of moments, as follows.

- The energy E_i of the pulse $s_i(l, t)$ is given by

$$E_i(l) = \int_{-\infty}^{\infty} s_i(l,\ t)\ \mathrm{d}t \tag{10.51}$$

- The pulse delay t_i follows from the first moment of of $s_i(l,\ t)$

$$t_i(l) = \frac{\int_{-\infty}^{\infty} t\ s_i(l,\ t)\ \mathrm{d}t}{E_i(l)} \tag{10.52}$$

- The width of the pulse that can be associated with σ_i is found from the second central moment

$$\sigma_i^2(l) = \frac{\int_{-\infty}^{\infty} [t - t_i(l)]^2\ s_i(l,\ t)\ \mathrm{d}t}{E_i(l)} \tag{10.53}$$

Higher central moments determine the pulse in even more detail; they are of minor importance in this context, because it has been shown in [11] that the impulse response of a long fiber with mode coupling can be adequately approximated by a Gaussian pulse shape. This approximation is valid provided that the fiber length is larger than the coupling length $L_c = 1/C$, where C is the coupling coefficient, which will be described later in this section. The corresponding impulse response and transfer characteristic are depicted in Figure 10.5(d).

We assume that the process of attenuation and mode conversion is that described by the following coupled power equation [12]:

$$\frac{\partial s_i(l,\ t)}{\partial l} + \frac{1}{v_i}\frac{\partial s_i(l,\ t)}{\partial t} = -2\alpha_i s_i(l,\ t) + \sum_j c_{ij}[s_j(l,\ t) - s_i(l,\ t)] \tag{10.54}$$

In this equation:

$\quad v_i \quad$ = group velocity in the ith channel
$\quad \alpha_i \quad$ = attenuation coefficient of the fields in the ith mode
$\quad c_{ij} \quad$ = coupling coefficient between the jth and the ith channel

In [12] this equation has been derived for certain non-uniform fibers, but equation (10.54) is so simple and has such an intuitive meaning that it can be postulated as a mathematical model directly, without derivation. Let us look at $s_i(l,\ t)$ and $s_i(l + \mathrm{d}l,\ t + \mathrm{d}t)$. If the function $s_i(l,\ t)$ is continuous and has continuous derivatives, then the power increase can be written as a first approximation

$$s_i(l + \mathrm{d}l,\ t + \mathrm{d}t) - s_i(l,\ t) = \frac{\partial s_i(l,\ t)}{\partial l}\ \mathrm{d}l + \frac{\partial s_i(l,\ t)}{\partial t}\ \mathrm{d}t \tag{10.55}$$

If a constant ratio v_i between $\mathrm{d}l$ and $\mathrm{d}t$ is assumed, or

$$v_i = \frac{\mathrm{d}l}{\mathrm{d}t} \tag{10.56}$$

then the increase is given by

$$\left[\frac{\partial s_i(l, t)}{\partial l} + \frac{1}{v_i}\frac{\partial s_i(l, t)}{\partial t}\right]dl \tag{10.57}$$

Thus, the left-hand side of equation (10.54) represents the power increase of the ith mode per unit length under the condition $v_i = dl/dt$.

If there is no mutual coupling or attenuation, then the right-hand side of equation (10.54) disappears. An observer who travels with the speed v_i along the fiber observes a constant power in the ith mode. The solution of equation (10.54) under these conditions is

$$s_i(l, t) = s_i(l - v_i t) \tag{10.58}$$

If there is no coupling, although there is attenuation, then the solution reads

$$s_i(l, t) = \exp(-2\alpha_i l) \, s_i(l - v_i t) \tag{10.59}$$

As a result of the coupling with the jth mode, the power in the ith mode increases. This increase is proportional to the difference between the power in the jth mode and ith mode. The proportional factor is called $c_{ij} = c_{ji}$.

When the mode conversion model resembles that described by equation (10.54), the following differential equations can be derived with the aid of equations (10.51)–(10.53):

$$\frac{dE_i(l)}{dl} = -2\alpha_i E_i(l) + \sum_j c_{ij}[E_j(l) - E_i(l)] \tag{10.60}$$

$$\frac{dt_i(l)}{dl} = \frac{1}{v_i} + \sum_j \frac{E_j(l)}{E_i(l)} \, c_{ij}[t_j(l) - t_i(l)] \tag{10.61}$$

$$\frac{d\sigma_i^2(l)}{dl} = \sum_j \frac{E_j}{E_i} \, c_{ij}[\sigma_j^2 - \sigma_i^2 + (t_j - t_i)^2] \tag{10.62}$$

Derivation of equations (10.60)–(10.62)
In the derivation use is made of the Fourier transform $S_i(l, \omega)$ of the signal $s_i(l, t)$ and of the first and second derivative of $S_i(l, \omega)$ with respect to ω, denoted as $S_i'(l, \omega)$ and $S_i''(l, \omega)$, respectively

$$S_i(l, \omega) = \int_{-\infty}^{\infty} s_i(l, t) \exp(-j\omega t) \, dt \tag{10.63}$$

$$S_i'(l, \omega) = -j\int_{-\infty}^{\infty} t \, s_i(l, t) \exp(-j\omega t) \, dt \tag{10.64}$$

$$S_i''(l, \omega) = -\int_{-\infty}^{\infty} t^2 \, s_i(l, t) \exp(-j\omega t) \, dt \tag{10.65}$$

For $\omega = 0$ and using equations (10.51)–(10.53), the above equations become

$$S_i(l, 0) = \int_{-\infty}^{\infty} s_i(l, t)\, dt = E_i(l) \tag{10.66}$$

$$S_i'(l, 0) = jE_i(l)t_i(l) \tag{10.67}$$

$$S_i''(l, 0) = -E_i(l)[\sigma_i^2(l) + t_i^2(l)] \tag{10.68}$$

Equations (10.66)–(10.68) are satisfied by

$$S_i(l, \omega) = E_i(l)\exp\left[-j\omega t_i(l) - \frac{\omega^2}{2}\sigma_i^2(l) + \cdots\right] \tag{10.69}$$

The higher-order terms in the exponent correspond to the higher-order central moments. In the case of Gaussian pulses they are zero.

The Fourier transform of equation (10.54) is written as

$$\left[\frac{1}{E_i(l)}\frac{dE_i(l)}{dl} - j\omega\frac{dt_i(l)}{dl} - \frac{\omega^2}{2}\frac{d\sigma_i^2(l)}{dl} + \cdots\right]S_i(l, \omega) + \frac{j\omega}{v_i}S_i(l, \omega)$$

$$= -2\alpha_i S_i(l, \omega) + \sum_j c_{ij}[S_j(l, \omega) - S_i(l, \omega)] \tag{10.70}$$

With $\omega = 0$, equation (10.70) equals equation (10.60). Differentiating equation (10.70) with respect to ω, substituting $\omega = 0$ and using equations (10.66), (10.67) and (10.60) leads to equation (10.61). Taking the second derivative of equation (10.70) with respect to ω, substituting $\omega = 0$, and using the equations (10.66)–(10.68), (10.60) and (10.61) leads to equation (10.62).

In the case of two coupled channels, equation (10.60) produces two first-order partial linear differential equations [13]

$$\frac{dE_1(l)}{dl} = 2\alpha_1 E_1(l) + C[E_2(l) - E_1(l)] \tag{10.71}$$

$$\frac{dE_2(l)}{dl} = -2\alpha_2 E_2(l) + C[E_1(l) - E_2(l)] \tag{10.72}$$

with $C = c_{12} = c_{21}$. The solution of this set of equations reads

$$E_1(l) = B_1(2\alpha_2 + C + p_1)\exp(p_1 l) + B_2 C\exp(p_2 l) \tag{10.73}$$

$$E_2(l) = B_1 C\exp(p_1 l) + B_2(2\alpha_1 + C + p_2)\exp(p_2 l) \tag{10.74}$$

with

$$p_{1,2} = -(\alpha_1 + \alpha_2 + C) \pm \sqrt{(\alpha_1 - \alpha_2)^2 + C^2} \tag{10.75}$$

the roots of the characteristic equation

$$\begin{vmatrix} -2\alpha_1 - C - p & C \\ C & -2\alpha_2 - C - p \end{vmatrix} = 0 \tag{10.76}$$

and

$$B_1 = \frac{(2\alpha_1 + C + p_2)\, E_1(0) - CE_2(0)}{(2\alpha_2 + C + p_1)(2\alpha_1 + C + p_2) - C^2} \tag{10.77}$$

$$B_2 = \frac{(2\alpha_2 + C + p_1)\, E_2(0) - CE_1(0)}{(2\alpha_2 + C + p_1)(2\alpha_1 + C + p_2) - C^2} \tag{10.78}$$

If $l \to \infty$, then the ratio between the energy of the different pulses approaches a limit; this results in a state of equilibrium

$$\lim_{l \to \infty} \frac{E_1(l)}{E_2(l)} = \frac{\alpha_2 - \alpha_1}{C} + \sqrt{\left(\frac{\alpha_2 - \alpha_1}{C}\right)^2 + 1} \tag{10.79}$$

In Figure 10.6 two examples of E_1 and E_2 as functions of Cl have been drawn. In Figure 10.6(a) it has been assumed that there is no attenuation. In Figure 10.6(b) it has been assumed that both channels have an attenuation that is equal to the coupling coefficient between the channels.

The general solution for the travel times is rather involved; therefore, we will restrict ourselves to the specific case where $\alpha_1 = \alpha_2 = \alpha$ and where the equilibrium condition for the energies is fulfilled from the beginning, i.e. $E_1/E_2 = 1$. The following set of equations must then be solved:

$$\frac{dt_1(l)}{dl} = \frac{1}{v_1} + C[t_2(l) - t_1(l)] \tag{10.80}$$

$$\frac{dt_2(l)}{dl} = \frac{1}{v_2} + C[t_1(l) - t_2(l)] \tag{10.81}$$

giving the solution

$$t_1(l) = t_1(0) + \frac{l}{2}\left(\frac{1}{v_1} + \frac{1}{v_2}\right) + \frac{1}{2}\left[t_2(0) - t_1(0)\right.$$
$$\left. - \frac{(1/v_2) - (1/v_1)}{2C}\right][1 - \exp(-2Cl)] \tag{10.82}$$

$$t_2(l) = t_2(0) + \frac{l}{2}\left(\frac{1}{v_1} + \frac{1}{v_2}\right) + \frac{1}{2}\left[t_1(0) - t_2(0)\right.$$
$$\left. - \frac{(1/v_1) - (1/v_2)}{2C}\right][1 - \exp(-2Cl)] \tag{10.83}$$

With the difference in travel times

$$T(l) \triangleq t_1(l) - t_2(l) \tag{10.84}$$

it follows from equations (10.82) and (10.83) that

$$T(l) = T(0) + \left[\frac{(1/v_1) - (1/v_2)}{2C} - T(0)\right][1 - \exp(-2Cl)] \tag{10.85}$$

In the equilibrium situation a constant difference in travel time is attained

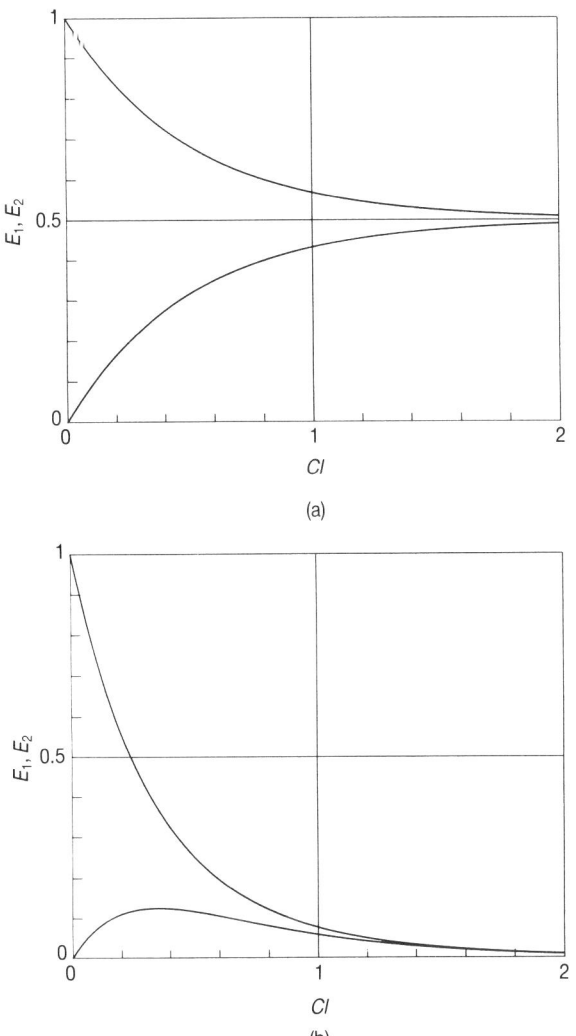

Figure 10.6 The pulse energies E_1 and E_2 as functions of Cl. (a) $\alpha_1 = \alpha_2 = 0$, $E_1(0) = 1$, $E_2(0) = 0$. (b) $\alpha_1 = \alpha_2 = C$, $E_1(0) = 1$, $E_2(0) = 0$.

$$\lim_{l \to \infty} T(l) = T(\infty) = \frac{1}{2C} \left(\frac{1}{v_1} - \frac{1}{v_2} \right) \tag{10.86}$$

The difference in travel time can be expressed in $T(0)$ and $T(\infty)$

$$T(l) = T(0) \exp(-2Cl) + T(\infty)[1 - \exp(-2Cl)] \tag{10.87}$$

If the light that enters the fiber comes directly from a modulated light source, such as an LED or a laser, then generally $T(0) = 0$. If, however, the light that is coupled into a fiber comes from another fiber of such a length that the equilibrium state is attained, then $T(0) \approx T(\infty)$. In Figure 10.7 the travel times t_1 and t_2 are shown as functions of Cl.

We now start to calculate the widths of the pulses travelling along the fibers. As before, for the sake of simplicity, it is assumed that $\alpha_1 = \alpha_2 = \alpha$ and $E_1 = E_2$. If the time difference from equation (10.87) is substituted in equation (10.62), the following two differential equations result:

$$\frac{d\sigma_1^2(l)}{dl} = C\sigma_2^2(l) - C\sigma_1^2(l)$$

$$+ C\{T(0)\exp(-2Cl) + T(\infty)[1 - \exp(-2Cl)]\}^2 \tag{10.88}$$

$$\frac{d\sigma_2^2(l)}{dl} = C\sigma_1^2(l) - C\sigma_2^2(l)$$

$$+ C\{T(0)\exp(-2Cl) + T(\infty)[1 - \exp(-2Cl)]\}^2 \tag{10.89}$$

The solution of these differential equations are

$$\sigma_{1,2}^2 = \frac{\sigma_1^2(0) + \sigma_2^2(0)}{2} \pm \frac{\sigma_1^2(0) - \sigma_2^2(0)}{2} \exp(-2Cl)$$

$$+ \frac{T^2(0)}{4}[1 - \exp(-4Cl)] + \frac{T(0)T(\infty)}{2}[1 - \exp(-2Cl)]^2$$

$$+ \frac{T^2(\infty)}{4}[4Cl + 4\exp(-2Cl) - \exp(-4Cl) - 3] \tag{10.90}$$

In general, the two modes are detected simultaneously by the photodiode, resulting in a single current pulse, which comprises the sum of the pulses $s_1(l, t)$ and $s_2(l, t)$ of the two modes. If the resulting pulse is denoted by $s_p(l, t)$, then

$$s_p(l, t) = s_1(l, t) + s_2(l, t). \tag{10.91}$$

With the definition from equations (10.51)–(10.53) and equation (10.84), the resulting travel time $t_p(l)$ and width $\sigma_p(l)$ can be found

$$t_p(l) = \frac{t_1(l) + t_2(l)}{2} \tag{10.92}$$

$$\sigma_p^2(l) = \frac{\sigma_1^2(l) + \sigma_2^2(l)}{2} + \frac{T^2(l)}{4} \tag{10.93}$$

If $T(0) = 0$ and if $\sigma_1(0) = \sigma_2(0) = \sigma(0)$, then it follows from equation (10.90) that

$$\sigma_{1,2}^2 = \sigma^2(0) + \frac{T^2(\infty)}{4}[4Cl + 4\exp(-2Cl) - \exp(-4Cl) - 3] \tag{10.94}$$

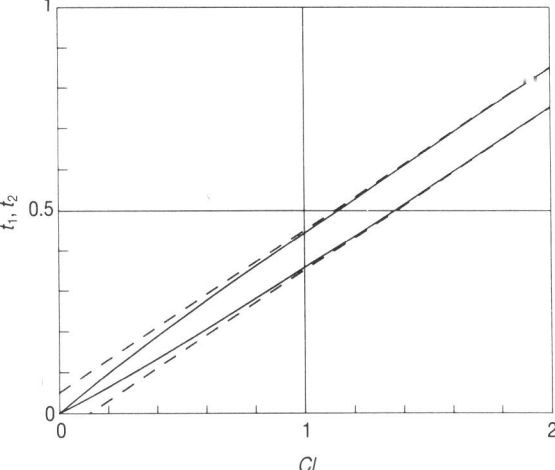

Figure 10.7 The travel times t_1 and t_2 as functions of Cl. Solid lines indicate $T(0) = 0$ and broken lines $T(0) = T(\infty)$.

and

$$\sigma_p^2(l) = \sigma^2(0) + \frac{T^2(\infty)}{4} [4Cl + 2\exp(-2Cl) - 2] \tag{10.95}$$

In Figure 10.8 a number of curves representing $\sigma_{1,2}/T(\infty)$ and $\sigma_p/T(\infty)$ as

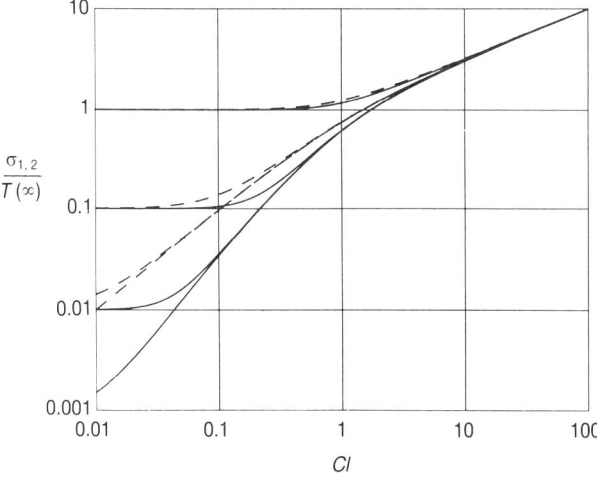

Figure 10.8 $\sigma_{1,2}/T(\infty)$ (solid lines) and $\sigma_p^2/T(\infty)$ (broken lines) as functions of Cl for the parameter values $\sigma(0)/T(\infty) = 0.001, 0.01, 0.1, 1.0$.

functions of Cl are depicted for different values of the parameter $\sigma(0)/T(\infty)$. In particular, when $T(0) = T(\infty)$ and when $\sigma_1(0) = \sigma_2(0) = \sigma(0)$, then

$$\sigma_{1,2}^2(l) = \sigma^2(0) + Cl\ T^2(\infty) \tag{10.96}$$

and

$$\sigma_{\mathrm{p}}^2(l) = \sigma^2(0) + ClT^2(\infty) + \frac{T^2(\infty)}{4} \tag{10.97}$$

Looking at these equations and Figure 10.8, the following conclusions can be drawn.

- When a steady-state (equilibrium) is introduced right from the beginning, the pulse width increments $\sigma_{1,2}$ and σ_{p} are roughly proportional to the square root of the length.
- When the input pulse width to the fiber is very small, then for lengths shorter than the coupling length $1/C$, the pulse width increment is proportional to the length. For lengths longer than the coupling length, the pulse width increment is proportional to the square root of the length increment, just like the case where the steady state is present from the very beginning.

We next consider the combined effect of mode coupling, and material and waveguide dispersion. This means that we shall combine the model introduced in this section with the results of Section 10.1.2. The addition of material and waveguide dispersion does not affect the travel times of the pulses as described by equations (10.80)–(10.83); however, the width of the pulses does change. The changes can be found with equation (10.40); by taking the derivative of σ_0^2 with respect to the length parameter l we obtain

$$\frac{\mathrm{d}\sigma_0^2}{\mathrm{d}l} = 2(\beta_0''B)^2 l \triangleq 2\mu l \tag{10.98}$$

Because the material and waveguide dispersion on one side and the mode coupling on the other side are independent, their combined effects can be described by adding the term $2\mu l$ to the right-hand side of equation (10.62) [14]. Let us again consider the case of two coupled modes; we then obtain the following set of equations:

$$\frac{\mathrm{d}\sigma_1^2}{\mathrm{d}l} = \frac{E_2}{E_1} C[\sigma_2^2 - \sigma_1^2 + T^2(l)] + 2\mu l$$

$$\frac{\mathrm{d}\sigma_2^2}{\mathrm{d}l} = \frac{E_1}{E_2} C[\sigma_1^2 - \sigma_2^2 + T^2(l)] + 2\mu l \tag{10.99}$$

Solving these equations for the special case $E_1 = E_2$, $t_1 = t_2 = 0$ and $\sigma_1(0) = \sigma_2(0) = \sigma(0)$ gives

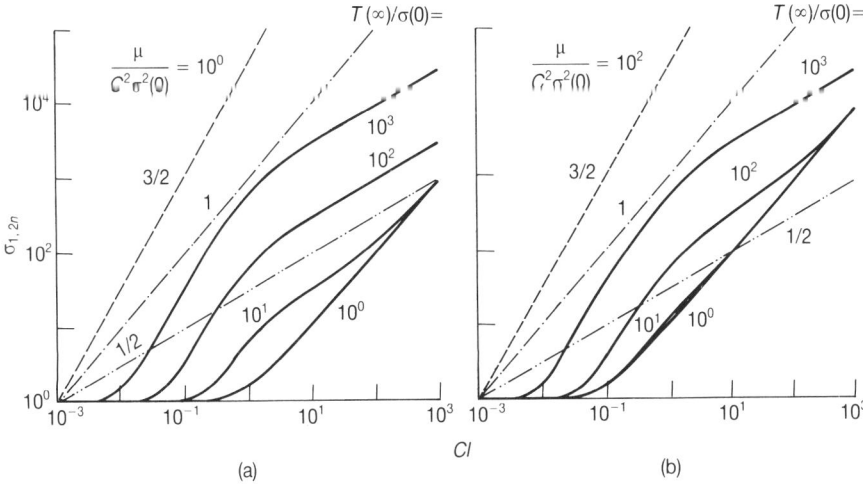

Figure 10.9 Normalized pulse width of a single mode versus the normalized fiber length, for various values of $T(\infty)/\sigma(0)$ and two fixed values of $\mu/[C^2\sigma^2(0)]$. (Source: reproduced with permission from *Archiv für Elektronik und Übertragungstechnik*, vol. 37, no. 11/12, 1983, pp. 399–401, 'Pulse broadening in optical fibers due to mode coupling and material dispersion', by W. van Etten and P. van de Mortel.)

$$\sigma_{1,2}^2(l) = \sigma^2(0) + \frac{T^2(\infty)}{4}$$

$$\times [4Cl + 4\exp(-2Cl) - \exp(-4Cl) - 3] + \frac{\mu}{C^2}(Cl)^2$$

$$(10.100)$$

In Figure 10.9 the function $\sigma_{1,2n} \triangleq \sigma_{1,2}(Cl)/\sigma(0)$ has been plotted on a double logarithmic scale for various values of $T(\infty)/\sigma(0)$ and fixed values of $\mu/[C^2\sigma^2(0)]$. The numbers on the broken lines represent the asymptotic values of the various slopes of the characteristics. Again, the pulses of the individual modes give a single current pulse in the photodiode. For the squared value of this pulse width, equation (10.93) can be substituted to find that

$$\sigma_p^2(l) = \sigma^2(0) + \frac{T^2(\infty)}{2}[2Cl - 1 + \exp(-2Cl)] + \frac{\mu}{C^2}(Cl)^2 \quad (10.101)$$

The function $\sigma_{pn} \triangleq \sigma_p(Cl)/\sigma(0)$ has been drawn in Figure 10.10 for different values of $T(\infty)/\sigma(0)$ and fixed values of $\mu/[C^2\sigma^2(0)]$; the function σ_{pn}^2 is approximated by

$$\sigma_{pn}^2 \approx 1 + \frac{T^2(\infty)}{2\sigma^2(0)}\left[2(Cl)^2 + \frac{4}{3}(Cl)^3\right] + \frac{\mu}{C^2\sigma^2(0)}(Cl)^2, \quad Cl < 1$$

$$(10.102)$$

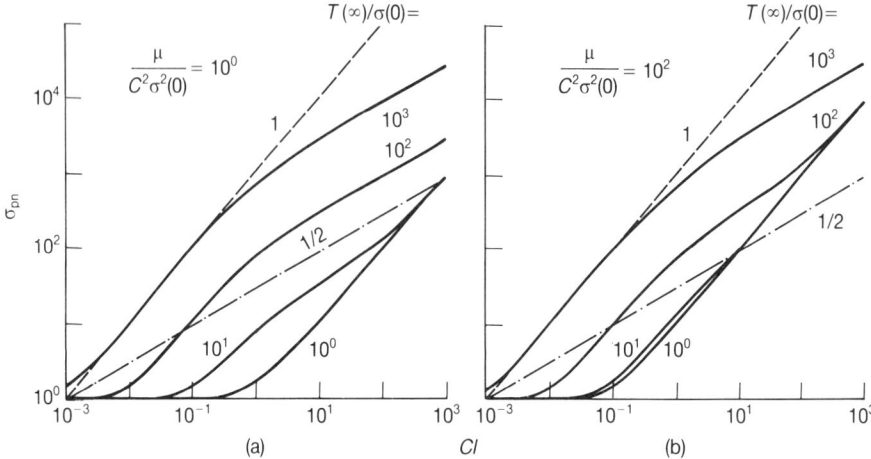

Figure 10.10 Normalized pulse width for two modes, as detected by a photodiode versus the normalized fiber length, for various values of $T(\infty)/\sigma(0)$ and two fixed values of $\mu/[C^2\sigma^2(0)]$. (Source: reproduced with permission from *Archiv für Elektronik und Übertragungstechnik*, vol. 37, no. 11/12, 1983, pp. 399–401, 'Pulse broadening in optical fibers due to mode coupling and material dispersion', by W. van Etten and P. van de Mortel.)

$$\sigma_{pn}^2 \approx 1 + \frac{T^2(\infty)}{2\sigma^2(0)}\,Cl + \frac{\mu}{C^2\sigma^2(0)}\,(Cl)^2, \qquad Cl > 1 \qquad (10.103)$$

Various characteristic parts can be discerned in the domain of the function σ_{pn}:

1. $Cl < 1$, but not $Cl \ll 1$. From equation (10.101) it follows that the pulse width σ_{pn} always shows a slope of unit in this part of the domain, when plotted on a double logarithmic scale.
2. $Cl > 1$, but $Cl < T^2(\infty)C^2/2\mu$ and $\mu/C^2 \ll T^2(\infty)/2$. The slope of σ_{pn} is 1/2 (see equation 10.102); however, for $Cl \gg T^2(\infty)C^2/2\mu$ the slope always returns to unity. This means that the material and waveguide dispersion are always dominant for long fibers.

As can be seen from Figure 10.10, the slope 1/2 may not appear at all. In that case, the material and waveguide dispersion have completely masked the mode mixing. If this does not occur, the pulse width increases proportionally with the square root of the length increment when the fiber length is longer than the coupling length. This suggests that the coupling coefficient C between the modes be increased. However, in this way, the optical power coupled to radiating and leaky modes increases. Narrowing the impulse response for a certain length of optical fiber causes the attenuation to increase too. There is therefore a trade-off between the two transmission parameters, in this respect. Mode coupling occurs spontaneously along the entire fiber, due to random imperfections in the fiber material and its geometry. The effect becomes more pronounced in splices, optical fiber connectors and bends (especially micro-bending in cables).

From this section, it becomes clear that the signal transfer via multimode optical fibers is difficult to predict, due to the following reasons:

1. It is usually not known exactly how the light source is coupled to the fiber since it is not known how strongly the different modes are excited.
2. The mode coupling coefficients are not known or, if they can be measured in the laboratory, these coefficients may differ under operating conditions.
3. The refractive index profile may deviate from the desired profile.

When designing communication systems in which multimode optical fibers serve as the transmission medium, it is important to measure as many of the transmission characteristics as possible; when the characteristics or parameters cannot be measured, the worst estimates for these should be used.

References

[1] A. K. Ghatak and K. Thyagarajan, *Contemporary Optics* (New York: Plenum Press, 1978).
[2] F. P. Kapron and D. B. Keck, 'Pulse transmission through a dielectric optical waveguide', *Applied Optics*, vol. 10, no. 7, July 1971, pp. 1519–23.
[3] A. Papoulis, *Probability, Random Variables and Stochastic Processes*, 2nd edn (New York: McGraw-Hill, 1984).
[4] S. D. Personick, 'Baseband linearity and equalization in fiber optic digital communication systems', *Bell System Technical Journal*, vol. 52, no. 7, September 1973, pp. 1175–94.
[5] W. van Etten, 'The ergodicity of laser light in connection with optical fibre transmission', *Optical and Quantum Electronics*, vol. 13, 1981, pp. 519–21.
[6] D. Marcuse, 'Pulse distortion in single-mode fibers', *Applied Optics*, vol. 19, no. 10, May 1980, pp. 1653–60.
[7] M. Miyagi and S. Nishida, 'Pulse spreading in a single-mode optical fiber due to third-order dispersion: effect of optical source bandwidth', *Applied Optics*, vol. 18, no. 13, July 1979, pp. 2237–40.
[8] P. J. B. Clarricoats and K. B. Chan, 'Electromagnetic-wave propagation along radially inhomogeneous dielectric cylinders', *Electronics Letters*, vol. 6, no. 22, October 1970, pp. 694–5.
[9] C. Yeh and G. Lindgren, 'Computing the propagation characteristics of radially stratified fibers: an efficient method', *Applied Optics*, vol. 16, no. 2, February 1977, pp. 483–93.
[10] S. D. Personick, *Optical Fiber Transmission Systems* (New York: Plenum Press, 1981).

[11] S. Kawakami and M. Ikeda, 'Transmission characteristics of a two-mode optical waveguide', *IEEE Journal of Quantum Electronics*, vol. QE-14, no. 8, August 1978, pp. 608–14.

[12] D. Marcuse, *Theory of Dielectric Optical Waveguides* (New York: Academic Press, 1974).

[13] S. Geckeler, 'Mode-mixing in optical fibres', *Archiv für Elektronik und Übertragungstechnik*, vol. 33, no. 1, January 1979, pp. 43–5.

[14] W. van Etten and P. van de Mortel, 'Pulse broadening in optical fibers due to mode coupling and material dispersion', *Archiv für Elektronik und Übertragungstechnik*, vol. 37, no. 11/12, November/December 1983, pp. 399–401.

Problems

10.1 Show that, in general, $E[m_o(t)]$ only becomes real if the integrand of equation (10.27) is symmetrical in τ_1 and τ_2.

10.2 Consider a transmission link with a silica single-mode fiber.
    Show that for an LED as the light source ($B \approx 10^{13}$ Hz) and a reasonable length (>1 km), equation (10.36) can be approximated by equation (10.38).

10.3 Show that for the link of Problem 10.2 and a laser as the light source ($B \approx 10^8$ Hz), equation (10.36) is approximated by equation (10.41).

10.4 Show that equation (10.90) is the solution of the set of equations (10.88) and (10.89).

10.5 Reason why pulse broadening due to material and waveguide dispersion always dominates pulse broadening due to mode mixing, in long fibers.

11

Power launching and coupling efficiency

11.1   The ray density of a Lambertian source in the phase space
11.2   Power launching from the source into a multimode fiber
11.3   Multimode fiber–fiber coupling
11.4   Coupling model for single-mode fibers
11.5   Power coupling from the source into single-mode fibers
11.6   Single-mode fiber–fiber coupling
11.7   Fiber–detector coupling
       References
       Problems

In optical fiber communication systems the coupling of optical power from the source to a fiber can be a major problem. In this chapter we shall consider the coupling of both a laser and an LED to a multimode as well as a single-mode fiber. Moreover, the coupling between two fibers with different parameters and geometrically misaligned is also considered. For multimode fibers the problems can be solved by means of ray optics. The propagation of light rays is described by means of the phase space. The important parameters that play a role in defining these coupling problems are numerical aperture, core radius and refractive index profile.

Light propagation in single-mode fibers is well described by electromagnetic fields. It has been shown in Section 5.3.1 that the field distribution of the LP_{01}

mode is reasonably well approximated by means of a Gaussian function. This reduces the coupling problem to a coupling between Gaussian beams.

This class of problems will be examined in this chapter and, for different configurations, the coupling efficiency or attenuation will be calculated as a function of the relevant parameters.

11.1 The ray density of a Lambertian source in the phase space

A Lambertian source is a light source that emits equal amounts of power per unit of projected area and per unit of solid angle from each part of its emitting surface. This optical power per unit of project area and per unit of solid angle is called the radiance L (units $W\,m^{-2}\,sr^{-1}$). The phase space is a mathematical tool to describe the radiation (in the geometrical optics context) that is emitted by a light source, or the radiation that is received by an optical component (for instance a fiber). We shall only consider surfaces for emitting and receiving light that can be described by two spatial coordinates (x and y), in which case the phase space has four coordinates, namely the two coordinates x and y, and two angle coordinates θ and ψ. In this four-dimensional space a function is defined that represents the density of the light rays from the source or receiving surface in question. This density is then identified as the radiance L. For a Lambertian source L is a constant by definition.

Let us consider Figure 11.1(a). In this diagram OD is part of a light ray emitted from point O of the source in the direction of point D. The emitting surface is in the x–y-plane, whereas the ray OD makes an angle θ with the z-axis. The projection of OD onto the x–z-plane is OC, whereas the angle between this projection and the z-axis is called θ_x. In the same way the projection of OD onto the y–z-plane is OA and the angle between OA and the z-axis is called θ_y. At a distance z from the origin, the x and y coordinates of point D are

$$x = BC = z\tan\theta_x \triangleq zu \tag{11.1a}$$

$$y = AB = z\tan\theta_y \triangleq zv \tag{11.1b}$$

$$z = z \tag{11.1c}$$

If the radiation has an angular symmetry, i.e. it only depends on θ, the angle between the ray OD and the z-axis, then the two variables u and v can be combined into a single variable w. This becomes clear when Figure 11.1(b) is considered, where the triangle OBD from Figure 11.1(a) is depicted. From Figure 11.1(b) it follows that

$$\cos\theta = \frac{1}{\sqrt{1+u^2+v^2}} \triangleq \frac{1}{\sqrt{1+w^2}} \tag{11.2a}$$

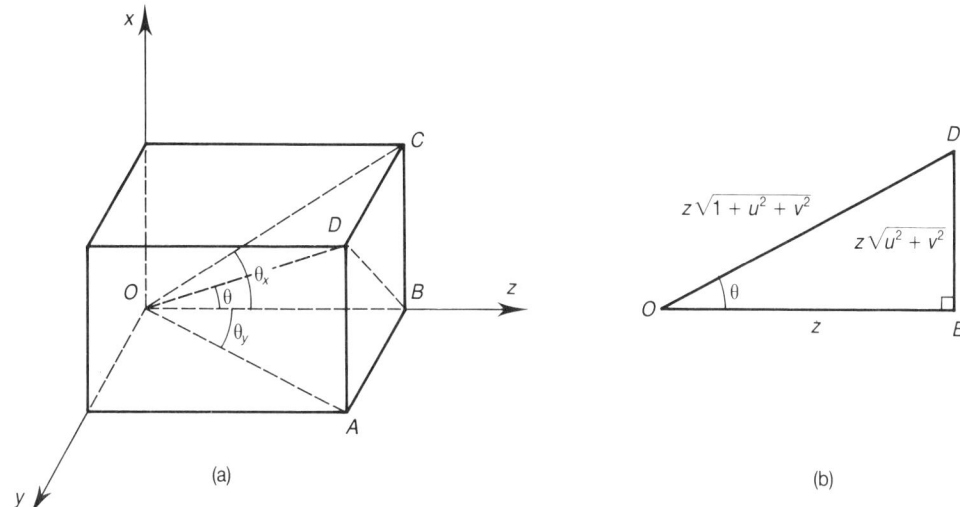

(a)   (b)

Figure 11.1 (a) Coordinates of the emitting surface and light ray *OD*. (b) Triangle *OBD*. (Source: reproduced with permission from *Applied Optics*, vol. 24, no. 7, April 1985, pp. 970–6, 'Loss in multimode fiber connections with a gap', by W. van Etten, W. Lambo and P. Simons.)

$$\tan \theta = \sqrt{u^2 + v^2} = w^2 \tag{11.2b}$$

The total power emitted by a light source is written as [1, p. 327]

$$P = \int_\Omega \int_S L \cos \theta \, d\Omega \, dS \tag{11.3}$$

where Ω is the solid angle over which the radiation extends and S the area of the emitting surface. The aim of this section is to determine how equation (11.3) can be written in terms of the variable w. Let us consider Figure 11.2(a) for this purpose. All the radiation leaving the emitter with an angle of $\{\theta, \theta + d\theta\}$ strikes the plane $z = 1$ in an annular region with radius w and width dw. The area of this region is $2\pi w \, dw$, but from the viewpoint of O, seen at an angle θ, the corresponding area is $2\pi w \cos \theta \, dw$ (see Figure 11.2b). The solid angle $d\Omega$ of this region, as seen from O, following from its definition, is

$$d\Omega = \frac{2\pi w \cos \theta \, dw}{1 + w^2} = \frac{2\pi w \, dw}{(1 + w^2)^{3/2}} \tag{11.4}$$

Inserting this into equation (11.3) gives

$$P = \int_w \int_S L \frac{2\pi w}{(1 + w^2)^2} \, dw \, dS \tag{11.5}$$

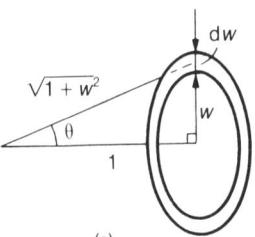
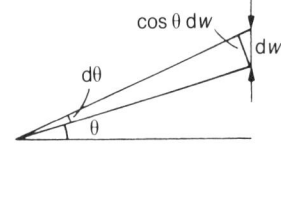

(a)   (b)

Figure 11.2 (a) Annular region illuminated by rays leaving the emitter with angle $\{\theta, \theta + d\theta\}$. (b) The annular region as seen from O. (Source: reproduced with permission from *Applied Optics*, vol. 24, no. 7, April 1985, pp. 970–6, 'Loss in multimode fiber connections with a gap', by W. van Etten, W. Lambo and P. Simons.)

The integration limits depend on the problem at hand. For this integration it can be seen that, besides the two coordinates x and y, one angle parameter w in the phase space suffices. The integrand of equation (11.5) can be interpreted as the density of light rays in the phase space.

Example 11.1

Let us consider an LED with a circular emitting surface and assume the LED to be a Lambertian source. When the radius of this surface is called b, then the total emitted power is written as

$$P = 2\pi L \int_0^\infty \frac{w}{(1 + w^2)^2}\, dw \int_0^b 2\pi r\, dr = 2\pi^2 b^2 L \int_0^\infty \frac{w\, dw}{(1 + w^2)^2} = \pi^2 b^2 L$$

(11.6)

As can be seen in this special case, the phase space reduces to two dimensions.

11.2 Power launching from the source into a multimode fiber

11.2.1 Launching from an LED into a multimode fiber

In this section we shall calculate the efficiency of power coupling from an LED into a fiber, where the fiber end-face is butt-jointed to the emitting surface of the LED. The radius of the fiber core is called a. In the preceding section the total power emitted by a circular LED of radius b has been calculated. Moreover, the density of rays in the phase space can be determined (see equation 11.5). We assume that $b > a$. If the bounds of the receiving device (in this case the fiber) can be determined in the phase space, then the power that is coupled into the fiber can be calculated with the aid of equation (11.5). In this case it again suffices to use two dimensions for the phase space, namely r and w, analogous to equation (11.6). Let us take the numerical aperture of the fiber $A(r)$, the limits then appear to be $r \in \{0, a\}$ and $w \in \{0, A(r)/\sqrt{1 - A^2(r)}\}$. The fiber is assumed to have a so-called power-law index profile (see also equation 1.2)

$$n(r) = n(0) \sqrt{1 - 2\Delta \left(\frac{r}{a}\right)^x}, \qquad 0 \le r \le a, \tag{11.7a}$$

$$n(r) = n(0) \sqrt{1 - 2\Delta}, \qquad r \ge a \tag{11.7b}$$

Excluding the leaky rays, the numerical aperture of such a fiber can be written as (see equations 6.68 and 6.70)

$$A(r) = \mathrm{NA} \sqrt{1 - \left(\frac{r}{a}\right)^x} \tag{11.8}$$

The power coupled into the fiber is found by integrating equation (11.5) with the limits given above

$$P = 2\pi L \int_0^a 2\pi r \left[\int_0^{T(r)} \frac{w}{(1 + w^2)^2} \, dw \right] dr$$

$$= 2\pi^2 (\mathrm{NA})^2 L \int_0^a r \left[1 - \left(\frac{r}{a}\right)^x \right] dr = \pi^2 a^2 L (\mathrm{NA})^2 \frac{x}{x + 2} \tag{11.9a}$$

with

$$T(r) \triangleq \frac{A(r)}{\sqrt{1 - A^2(r)}} \tag{11.9b}$$

From this result both the power in a step index fiber as well as in a graded index fiber can be obtained. For a step index fiber, $x \to \infty$ is taken, so that

$$P_s = \pi^2 a^2 L (\mathrm{NA})^2 \tag{11.10}$$

and the efficiency of coupling an LED to a step index fiber becomes

$$\eta_s = \frac{a^2 (\mathrm{NA})^2}{b^2} \tag{11.11}$$

In the case of a fiber with a parabolic profile, we take $x = 2$ and

$$P_p = \frac{\pi^2 a^2 L (\mathrm{NA})^2}{2} \tag{11.12}$$

This represents half the power that is coupled into a step index fiber with the same core radius and the same central numerical aperture. Consequently, the efficiency in this case becomes

$$\eta_p = \frac{a^2(\text{NA})^2}{2b^2} \tag{11.13}$$

which is half of that for a step index fiber.

Example 11.2

Let us consider the situation where light of an LED is coupled into a step index fiber with NA = 0.2 and $a = b$. It follows that $\eta = 4$ percent; therefore, the loss is 14 dB. For a fiber with a parabolic profile and the same central numerical aperture the efficiency becomes $\eta = 2$ percent and the loss 17 dB.

If the radius of the emitting surface of the LED is larger than the core radius of the fiber the coupling loss is even greater, as shown by equations (11.11) and (11.13). In that case no coupling improvement can be achieved by using optical appliances, such as a lens [2]. This result also follows from the radiance conservation law [3, p. 189]. The law states that when making an image with an optical system, the radiance of the image is not greater than the radiance of the original, provided that the refractive index of the medium of the image equals the refractive index of the medium of the original. When the radius of the source or its NA is smaller than that of the fiber, it follows from equations (11.10) and (11.12) that it makes sense to use optical systems when enlarging these parameters in order to fill completely both the fiber core and its NA.

11.2.2 Coupling a laser diode to a multimode fiber

The coupling of coherent laser light into a fiber differs substantially from the coupling of incoherent light from an LED into a fiber. We can see that an LED behaves like a Lambertian source in Figure 11.3(a). The radiation from a laser diode is quite different: at some distance from the source all light rays seem to be emitted from a single point (see Figure 11.3b).

The light from a laser diode operating in the fundamental transversal mode is, by approximation, described by means of a Gaussian beam with irradiance ([4, 5] see also Appendix 3)

$$S = S_0 \exp\left[-2\left(\frac{x}{w_x}\right)^2 - 2\left(\frac{y}{w_y}\right)^2\right] \tag{11.14}$$

with $w_{x,y}$ the values of x and y, respectively, where the irradiance drops a factor of e^{-2} with respect to the on-axis value. The parameters w_x and w_y are functions of z

$$w_{x,y} = w_{0x,0y}\sqrt{1 + \left(\frac{\lambda z}{\pi w_{0x,0y}^2}\right)^2} \approx \frac{\lambda z}{\pi w_{0x,0y}} \tag{11.15}$$

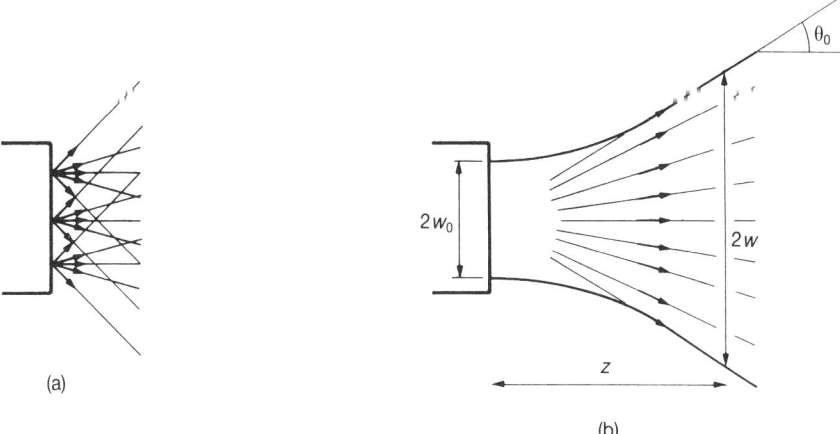

Figure 11.3 Radiation from: (a) an LED; (b) a laser diode.

The latter approximation only holds for $z > z_0$ (see equation A3.18). The parameters w_{0x} and w_{0y} have, in general, different values for semiconductor lasers. At some distance from the laser diode, the ray can be described by its radiant intensity ([6] and Appendix 3)

$$I = I_o \exp\left[-2\left(\frac{\theta_x}{\theta_{0x}}\right)^2 - 2\left(\frac{\theta_y}{\theta_{0y}}\right)^2\right]$$

(11.16)

The angles θ_{0x} and θ_{0y}, where the radiant intensity has been decreased by a factor of e^{-2} with respect to the maximum value on the z-axis, are derived from equation (11.15)

$$\theta_{0x,0y} = \arctan\frac{\lambda}{\pi w_{0x,0y}} \approx \frac{\lambda}{\pi w_{0x,0y}}$$

(11.17)

At $z = 0$ the beam is as narrow as possible and this smallest cross-section is called the waist of the beam. When making an image of the emitting facet of the laser diode, a second waist is created which can be larger or smaller than the first one [5]; in compliance with equation (11.17), the characteristic angle θ_0 can be smaller or larger. There is thus a trade-off between the characteristic angle θ_0 and the minimum beam width $2w_0$. As a rule, the dimensions of a laser facet are smaller than the core diameter of the fiber. A magnified image of the laser facet at the end-face of a fiber decreases the characteristic angle of the beam (which is often larger than the numerical aperture of the fiber), in such a way that it improves the power launching efficiency. If the waist of a laser beam coincides with the end-face of a fiber, by butt jointing or by imaging, the launching efficiency is not determined by the near-field distribution (if the waist is smaller than the fiber core diameter). Light acceptance by the fiber is limited in the far-field distribution, as described by equation (11.16). Provided that its conditions are fulfilled and the axis of symmetry of the beam coincides with the fiber axis, the launching efficiency can be written as

$$
\eta = \frac{\displaystyle\int_0^{\theta_n} \exp\left[-2\left(\frac{\theta_x}{\theta_{0x}}\right)^2\right] d\theta_x \ \int_0^{\theta_n} \exp\left[-2\left(\frac{\theta_y}{\theta_{0y}}\right)^2\right] d\theta_y}{\displaystyle\int_0^{\pi/2} \exp\left[-2\left(\frac{\theta_x}{\theta_{0x}}\right)^2\right] d\theta_x \ \int_0^{\pi/2} \exp\left[-2\left(\frac{\theta_y}{\theta_{0y}}\right)^2\right] d\theta_y}
$$

$$
\triangleq H(\text{NA}, \theta_{0x}) \, H(\text{NA}, \theta_{0y})
\tag{11.18}
$$

with

$$
\theta_n = \arcsin \text{NA}
\tag{11.19}
$$

In most practical situations the following approximation is allowed:

$$
\int_0^{\pi/2} \exp\left[-2\left(\frac{\theta}{\theta_0}\right)^2\right] d\theta \approx \int_0^{\infty} \exp\left[-2\left(\frac{\theta}{\theta_0}\right)^2\right] d\theta = \sqrt{2\pi}\,\theta_0
\tag{11.20}
$$

so that

$$
\eta \approx \text{erf}\left(\sqrt{2}\,\frac{\theta_n}{\theta_{0x}}\right) \text{erf}\left(\sqrt{2}\,\frac{\theta_n}{\theta_{0y}}\right)
\tag{11.21}
$$

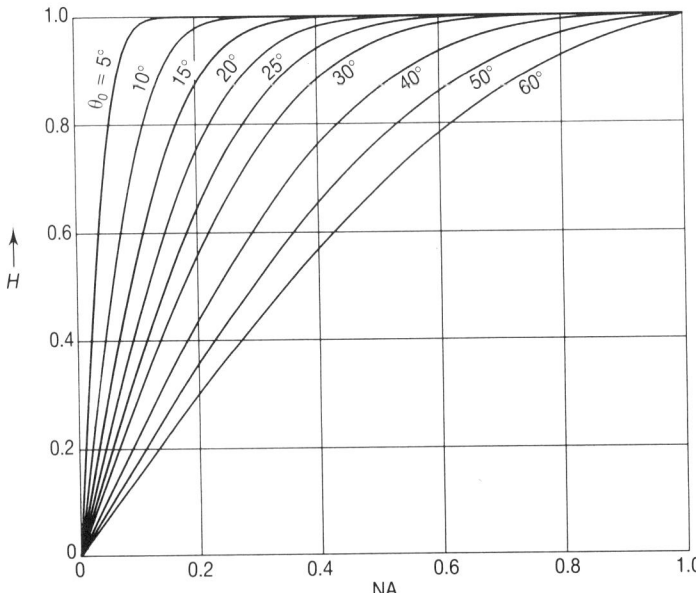

Figure 11.4 The function *H*, which is used to calculate the coupling efficiency from laser diode to multimode fiber.

Thus, the launching efficiency is approximated by the product of two error functions. In Figure 11.4 the function H of equation (11.18) is depicted as a function of NA, the numerical aperture of the fiber, with θ_0 as a parameter.

In order to determine the launching efficiency, Figure 11.4 has to be used twice. Firstly, to evaluate H with θ_{0x} as a parameter, and secondly with θ_{0y} as a parameter. The efficiency is given by the product of the two values obtained for H. The values of θ_{0x} and θ_{0y} are found on the data sheet from the laser diode supplier; one value of θ_0 is given for the plane of the p–n junction and the other for the plane perpendicular to the p–n junction. These two values can differ substantially.

Example 11.3
Let us consider the following practical values: NA $= 0.2$, $\theta_{0x} = 40°$ and $\theta_{0y} = 15°$. By means of Figure 11.4 it can be found that $H_x = 0.43$ and $H_y = 0.87$. The efficiency is therefore: $\eta = H_x H_y = 37$ percent.

From Example 11.3 it follows that a laser–fiber coupling produces a much greater efficiency than an LED–fiber coupling (compare Example 11.3 with Example 11.2). In Example 11.3 the efficiency can be improved further by using a cylindrical lens. Such an improvement is impossible for the LED–fiber coupling when the emitting area of the LED is larger than the cross-section of the fiber core (see Section 11.2.1), which is usually the case. The lens can be integrated with the fiber by making the fiber ends spherical, by fusing or etching the end-faces [7].

11.3 Multimode fiber–fiber coupling

In this section we shall calculate the loss that occurs when joining two fibers, as in a connector (see Chapter 15). We will restrict ourselves to a butt joint connection. Losses in a fiber connection fall into two categories:

1. Intrinsic losses are caused by different numerical apertures, diameters or index profiles of the fibers being joined. Fresnel reflection losses also belong to this category. At each glass–air interface 4 percent of the optical power is reflected, which results in a loss of 0.14 dB. Intrinsic loss can be avoided by matching the relevant parameters of the fibers being joined.
2. Extrinsic losses are caused by the mechanical misalignment of fibers with identical parameters; these losses are minimized by aligning the end-faces precisely.

In this section we shall show how to calculate the intrinsic and extrinsic losses for some typical misalignments. The collection is not complete, but the configurations presented and the methods used are sufficient to handle other configurations easily. In all cases we assume that the emitting fiber behaves like a Lambertian source with a limited (local) numerical aperture. Moreover, in these investigations, only guided rays will be considered and we will restrict ourselves to fibers with a power-law index profile (see equation 11.7).

11.3.1 Intrinsic losses

In this section we assume that the fibers to be joined are perfectly aligned, i.e. there are no extrinsic losses. As a starting point we take equation (11.9a); the emitting fiber is called fiber 1, while the receiving fiber is called fiber 2.

Mismatch of core diameters
Firstly, we take: $NA_1 = NA_2$, $x_1 = x_2$ and $a_1 \neq a_2$. If the core diameter of the receiving fiber is larger than the core diameter of the emitting fiber, then all the light emitted by fiber 1 is accepted by fiber 2. In the case of $a_1 > a_2$, some of the light emitted is not accepted. The power in both the emitting and the receiving fibers is described by equation (11.9a), because each point on the receiving fiber is illuminated by a point on the emitting fiber with a larger numerical aperture. The coupling loss is thus written as

$$
L_d =
\begin{cases}
-10 \log \left(\dfrac{a_2}{a_1} \right)^2, & a_2 \leq a_1 \\[2ex]
0, & a_2 \geq a_1
\end{cases}
\tag{11.22}
$$

It is easy to see that if the core diameter of the transmitting fiber is 1 percent smaller than the core diameter of the receiving fiber, it introduces a loss of 0.087 dB compared to ideally matched fibers, when $a_1 = a_2$.

Mismatch of numerical apertures
If $a_1 = a_2$, $x_1 = x_2$ and $NA_1 \neq NA_2$, it follows from equation (11.9a) that

$$
L_n =
\begin{cases}
-10 \log \left(\dfrac{NA_2}{NA_1} \right)^2, & NA_2 \leq NA_1 \\[2ex]
0, & NA_2 \geq NA_1
\end{cases}
\tag{11.23}
$$

Since equation (11.23) equals equation (11.22) it can be concluded that if the numerical aperture of the receiving fiber is 1 percent smaller than that of the transmitting fiber there is again a loss of 0.087 dB.

Mismatch of index profiles
If $a_1 = a_2$, $NA_1 = NA_2$ and $x_1 \neq x_2$, it follows from equation (11.9a) that

$$
L_x = \begin{cases} -10 \log \dfrac{x_2(x_1 + 2)}{x_1(x_2 + 2)}, & x_2 \leq x_1 \\ \\ 0, & x_2 \geq x_1 \end{cases} \tag{11.24}
$$

If a step index fiber is used as the emitting fiber $(x_1 \to \infty)$ and a parabolic index fiber is used as the receiving fiber $(x_2 = 2)$, a loss of 3 dB is introduced. An emitting fiber with $x_1 = 2.2$ and a receiving fiber with $x_2 = 2$ leads to a loss of 0.2 dB.

Combination of mismatches
When combining mismatches $a_2 > a_1$, $NA_2 > NA_1$ and $x_2 > x_1$, the loss is 0 dB. The combination of $a_2 < a_1$, $NA_2 < NA_1$ and $x_2 < x_1$ gives

$$
L_c = -10 \log \frac{(a_2 NA_2)^2(x_1 + 2)x_2}{(a_1 NA_1)^2(x_2 + 2)x_1} \tag{11.25}
$$

All other possible combinations lead to more complicated calculations and solving them requires methods that will be dealt with in the next section. Reference [8] presents the distributions of losses that are found when, with certain distributions of the parameters, random combinations of mismatches occur with a long series of connections.

11.3.2 Extrinsic losses

The mechanical misalignments, which are the causes of the extrinsic losses, are radial offset, separation of the end-faces, angular misalignment and combinations of all these misalignments.

Radial offset of two step index fibers
Here we consider a situation where the fiber end-faces are in contact with each other and both axes are moved a distance of $2d$ radially. Contrary to the literature [9], we shall use the description of the density of light rays in the phase space as a basis for our calculations (see Section 11.1). These calculations are difficult to perform for the general case of a power-law index profile according to equation (11.7); therefore, we shall separately investigate the connection between two step index and two graded index fibers with a parabolic profile.

Starting with step index fibers, because the cores of the emitting and the receiving fibers overlap partially and each point of the emitting fiber has the same numerical aperture as the matching point on the receiving fiber, the problem is quite simple. The coupling efficiency is determined by the ratio of the overlapping area of two circles with radius a and centers $2d$ apart, and the

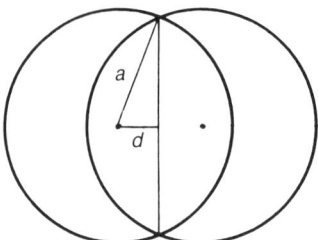

Figure 11.5 Two fibers with a radial separation of 2d.

total area of a circle with radius a (see Figure 11.5). The area of the overlap is

$$O = 2a^2 \arccos\left(\frac{d}{a}\right) - 2d\sqrt{a^2 - d^2} \tag{11.26}$$

and the efficiency becomes

$$\eta = \frac{2}{\pi}\left(\arccos\left(\frac{d}{a}\right) - \frac{d}{a}\sqrt{1 - \left(\frac{d}{a}\right)^2}\right) \tag{11.27}$$

It can be seen that the efficiency is independent of the numerical aperture. In Figure 11.6(a) the efficiency according to equation (11.27) is shown as a function of the normalized shift d/a. By means of a series expansion, an approximation of equation (11.27) is found for small values of d/a

$$\eta \approx 1 - \frac{4}{\pi}\frac{d}{a}, \qquad \text{for } \frac{d}{a} \ll 1 \tag{11.28}$$

This approximation is shown in Figure 11.6(a) as a dotted line. In Figure 11.6(b) the loss from equation (11.27) is shown along with the loss from equation (11.28), which is shown as a dotted line.

Radial offset of two graded index fibers with parabolic index profile
Unlike the coupling of two step index fibers, in this case each point on the end-face of the emitting fiber corresponds with a point on the receiving fiber of which the greatest acceptance angle differs from the greatest emitting angle of the corresponding contact point of the emitting fiber. Thus, the loss has two origins: one part of the emitting fiber emits light that does not strike the core of the receiving fiber; and, in the overlapping part of the cores, some of the emitted light rays have an emitting angle larger than the greatest acceptance angle of the receiving fiber. Based on these facts we can expect a larger attenuation than that calculated for step index fibers. The density of light rays in the phase space of the emitting fiber is

$$F_1(r, w) = Lp_a(r)\, p_{N(r)}(w)\, \frac{2\pi w}{(1 + w^2)^2} \tag{11.29}$$

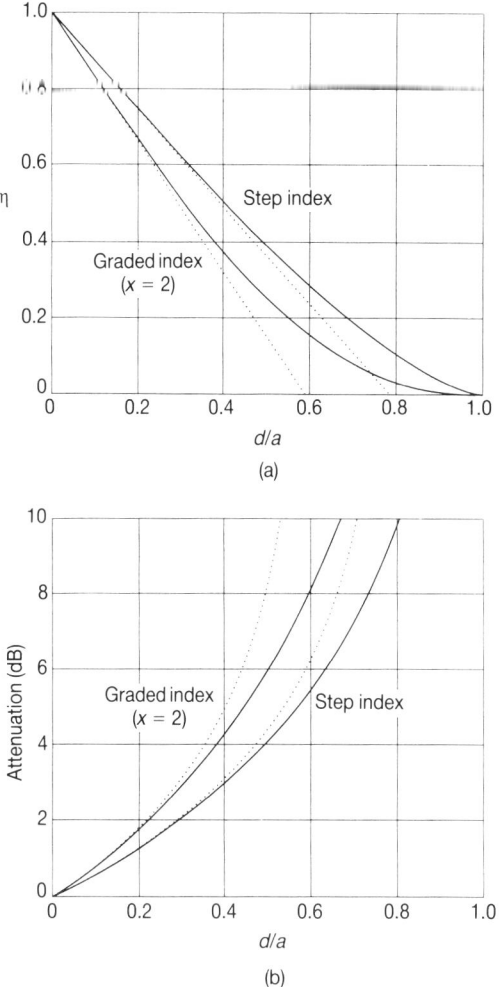

Figure 11.6 Coupling of two identical multimode fibers with a radial separation of $2d$. (a) Efficiency of the joint. (b) Attenuation (in dB).

where

$$p_a(\xi) \overset{\triangle}{=} \begin{cases} 1, & \text{for } 0 \le \xi \le a \\ 0, & \text{for other values of } a, \end{cases} \tag{11.30}$$

and $N(r)$ is explained below. Since the fibers are in contact, we have to find the greatest emission angle of the transmitting fiber. This angle is determined by the greatest angle that a ray inside the transmitting fiber can make with the axis and by the refractive index of the receiving fiber at the contact point under consideration. From equations (6.66) and (6.68) it follows that inside the transmitting fiber

$$\sin \theta_{1,\max} = \frac{\sqrt{n^2(r) - n^2(a)}}{n(r)} \tag{11.31}$$

This means that the greatest emitting angle inside the receiving fiber is given by

$$\sin \theta_{2,\max} = \frac{\sqrt{n^2(r) - n^2(a)}}{n(r')} \approx \sqrt{2\Delta} \, \sqrt{1 - (r/a)^2} \tag{11.32}$$

By means of this approximation we can find that

$$N(r) = \frac{\sqrt{2\Delta} \, \sqrt{1 - (r/a)^2}}{\sqrt{1 - 2\Delta[1 - (r/a)^2]}} \tag{11.33}$$

$N(r)$ can be interpreted as the tangent of the greatest angle $\theta_{\max}$ that a light ray inside the fibers can make with the axis.

In equation (11.29) the first p-function describes the spatial limits of the emitting end-face of the fiber, whereas the second p-function describes the limits of the emission angle. The receiving fiber also has its limits in the phase space, so that for this fiber end-face we get (see Figure 11.7)

$$F_2(r, w) = L p_a(r) \, p_a(r') \, p_{N(r)}(w) \, p_{N(r')}(w) \, \frac{2\pi w}{(1 + w^2)^2} \tag{11.34}$$

The power accepted by the receiving fiber is found by integrating equation (11.34) over r and w.

In Figure 11.7 the shaded area I represents the inequality $r < r'$ and, consequently, $N(r) > N(r')$. In the shaded area II the two inequalities are reversed. From this we can conclude that it is sufficient to integrate only over the shaded area II, with $p_{N(r)}(w)$ as the limit of w. This integration leads to the power accepted by the receiving fiber

$$P_2 = 8\pi L \int_{\phi=0}^{\arccos d/a} d\phi \int_{r=d/\cos\phi}^{a} r \, dr \int_0^{N(r)} \frac{w}{(1 + w^2)^2} \, dw \tag{11.35}$$

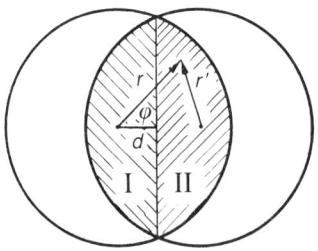

Figure 11.7 The integration area for calculating the coupling efficiency of two radially displaced, graded index fibers.

Elaborating equation (11.35) gives

$$P_2 = 2\pi a^2 2\Delta L\left(\frac{1}{2}\arccos\frac{d}{a} - \frac{d}{a}\sqrt{1 - \left(\frac{d}{a}\right)^2}\left\{1 - \frac{1}{6}\left[1 + 2\left(\frac{d}{a}\right)^2\right]\right\}\right)$$

(11.36)

The efficiency of the connection is found by inserting $d = 0$ in equation (11.36) and then dividing equation (11.36) by the result of that insertion

$$\eta = \frac{4}{\pi}\left\{\frac{1}{2}\arccos\frac{d}{a} - \frac{d}{a}\sqrt{1 - \left(\frac{d}{a}\right)^2}\left[\frac{5}{6} - \frac{1}{3}\left(\frac{d}{a}\right)^2\right]\right\}$$

(11.37)

The efficiency and the corresponding attenuation are drawn in Figure 11.6 as functions of the normalized offset d/a. By means of a series expansion the efficiency can be approximated for small values of the normalized offset

$$\eta \approx 1 - \frac{16}{3\pi}\frac{d}{a}, \qquad \text{for}\ \frac{d}{a} \ll 1$$

(11.38)

This approximation is shown in Figure 11.6 as a dotted line.

End separation of two step index fibers

In this section we calculate the loss that occurs when the end-faces of two step index fibers are separated, although their axes are aligned. It is assumed that the gap is filled with air ($n = 1$). We shall subsequently put a matching liquid with an arbitrary refractive index in the gap. The distance between the parallel end-faces is called z (see Figure 11.8).

Once again, this problem will be treated as one of light rays in the phase space. The coupling between step index fibers will be analyzed separately from the coupling between graded index fibers, because a common treatment could lead to integrals that are difficult to solve [10]. The density of light rays at the end-face of the emitting fiber reads

$$F_1(x, y, u, v) = Lp_a(\sqrt{x^2 + y^2})\ p_T(w)\frac{2\pi w}{(1 + w^2)^2}$$

(11.39)

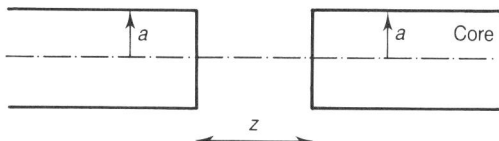

Figure 11.8 Two fibers of core radius a and a distance of z between the end-faces. The fiber axes are assumed to be aligned. (Source: reproduced with permission from *Applied Optics*, vol. 24, no. 7, April 1985, pp. 970–6, 'Loss in multimode fiber connections with a gap', by W. van Etten, W. Lambo and P. Simons.)

From Figure 11.1 and equation (11.1) it follows that a light ray that is described by the coordinates (x, y, u, v) in the phase space of an emitting fiber can be described in the phase space of the receiving fiber by means of the coordinates $(x + uz, y + vz, u, v)$. With this in mind we can write the density of the light rays in the phase space of the receiving fiber as

$$F_2(x, y, u, v) = F_1(x + uz, y + vz, u, v) \, p_a(\sqrt{x^2 + y^2}) \, p_T(w)$$

$$= 2\pi L p_a (\sqrt{x^2 + y^2})$$

$$\times p_a[\sqrt{(x + uz)^2 + (y + vz)^2}] \, p_T(w) \, \frac{w}{(1 + w^2)^2}$$

$$(11.40)$$

Integrating equation (11.40) determines the power that is accepted by the receiving fiber

$$P_2 = 2\pi L \int_x \int_y \int_w p_a(\sqrt{x^2 + y^2}) \, p_a[\sqrt{(x + uz)^2 + (y + vz)^2}]$$

$$\times p_T(w) \, \frac{w}{(1 + w^2)^2} \, \mathrm{d}x \, \mathrm{d}y \, \mathrm{d}w$$

$$= 2\pi L \int_w \left\{ \int_x \int_y p_a(\sqrt{x^2 + y^2}) \, p_a[\sqrt{(x + uz)^2 + (y + uv)^2}] \, \mathrm{d}x \, \mathrm{d}y \right\}$$

$$\times p_T(w) \, \frac{w}{(1 + w^2)^2} \, \mathrm{d}w \qquad (11.41)$$

A closer inspection of the double integral between brackets in this equation leads to the conclusion that this integral represents the overlap area of two circles of radius a, whose origins are moved a distance uz in the x-direction and vz in the y-direction. This means that the combined shift of the origins is wz. Subsequently, the area of overlap can be calculated easily.

For the power received we find that

$$P_2 = 2\pi L \int_w 2a^2 \left[\arccos\left(\frac{zw}{2a}\right) - \frac{zw}{2a} \sqrt{1 - \left(\frac{zw}{2a}\right)^2} \right]$$

$$\times p_a(zw) \, p_T(w) \, \frac{w}{(1 + w^2)^2} \, \mathrm{d}w$$

$$= 2\pi L \int_0^W 2a^2 \left[\arccos\left(\frac{zw}{2a}\right) - \frac{zw}{2a} \sqrt{1 - \left(\frac{zw}{2a}\right)^2} \right] \frac{w}{(1 + w^2)^2} \, \mathrm{d}w$$

$$(11.42)$$

with

$$W \triangleq \min(2a/z, T) \qquad (11.43)$$

The power emitted is calculated by substituting $z = 0$ into equation (11.40) to give

$$P_1 = \pi^2 La^2(\mathrm{NA})^2 \tag{11 44}$$

Ultimately, the coupling efficiency is found by dividing equation (11.42) by equation (11.44). In Figure 11.9(a) the efficiency is depicted as a function of the

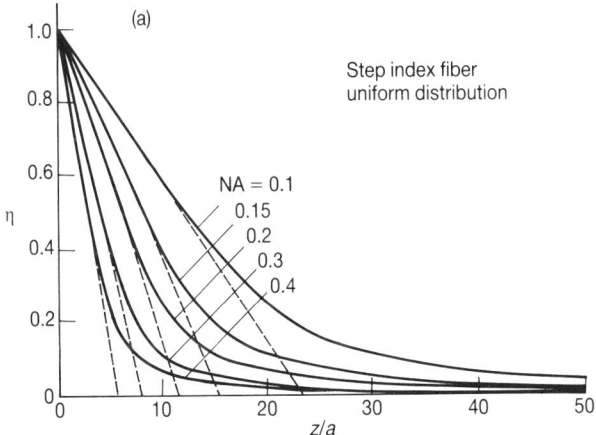

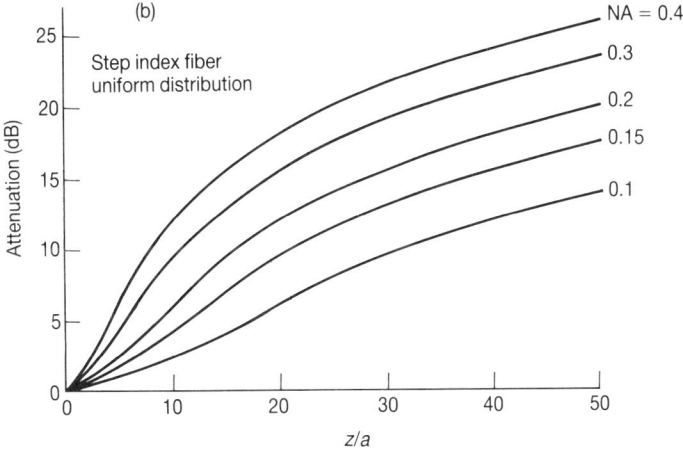

Figure 11.9 Coupling of two identical step index fibers with a gap. (a) Efficiency as a function of z/a with the numerical aperture NA as a parameter. (b) Attenuation (in dB) as a function of z/a. (Source: reproduced with permission from *Applied Optics*, vol. 24, no. 7, April 1985, pp. 970–6, 'Loss in multimode fiber connections with a gap', by W. van Etten, W. Lambo and P. Simons.)

normalized end separation z/a. Moreover, Figure 11.9(b) shows the corresponding attenuation of the coupling. The results in Figure 11.9 were obtained by evaluating equation (11.42) numerically. Based on a series expansion, the efficiency for small values of z/a can be approximated analytically by

$$\eta \approx 1 - \frac{z}{a} \frac{2}{\pi(\text{NA})^2} \left[\arcsin(\text{NA}) - \text{NA}\sqrt{1 - (\text{NA})^2} \right], \qquad \text{for } \frac{z}{a} \ll 1$$

(11.45)

For small values of the numerical aperture, the last expansion can be simplified still further. It is easy to see that

$$\eta \approx 1 - \frac{4}{3\pi} \text{NA} \frac{z}{a}, \qquad \frac{z}{a} \ll 1, \text{NA} \ll 1$$

(11.46)

In Figure 11.9 the approximation is shown as dashed lines.

End separation of two graded index fibers with a parabolic index profile

A similar procedure to that for coupling two step index fibers with end separation can be used to calculate the efficiency and attenuation when coupling two graded index fibers with end separation. There is just one difference, namely the numerical aperture of a graded index fiber is a function of the distance r to the core axis, so that the integration limit of w depends on r. The density of light rays in the phase space of the emitting fiber is therefore given by

$$F_1(x, y, u, v) = 2\pi L p_a(r) \, p_{T(r)}(w) \frac{w}{(1 + w^2)^2}$$

(11.47)

The density of light rays in the phase space of the receiving fiber becomes

$$F_2(x, y, u, v) = F_1(x + uz, y + vz, u, v) \, p_a(\sqrt{x^2 + y^2}) \, p_{T(r')}(w)$$

$$= 2\pi L p_a(r) \, p_a(r') \, p_{T(r)}(w) \, p_{T(r')}(w) \frac{w}{(1 + w^2)^2} \quad (11.48)$$

with

$$r' \triangleq \sqrt{(x + uz)^2 + (y + vz)^2}$$

(11.49)

Integrating equation (11.48) determines the power that is accepted by the receiving fiber

$$P_2 = 2\pi L \int_x \int_y \int_w p_a(r) \, p_a(r') \, p_{T(r)}(w) \, p_{T(r')}(w) \frac{w}{(1 + w^2)^2} \, dw \quad (11.50)$$

The integration over x and y represents the overlap of two circles of radius a, with their centers separated by wz. However, a complication now occurs: the integration limit of w is a function of the coordinates x and y. In order to find

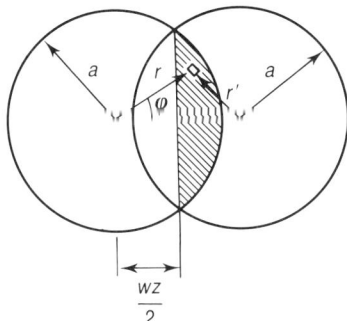

Figure 11.10 Two circles with different centers *wz* apart that are used to determine the integration limits of equation (11.46). (Source: reproduced with permission from *Applied Optics*, vol. 24, no. 7, April 1985, pp. 970–6, 'Loss in multimode fiber connections with a gap', by W. van Etten, W. Lambo and P. Simons.)

the correct integration limits let us consider Figure 11.10, where the two circles with the shifted centers are shown.

Because of the symmetry, the integration over the shaded area gives $P_2/2$, whilst using the polar coordinates r and ϕ, instead of x and y, the integration limits become

$$0 \le \phi \le \arccos\frac{wz}{2a} \le \frac{\pi}{2} \tag{11.51}$$

$$0 \le \frac{wz}{2\cos\phi} \le r \le a \tag{11.52}$$

In the shaded area $r' < r$, consequently $A(r) < A(r')$. This means that the fourth p-function in equation (11.50) becomes redundant. The third p-function leads to the following integration limit with respect to the angle parameter

$$w \le \sqrt{\frac{(\mathrm{NA})^2(a^2 - r^2)}{a^2 - (\mathrm{NA})^2(a^2 - r^2)}} \tag{11.53}$$

From equation (11.52) it follows that $w < 2r\cos(\phi)/z$, whereas from Figure 11.10 it can be seen that $w \le 2a/z$, an inequality that is produced automatically when the first is satisfied. The requirement of $w < 2a\cos(\phi)/z$, which follows from equation (11.51), is also redundant, because $r \le a$. In view of these considerations, equation (11.50) can be rewritten as

$$
\begin{aligned}
P_2 &= 8\pi L \int_0^a r\,\mathrm{d}r \int_0^{\pi/2} \mathrm{d}\phi \int_0^{W_g} \frac{w}{(1 + w^2)^2}\,\mathrm{d}w \\
&= \pi^2 L a^2 - 4\pi L \int_0^a r\,\mathrm{d}r \int_0^{\pi/2} \mathrm{d}\phi\, \frac{1}{1 + W_g^2}
\end{aligned}
\tag{11.54}
$$

with

$$W_g \triangleq \min\left\{\frac{2r}{z}\cos\phi, \; T(r)\right\}$$ (11.55)

The power emitted by fiber 1 can be written as $\pi^2 L a^2 (NA)^2/2$; thus the coupling efficiency is found by dividing equation (11.54) by this expression.

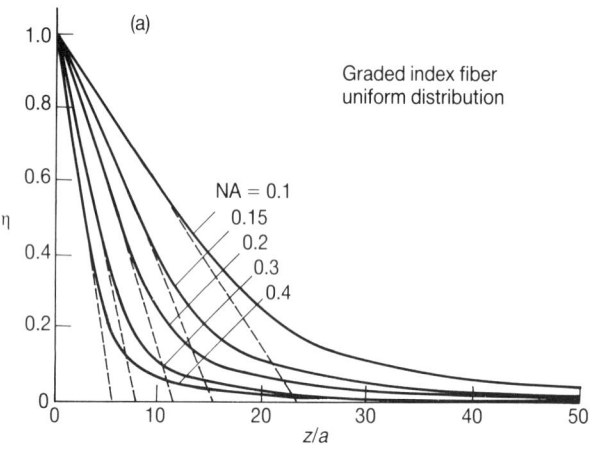

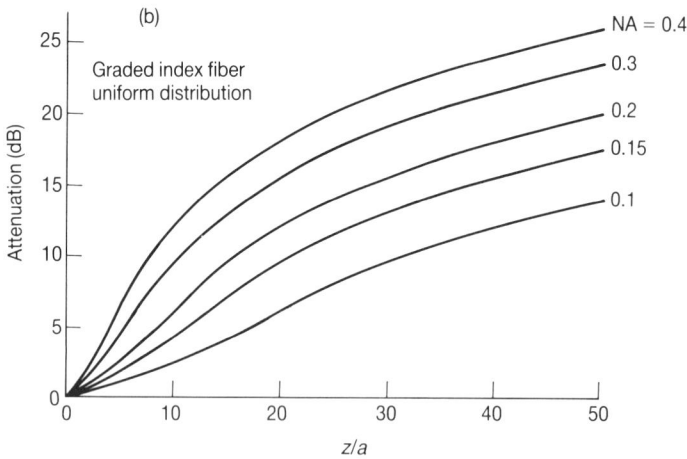

Figure 11.11 Coupling between two graded index fibers ($x = 2$) with a gap. (a) Efficiency as a function of z/a with the numerical aperture NA as parameter. (b) Attenuation (in dB) as a function of z/a. (Source: reproduced with permission from *Applied Optics*, vol. 24, no. 7, April 1985, pp. 970–6, 'Loss in multimode fiber connections with a gap', by W. van Etten, W. Lambo and P. Simons.)

Computation can evaluate the twofold integral of equation (11.54) numerically and this has been done in Figure 11.11(a) where the efficiency is given as a function of the normalized separation whilst Figure 11.11(b) shows the corresponding attenuation. In [10] the efficiency for small numerical apertures is approximated as

$$\eta \approx 1 - \frac{1}{2} \mathrm{NA} \frac{z}{a}, \qquad \frac{z}{a} \ll 1, \mathrm{NA} \ll 1 \tag{11.56}$$

This approximation is shown in Figure 11.11(a) as dashed lines. In [10] the efficiency and attenuation have been calculated for the so-called steady-state power distribution in the fibers. The theory so far has been based on the assumption that the gap between the fiber end-faces is filled with air ($n = 1$). The equations can be changed easily for gaps filled with other media, e.g. a matching liquid. From the definition of the numerical aperture (see equation 6.66) it is easy to see that a fiber accepts light at its end-face from a dielectric with a maximum acceptance angle given by $\sin \theta'_{\max} = (1/n) \sin \theta_{\max}$. Figures 11.9 and 11.11 show that a matching liquid reduces losses.

Interference in fiber connections
In a fiber connection with a gap between the end-faces of the fibers, the intrinsic Fresnel loss has a special effect. The parallel fiber end-faces, together with the gap, form a Fabry–Perot cavity. An oscillating loss, as a function of the end separation, is superimposed on the loss described in this section, due to the wavelength-dependent transfer of the cavity [11, 12]. We shall now analyze this situation.

Figure 11.12 shows the cores of the two fibers, with the multiple reflection from a ray on the end-faces. Contrary to what the figure suggests, the arrows are not the directions of the fields, but the directions of the Poynting vectors associated with the fields, which are assumed to be local plane waves.

For the sake of simplicity, we consider only identical step index fibers. The interference effect using graded index fibers differs only slightly, as the refractive index variation of a graded index fiber is in the order of 1 percent; the attenuation has already been described.

It is our intention to incorporate the interference effects; therefore, the calculations have to be based on the electric field. E_0 is the complex amplitude of the field just before the ray strikes the glass–air interface of the transmitting fiber (fiber 1); E_1 is the complex amplitude of the field just beyond this glass–air interface, and so on. It is easy to see that

$$E_2 = tt' \, E_0 \exp\left(\mathrm{j} \frac{\delta}{2}\right) \tag{11.57}$$

and

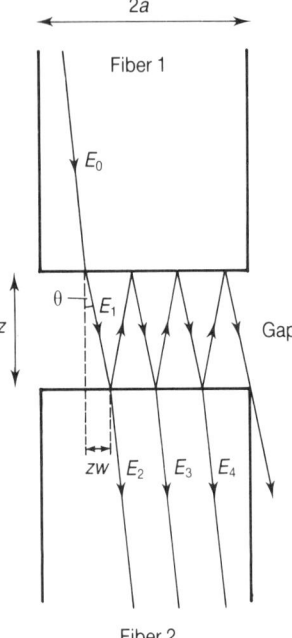

Figure 11.12 Multiple reflections of a light ray in a gap between the end-faces of two fibers. (Source: reproduced with permission from *Applied Optics*, vol. 26, no. 7, April 1987, pp. 1158–61, 'Attenuation and interference in multimode fiber connections', by F. Budé and W. van Etten.)

$$E_3 = tr^2t' \, E_0 \exp\left(j\,\frac{\delta}{2}\right)\exp\left(j\delta\right), \tag{11.58}$$

where t is the field transmission coefficient from glass to air, t' is the field transmission coefficient from air to glass and r is the field reflection coefficient of an air–glass interface. Moreover [3]

$$\delta \stackrel{\triangle}{=} \frac{4\pi}{\lambda_0}\,nz\cos\theta \tag{11.59}$$

where n is the refractive index in the gap (assumed to be unity) and λ_0 is the wavelength of the light in free space (assumed to be equal to the wavelength in air).

The effects of the field components after E_3 are neglected because of their small amplitudes, which are proportional to their respective powers of r. With these approximations in mind, Figure 11.12 can be decomposed as is shown in Figure 11.13. With this figure and the previous method, the complex amplitude of the total field in the receiving fiber (fiber 2) reads

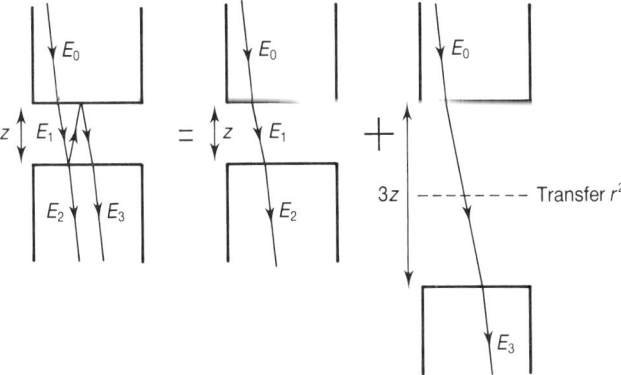

Figure 11.13 Decomposition of a transmitted light ray into its two main components. (Source: reproduced with permission from *Applied Optics*, vol. 26, no. 7, April 1987, pp. 1158–61, 'Attenuation and interference in multimode fiber connections', by F. Budé and W. van Etten.)

$$E_t = E_0 tt' \, p_a(\sqrt{x^2 + y^2}) \, p_a[\sqrt{(x + uz)^2 + (y + vz)^2}]$$

$$\times \exp\left(j\frac{\delta}{2}\right) p_T(w)$$

$$+ E_0 tt' r^2 \, p_a(\sqrt{x^2 + y^2}) \, p_a[\sqrt{(x + 3uz)^2 + (y + 3vz)^2}]$$

$$\times \exp\left(j\frac{\delta}{2}\right) \exp(j\delta) \, p_T(w) \tag{11.60}$$

and analogously with equation (11.9b) we obtain

$$T \triangleq \frac{\mathrm{NA}}{\sqrt{1 - (\mathrm{NA})^2}} \tag{11.61}$$

where NA is the numerical aperture of the fibers and u, v and w are already defined. The radiance L is now given by

$$L = \alpha \int_0^\infty E_t E_t^* G(v) \, \mathrm{d}v \tag{11.62}$$

where α is a constant that is of no interest to this problem and E_t^* is the complex conjugate of E_t. The spectral density $G(v)$ of the source (with v the optical frequency) is assumed to be Gaussian

$$G(v) = \exp\left[-\frac{1}{2}\left(\frac{v - v_0}{\Delta v}\right)^2\right] \tag{11.63}$$

Equation (11.60) can be used to find the product of E_t and E_t^*

$$E_1 E_1^* = E_0^2(tt')^2\, p_a(\sqrt{x^2 + y^2})\, p_a[\sqrt{(x + uz)^2 + (y + vz)^2}]\, p_T(w)$$

$$+ E_0^2(tt')^2 r^4\, p_a(\sqrt{x^2 + y^2})$$

$$\times\, p_a[\sqrt{(x + 3uz)^2 + (y + 3vz)^2}]\, p_T(w)$$

$$+ E_0^2(tt')^2 r^2\, p_a(\sqrt{x^2 + y^2})\, p_a[\sqrt{(x + uz)^2 + (y + vz)^2}]$$

$$\times\, p_a[\sqrt{(x + 3uz)^2 + (y + 3vz)^2}]\, p_T(w)\, 2\cos\delta \qquad (11.64)$$

We shall assume below that the transmission and reflection coefficients are independent of the incident angle of the ray, i.e. independent of w. For small angles this is a reasonable assumption (see [3]). Inserting equation (11.64) in equation (11.62) and the result in its turn in equation (11.5) is no problem for the first two terms of equation (11.64). These terms are independent of v and the integrations over w and S have the same form as in equation (11.41). The third term is complicated, as δ depends on the optical frequency. Thus, δ can be written in the form

$$\delta \triangleq lv \qquad (11.65)$$

where l is a quantity independent of v. Inserting the third term of equation (11.64) in equation (11.62) gives an integral of the form

$$I = \int_0^\infty \exp\left[-\frac{1}{2}\left(\frac{v - v_0}{\Delta v}\right)^2\right] \cos(lv)\, dv \qquad (11.66)$$

The lower integration limit can be extended to $-\infty$, because the exponential function in the integrand becomes negligible for negative values of v. If, moreover, $v - v_0$ is replaced by a new variable, then the integral can be found in [13, p. 480]

$$I = \Delta v\, \sqrt{2\pi}\, \cos(lv_0) \exp\left[-\frac{l^2(\Delta v)^2}{2}\right] \qquad (11.67)$$

After combining this equation with equation (11.41), we find that the power accepted by the receiving fiber is given by

$$P_2 = 2\pi\alpha E_0^2(tt')^2 \left\{ \int_0^{W_1} 2a^2\left[\arccos\left(\frac{\zeta w}{2}\right) - \left(\frac{\zeta w}{2}\right)\sqrt{1 - \left(\frac{\zeta w}{2}\right)^2}\right] \right.$$

$$\times\, \frac{w}{(1 + w^2)^2}\, dw + r^4 \int_0^{W_3} 2a^2\left[\arccos\left(\frac{3\zeta w}{2}\right) - \left(\frac{3\zeta w}{2}\right)\right.$$

$$\times\, \sqrt{1 - \left(\frac{3\zeta w}{2}\right)^2}\, \frac{w}{(1 + w^2)^2}\, dw + r^2 \int_0^{W_3} 2a^2\left[\arccos\left(\frac{3\zeta w}{2}\right) - \left(\frac{3\zeta w}{2}\right)\right.$$

$$\left.\times\, \sqrt{1 - \left(\frac{3\zeta w}{2}\right)^2}\, \frac{w}{(1 + w^2)^2}\, 2\cos(lv_0)\exp\left[-\frac{l^2(\Delta v)^2}{2}\right] dw \right\} \qquad (11.68)$$

with

$$W_1 \overset{\triangle}{=} \min\left(T, \frac{2}{\zeta}\right) \tag{11.69}$$

$$W_3 \overset{\triangle}{=} \min\left(T, \frac{2}{3\zeta}\right) \tag{11.70}$$

$$l \overset{\triangle}{=} \frac{4\pi}{c} nz \frac{1}{\sqrt{1+w^2}} \tag{11.71}$$

Since the second term of equation (11.68) is smaller than the third term by a factor r^2 and smaller than the first one by a factor r^4, it is neglected. The first term of equation (11.68) can be recognized in equation (11.41), whilst the third term represents the oscillating characteristic that was reported in [11] and [12]. Moreover, it follows that the oscillation becomes very small for end separations exceeding 50 μm, even when using a laser as the light source. However, as seen from equation (11.46), the integral of the first term of equation (11.68) can then be approximated for standard multimode fibers, resulting in the efficiency

$$\eta = (tt')^2 \left(1 - \frac{4\mathrm{NA}\zeta}{3\pi}\right) + \frac{8}{\pi(\mathrm{NA})^2} (tt')^2 r^2$$

$$\times \int_0^{W_3} \left[\arccos\left(\frac{3\zeta w}{2}\right) - \left(\frac{3\zeta w}{2}\right)\sqrt{1 - \left(\frac{3\zeta w}{2}\right)^2}\right] \frac{w}{(1+w^2)^2}$$

$$\times \cos\left(l v_0\right) \exp\left[-\frac{l^2(\Delta v)^2}{2}\right] \mathrm{d}w \tag{11.72}$$

Except for the factor $(tt')^2$ due to the transmission through two glass–air interfaces, the first term of this equation describes the attenuation due to the gap between the two fiber end-faces, whereas the second term in this equation accounts for the oscillations in transfer efficiency as a function of the end separation distance z. This latter term has been calculated approximately in [11]. The integral of the second term was evaluated numerically for two cases: an LED with $\lambda_0 = 0.82 \mu$m and $\Delta\lambda = 12$ nm; as well as a laser with $\lambda_0 = 0.82 \mu$m and $\Delta\lambda = 0.45 \times 10^{-7} \mu$m. In both cases $a = 25 \mu$m and $\mathrm{NA} = 0.2$. Figure 11.14 shows η as a function of the end separation for an LED, whereas Figure 11.15 represents its calculated values for a laser, when $tt' = 0.96$ and $r = 0.2$.

Some interesting remarks can be made with respect to the interpretation of these figures.

- The mean value of the oscillating phenomenon gives $\eta = 0.92$ for $z = 0$. This corresponds to a Fresnel reflection loss of twice that for a glass–air interface.
- Near $z = 0$ the amplitude of the oscillation is approximately 8 percent; therefore, at $z = 0$ no discontinuity can be expected. A detailed examination of the amplitude shows that it is 0.5 percent lower than calculated. This

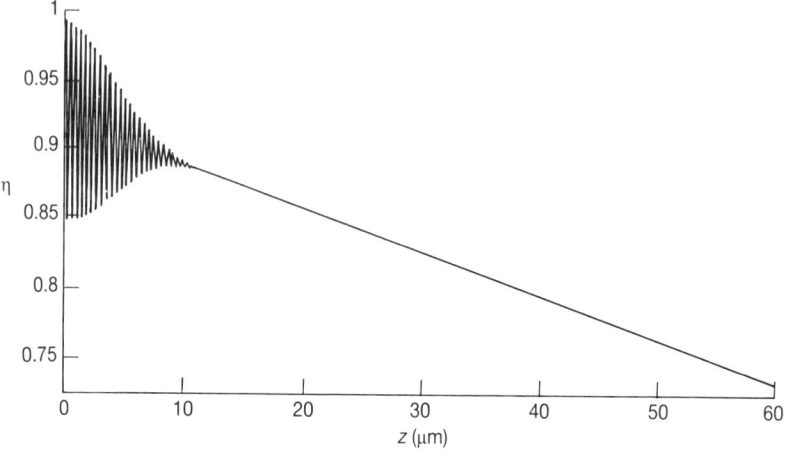

Figure 11.14 Power transfer between two separated fibers as a function of the end-face distance. LED source, NA = 0.2, $a = 25\ \mu m$, $\lambda_0 = 0.82\ \mu m$, $\Delta\lambda = 12$ nm. (Source: reproduced with permission from *Applied Optics*, vol. 26, no. 7, April 1987, pp. 1158–61, 'Attenuation and interference in multimode fiber connections', by F. Budé and W. van Etten.)

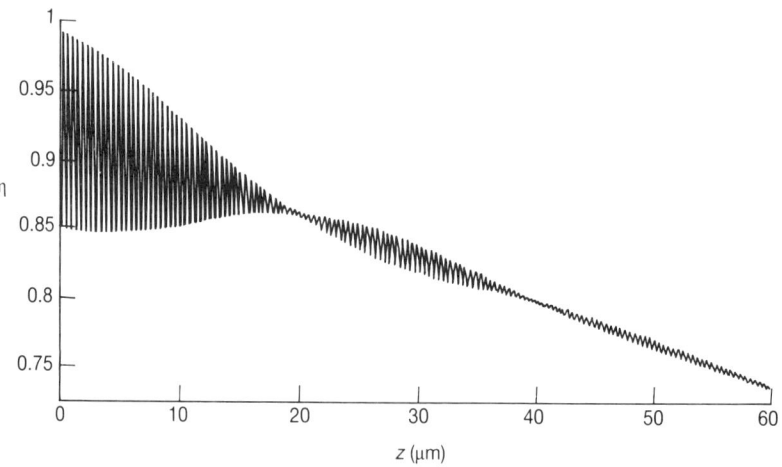

Figure 11.15 Power transfer between two separated fibers as a function of the end-face distance. Laser source, NA = 0.2, $a = 25\ \mu m$, $\Delta\lambda = 0.45 \times 10^{-7}\ \mu m$. (Source: reproduced with permission from *Applied Optics*, vol. 26, no. 7, April 1987, pp. 1158–61, 'Attenuation and interference in multimode fiber connections', by F. Budé and W. van Etten.)

difference stems from the fact that only one reflection has been taken into account, in our approximation.
● The periodic length of the oscillating phenomenon is roughly half the wavelength and it agrees with that for a Fabry–Perot cavity.

Except for the interference effects, when a correction factor of $(tt')^2$ is used with respect to the results of Figure 11.9, the results of Figures 11.14 and 11.15 arrive at the same attenuation value for all separation distances. In this approach, it can be seen that the interference effects are superimposed on the attenuation. The finite area of a fiber core has little impact on the interference (compare these results with those of [11]). With a laser as the light source, the interference is practically independent of the laser linewidth.

Angular misalignment between fibers

The problem of angular misalignment between the axes of the fibers being joined is not unambiguously defined; this becomes clear when Figure 11.16 is considered. In Figure 11.16(a) the end-planes of the fibers are perpendicular to the fiber axes. The fiber cores are prevented from being in contact with each other by the presence of the cladding and a coating. In this configuration, the coupling efficiency depends largely on the ratio of the core diameter to the outer fiber diameter. In Figure 11.16(b) the configuration is better defined, but this seldom occurs in a connector, although it may be a satisfactory model for a splice (a splice is a permanent joint between two fibers). For the case given by Figure 11.16(a), no analytical solution is known; such a situation can only be analyzed by ray tracing. This computer technique proceeds as follows. The phase space of the emitting fiber is filled with a uniform lattice of light rays and, by means of a computer program, the trajectory of each light ray is calculated and the number of rays entering the core of the receiving fiber as a guided ray is recorded. From the ratio of this number and the total number of rays in the lattice, the efficiency of the coupling is determined. The analytical solution for the configuration given in Figure 11.16(b) is given below. Some approximations have to be made; however, these cause only negligible deviations from the exact

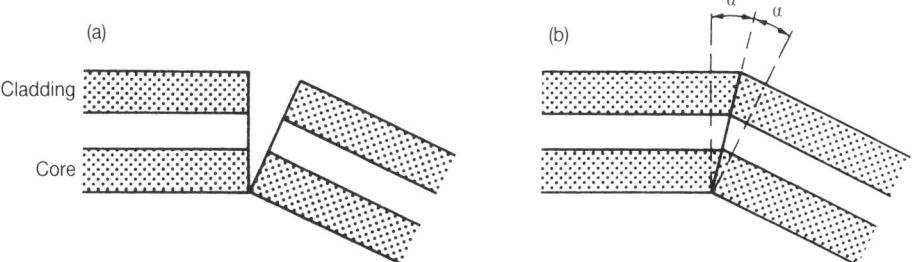

Figure 11.16 Angular misalignment of two fibers: (a) with a gap; (b) end-faces in contact.

solution, especially for small values of the numerical aperture. Again, it is assumed that fibers of the same type are being joined (both fibers are step index or graded index, with the same core radius, numerical aperture, etc.). Moreover, it is assumed that the fiber axes intersect, so that the core–cladding interfaces of the fibers are in line with each other. The radiation and acceptance of power is described by equation (11.3). The integration over S has to be taken for the entire fiber core contact plane A (see Figure 11.17).

We can describe the problem by the overlap of two solid angles. The first solid angle is the one over which the radiation of the transmitting fiber extends; the second is the angle where the radiation is accepted by the receiving fiber (see Figure 11.18). The phase space, which was introduced in Section 11.1, is redefined here with respect to the angle parameter. For this purpose, we will consider Figure 11.19 and define

$$t \stackrel{\triangle}{=} \sin \theta \tag{11.73}$$

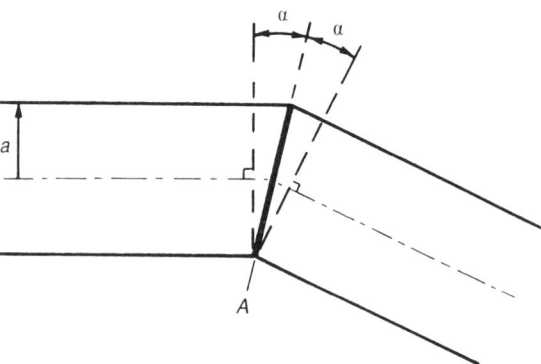

Figure 11.17 Parameters of two fibers with angular misalignment.

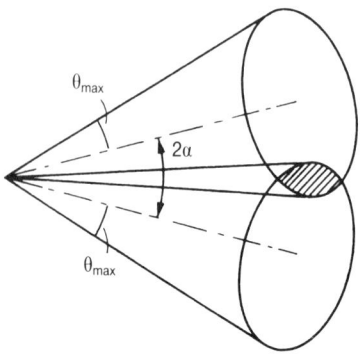

Figure 11.18 Overlap area (shaded) of the radiation and acceptance cones.

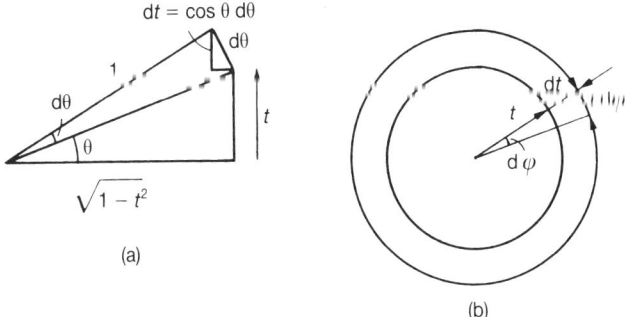

Figure 11.19 (a) Definition of the angle parameter t. (b) Annular region comprising $d\Omega$.

From Figure 11.19 it can be seen that the angle intervals $\{\theta, \theta + d\theta\}$ and $\{\phi, \phi + d\phi\}$ correspond to a solid angle

$$d\Omega = t \, d\phi \, d\theta = t \, d\phi \, \frac{dt}{\cos \theta} \tag{11.74}$$

Substituting this result in equation (11.3) yields

$$P = \int_t \int_\phi \int_A L \cos{(\theta)} t \, d\phi \, \frac{dt}{\cos \theta} \, dS = \int_t \int_\phi \int_A L t \, dt \, d\phi \, dS \tag{11.75}$$

Equation (11.75) has a very simple interpretation. Except for the integration over A, the integration over t and ϕ can be seen as the overlap of two circles with radius

$$M(r) \triangleq \sin \theta_{max} = \begin{cases} \sqrt{2\Delta}, & \text{for step index fibers} \\ & \qquad\qquad\qquad\qquad (11.76a) \\ \sqrt{2\Delta} \cdot \sqrt{\dfrac{1 - (r/a)^2}{1 - 2\Delta(r/a)^2}}, & \text{for parabolic index fibers} \end{cases}$$

$$(11.76b)$$

The origins have been separated by

$$\gamma \triangleq \sin \alpha \tag{11.77}$$

Using these definitions the power in the receiving fiber can be written as

$$P_2 = \int_A L \left[2M^2(r) \arccos \frac{\gamma}{M(r)} - 2\gamma \sqrt{M^2(r) - \gamma^2} \right] dS \tag{11.78}$$

Using the knowledge available about the overlap area of the two circles of equal diameter (see equation 11.26), integrating over the contact plane A with a radiance L becomes the same as integrating over a plane perpendicular to the fiber axis with a radiance of

$$L' \triangleq \frac{L}{\cos \alpha} \tag{11.79}$$

For step index fibers the problem is nearly solved, since $M(r)$ is a constant and integrating over S simply gives πa^2. After dividing by the well-known value for the power in a transmitting fiber $P_1 = \pi^2 a^2 L' 2\Delta$ (see equation 11.10), we obtain

$$\eta = \frac{2}{\pi} \left(\arccos \frac{\gamma}{\sqrt{2\Delta}} - \frac{\gamma}{\sqrt{2\Delta}} \sqrt{1 - \frac{\gamma^2}{2\Delta}} \right) \tag{11.80}$$

For graded index fibers with a parabolic index profile, two differences arise when compared with step index fibers. Firstly, the value $M(r)$ depends on r (see equation 11.76b) and, secondly, integrating over S does not cover the whole plane of A. This can be seen from Figure 11.18 and the larger the value of r, the smaller the value of $M(r)$, according to equation (11.76b). This means that for a larger r the spread of the cones is smaller, whereas the angle α remains constant. In this way, the cones do not overlap any more beyond a certain value of r. For a fixed value of α, this puts a limit of $r_{max} < a$ on the integration over r. Substituting equation (11.76b) in equation (11.78) leads to a complicated integral which has no simple analytical solution; therefore we introduce the approximation

$$M(r) \approx \sqrt{2\Delta} \sqrt{1 - (r/a)^2} \tag{11.81}$$

From Figure 11.18 it can be seen that r_{max} follows from the equation

$$\sin \alpha = \sin \theta_{max} \approx \sqrt{2\Delta} \sqrt{1 - (r/a)^2} \tag{11.82}$$

Solving this equation yields

$$r_{max} \approx a \sqrt{1 - \frac{\gamma^2}{2\Delta}} \tag{11.83}$$

Equation (11.78) now becomes

$$P_2 = 4\pi L' \int_0^{r_{max}} \left\{ 2\Delta \left[1 - \left(\frac{r}{a} \right)^2 \right] \arccos \frac{\gamma}{\sqrt{2\Delta} \sqrt{1 - (r/a)^2}} \right.$$
$$\left. - \gamma \sqrt{2\Delta} \sqrt{\frac{2\Delta - \gamma^2}{2\Delta} - \left(\frac{r}{a} \right)^2} \right\} r \, dr \tag{11.84a}$$

Substituting $\gamma / [\sqrt{2\Delta} \sqrt{1 - (r/a)^2}]$ by x, the first term of the integrand takes the form of $\arccos(x)/x^5$, the solution for which can be found in [13, p. 88]. The second term of the integrand is easily solved with standard integration methods. After solving the integral we have to divide it by the power in the transmitting fiber $P_1 = \pi^2 a^2 L' \Delta$ in order to get the efficiency

$$\eta = \frac{4}{\pi} \left[\frac{1}{2} \arccos \left(\frac{\gamma}{\sqrt{2\Delta}} \right) - \frac{\gamma}{\sqrt{2\Delta}} \sqrt{1 - \frac{\gamma^2}{2\Delta} \left(\frac{5}{6} - \frac{1}{3} \frac{\gamma^2}{2\Delta} \right)} \right] \qquad (11.85)$$

Comparing equations (11.80) and (11.85) with equations (11.27) and (11.37), it can be seen that the expressions for the efficiencies of angular misalignments are identical to those found for the radial offset. This requires the normalized shift d/a in the latter equations to be replaced by the normalized angular misalignment $\sin \alpha / \sin \theta_{max}$. Figure 11.6 can therefore also be used for angular misalignment, if such a replacement is made. The same holds for the approximations in equations (11.28) and (11.38).

Notwithstanding the fact that the results presented in [14] look different at first sight, the efficiencies are identical to those derived earlier; this becomes clear after making the right substitutions and rearranging the equations.

Combinations of the misalignments
No analytical solutions are known for combinations of extrinsic mismatches; however, ray tracing is a good way of treating such cases. Reference [15] includes many solutions with this method, and also gives some simple, experimentally derived expressions.

It can be concluded from this section that end separation leads to the smallest of the extrinsic losses investigated. In contrast to the other losses, end-separation losses depend on the numerical aperture of the fibers. Radial offset and angular misalignment have equal effects on the loss, and they lead to greater losses than end separation. In general, an angular alignment in a connection is easier to achieve than a radial one; the latter gives the most problems.

11.4 Coupling model for single-mode fibers

Light propagation in single-mode step index fibers can be described simply by means of electromagnetic fields (see Chapter 5.) Although we shall restrict the examination to single-mode step index fibers, the results can just as easily apply to other single-mode fiber index profiles (see [16]). Thus the analysis has to start by investigating the coupling of an arbitrary incident field to the mode fields of a fiber. In general, this problem can be split up into two parts: firstly, the transmission of an incident field through the interface between the medium where the incident field occurs and the fiber; and secondly, the excitation of the fiber's mode fields by this transmitted field. The first part will include Fresnel reflection and refraction. The laws describing these effects for plane waves are found in [3] and for Gaussian beams in Appendix 3. The second aspect will be dealt with below.

We shall call the electric field vector of the nth fiber mode $\mathbf{E}_{fn}$, while the corresponding magnetic field vector is represented by $\mathbf{H}_{fn}$. It is well known (see [17]) that the fiber modes are all orthogonal. This holds for both guided modes and radiation modes. Moreover, we normalize the mode fields so that they are orthonormal, as follows:

$$\int\int (\mathbf{E}_{\mathrm{f}n} \times \mathbf{H}_{\mathrm{f}n}^*).\mathbf{a}_z \, \mathrm{d}s = \delta_{nm} \triangleq \begin{cases} 1, & \text{for } n = m \\ \\ 0, & \text{for } n \neq m \end{cases} \tag{11.86}$$

where δ_{nm} is the Kronecker delta and $\mathbf{a}_z$ the unit vector in the z-direction or the direction of the fiber axis. Integration has to be done across the entire fiber end-face.

The transmitted incident field can now be written as the weighted sum of the fiber mode fields, as follows

$$\mathbf{E}_{\mathrm{i}} = \sum_k c_k \mathbf{E}_{\mathrm{f}k} \tag{11.87a}$$

$$\mathbf{H}_{\mathrm{i}} = \sum_k c_k \mathbf{H}_{\mathrm{f}k} \tag{11.87b}$$

The summation extends over all guided modes, and also includes integration over the continuance of radiation modes. In order to determine the coefficients c_k we invoke the orthonormal condition given by equation (11.86)

$$\int\int (\mathbf{E}_{\mathrm{i}} \times \mathbf{H}_{\mathrm{f}k}^*).\mathbf{a}_z \, \mathrm{d}s = \int\int \sum_k c_k (\mathbf{E}_{\mathrm{f}k} \times \mathbf{H}_{\mathrm{f}k}^*).\mathbf{a}_z \, \mathrm{d}s$$

$$= \sum_k c_k \int\int (\mathbf{E}_{\mathrm{f}k} \times \mathbf{H}_{\mathrm{f}k}^*).\mathbf{a}_z \, \mathrm{d}s = c_k \tag{11.88}$$

Of course, we are primarily interested in the guided mode of the single-mode fiber, therefore when making $k = 0$ for this mode we can obtain its coupling coefficient c_0 by introducing this value into equation (11.88) and solving the integral. The power that is carried by this mode is then developed from the Poynting theorem [18]

$$P_0 = \frac{1}{2} \mathrm{Re} \left\{ c_0 c_0^* \int\int (\mathbf{E}_{\mathrm{f}0} \times \mathbf{H}_{\mathrm{f}0}^*).\mathbf{a}_z \, \mathrm{d}s \right\} = \frac{1}{2} |c_0|^2 \tag{11.89}$$

Normalizing the transmitted incident field

$$\int\int (\mathbf{E}_{\mathrm{i}} \times \mathbf{H}_{\mathrm{i}}^*).\mathbf{a}_z \, \mathrm{d}s = 1 \tag{11.90}$$

allows the incident power to be written as

$$P_{\mathrm{i}} = \frac{1}{2} \mathrm{Re} \left\{ \int\int (\mathbf{E}_{\mathrm{i}} \times \mathbf{H}_{\mathrm{i}}^*).\mathbf{a}_z \, \mathrm{d}s \right\} = \frac{1}{2} \tag{11.91}$$

Consequently, the power coupling from the incident beam to the guided mode is simply written as

$$\eta = |c_0|^2 \tag{11.92}$$

Let us now assume that the incident field consists of a Gaussian beam that is polarized linearly and transversely (see Appendix 3)

$$E_{ix} = E_i \exp\left(-\frac{r'}{w^2_0}\right) \tag{11.93}$$

with $2w_0$ the width of the waist of this Gaussian beam. From Chapter 3 it follows that the guided modes of weakly guiding fibers are linearly and nearly transversely polarized too. For a single-mode fiber we are more interested in the LP_{01} mode; the transverse magnetic field of this mode is given by

$$H_{f0,y} = \begin{cases} E_0 \sqrt{\dfrac{\varepsilon_1}{\mu}} \dfrac{J_0[u(r/a)]}{J_0(u)}, & r \leq a \\[4mm] E_0 \sqrt{\dfrac{\varepsilon_2}{\mu}} \dfrac{K_0[w(r/a)]}{K_0(w)}, & r \geq a \end{cases} \tag{11.94}$$

where the various parameters in this equation have been defined in Chapter 3.

Substituting equations (11.93) and (11.94) in (11.88) and (11.89), we obtain the power coupling efficiency for a Gaussian beam when it is exciting a single-mode fiber, provided that the correct normalization is introduced. The integration of equation (11.88) had to be performed numerically, and so it is seen that η is a function of the Gaussian beam parameter w_0 and this function has a maximum value. Figure 11.20 shows this maximum value of η as a function of the normalized frequency v (see Chapter 3, equation 3.54).

As shown in Figure 11.20, the coupling efficiency is quite high over a wide

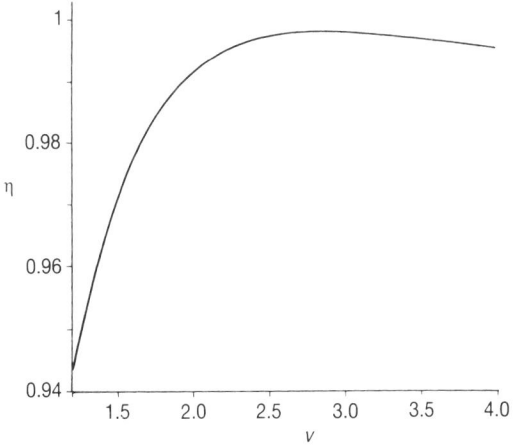

Figure 11.20 Maximum power coupling from a Gaussian beam to a single-mode fiber as a function of the normalized frequency v.

range of v-values. From $v = 1.2$ onwards, the efficiency is greater than 94 percent. At $v = 2.405$, the cut-off frequency of the next higher mode, the efficiency is better than 99 percent. It can be concluded from this figure that the field of the LP_{01} mode is reasonably well approximated by a Gaussian beam for most practical values of v, as far as the coupling is concerned. The optimal value of the beam parameter w_0, normalized with the core radius of the fiber, can be closely approximated by the empirical expression [16]

$$\frac{w_0}{a} = 0.65 + \frac{1.619}{\sqrt{v^3}} + \frac{2.879}{v^6} \tag{11.95}$$

Once these results have been established, problems of joining single-mode fibers are reduced simply to coupling Gaussian beams [19].

If we restrict our considerations to linear transverse fields, which is a reasonable assumption for Gaussian beams emitted by a semiconductor laser and for weakly guiding fibers, the power coupling efficiency η can be computed by just using electric fields. From equations (5.14) and (5.15) it follows that

$$H_{f0,y} = E_{f0,x} \sqrt{\frac{\varepsilon}{\mu}} \tag{11.96}$$

or similarly

$$\mathbf{H}_{f0} = \frac{\mathbf{a}_z \times \mathbf{E}_{f0}}{Z_f} \tag{11.97}$$

where

$$Z_f \overset{\triangle}{=} \sqrt{\frac{\mu}{\varepsilon}} \tag{11.98}$$

is the field impedance of the fiber. The same procedure can be used for the incident field; equations (11.89) and (11.91) are then changed into

$$P_0 = \frac{1}{2} \text{Re} \left\{ c_0 c_0^* \frac{1}{Z_f} \int\int |E_{f0}|^2 \, ds \right\} \tag{11.99}$$

and

$$P_i = \frac{1}{2} \text{Re} \left\{ \frac{1}{Z_f} \int\int |E_{it}|^2 \, ds \right\} \tag{11.100}$$

without changing the power coupling efficiency η. In equation (11.100) E_{it} is the transverse component of $\mathbf{E}_i$. Since the field impedance Z_f plays no role in calculating η, in this case, it is more convenient to normalize the electric fields in the following way

$$\int\int |E_i|^2 \, ds = \int\int E_i E_i^* \, ds = 1 \tag{11.101}$$

and

$$\int\int |E_{fo}|^2 \, ds = \int\int E_{f0}.E_{f0}^* \, ds = 1 \tag{11.102}$$

Consequently, the coupling coefficient c_0 changes into

$$c_0 = \int\int E_{it}.E_{f0}^* \, ds \tag{11.103}$$

while the power coupling η remains the same as that given by equation (11.92). In the following sections, we shall use the latter normalization method.

11.5 Power coupling from the source into single-mode fibers

11.5.1 Launching from a laser into a single-mode fiber

Once we have developed the model described in the preceding section, and using the results from Appendix 3 for Gaussian beams, calculating the efficiency of power coupling in different situations becomes quite straightforward. In this section we consider semiconductor lasers emitting light in the fundamental transverse and lateral mode, when the laser beam conforms to a Gaussian beam in the following manner (see also Appendix 3)

$$E_i = \frac{\sqrt[4]{2}}{\sqrt[4]{\pi}\,\sqrt{w_x}} \exp\left(-\frac{x^2}{w_x^2} - jk\frac{x^2}{2R_x}\right)$$

$$\times \frac{\sqrt[4]{2}}{\sqrt[4]{\pi}\,\sqrt{w_y}} \exp\left(-\frac{y^2}{w_y^2} - jk\frac{y^2}{2R_y}\right) \tag{11.104}$$

We therefore assume that the axis of symmetry for this beam is along the line of the fiber axis. The phase components jkz and $\phi(z)$ do not affect this problem, because they are independent of x and y. Describing this field as a product of two functions – one being a function of x and the other a function of y – we then have the general case of a laser beam with an elliptical beam cross-section. The fiber field can also be approximated by a Gaussian field distribution, but with a circular cross-section

$$E_{f0} = \frac{\sqrt[4]{2}}{\sqrt[4]{\pi}\,\sqrt{w_f}} \exp\left(-\frac{x^2}{w_f^2}\right) \frac{\sqrt[4]{2}}{\sqrt[4]{\pi}\,\sqrt{w_f}} \exp\left(-\frac{y^2}{w_f^2}\right) \tag{11.105}$$

Using (see [13])

$$\int_{-\infty}^{\infty} \exp(-ax^2) \, dx = \sqrt{\frac{\pi}{a}} \tag{11.106}$$

it is easy to verify that both the laser and the fiber fields satisfy the normalization conditions given by equations (11.101) and (11.102). Although we are considering the transmission of an incident field through the interface between the medium where the laser beam originates on the one hand, and the transmission through the fiber on the other hand as separate problems, we must emphasize that the quotient k/R is independent of the medium, whilst the waist parameter w does not change at such an interface. Since both the fiber and the laser field are given as products of two functions, the field coupling coefficient c_0 can be written likewise

$$c_0 = c_x c_y \tag{11.107}$$

with

$$
c_x = \frac{\sqrt{2}}{\sqrt{\pi w_x w_f}} \int_{-\infty}^{\infty} \exp\left(-\frac{x^2}{w_x^2} - jk\frac{x^2}{2R_x}\right) \exp\left(-\frac{x^2}{w_f^2}\right) dx
$$

$$
= \frac{\sqrt{2}}{\sqrt{\dfrac{w_f}{w_x} + \dfrac{w_x}{w_f} + jk\dfrac{w_x w_f}{2R_x}}} \tag{11.108}
$$

The coefficient c_y can be found by replacing the subscript x in equation (11.108) by the subscript y; the power coupling efficiency then becomes

$$
\eta = |c_0|^2 = |c_x c_y|^2 = |c_x|^2 |c_y|^2
$$

$$
= 2 \Bigg/ \left[\sqrt{\left(\frac{w_f}{w_x} + \frac{w_x}{w_f}\right)^2 + \frac{k^2 w_x^2 w_f^2}{4R_x^2}} \right]
$$

$$
\times 2 \Bigg/ \left[\sqrt{\left(\frac{w_f}{w_y} + \frac{w_y}{w_f}\right)^2 + \frac{k^2 w_y^2 w_f^2}{4R_y^2}} \right] \tag{11.109}
$$

The greatest coupling efficiency is achieved when the denominator of equation (11.109) is minimal. This is the case if $R_x \to \infty$ and $R_y \to \infty$, i.e. if the laser beam has a waist at the end-face of the fiber. Moreover, the condition $w_x = w_y = w_f$ has to be fulfilled, which means that the waist of the laser bundle must be circular and with a diameter similar to that of the fiber field. In that case the efficiency is unity.

11.5.2 Launching from an LED into a single-mode fiber

The incident field described here is not a coherent Gaussian beam, but an incoherent source radiating in all the directions of a hemisphere, rather like a Lambertian source. We assume that the radiating area of the LED is much larger than the parameter w_f of the fiber field. Moreover, we assume that the LED is butt-jointed to the fiber. Radiation is modelled as a superposition of coherent plane waves, each having an angle θ with the axis of the fiber (see Figure 11.21). There is no phase relationship between the beams with angles $\theta = \theta_1$ and $\theta = \theta_2 \neq \theta_1$. Consequently, the total power is found by integrating the power in all individual beams $0 < \theta < \pi/2$ [17]. When making the calculations we combine the method developed in Section 11.1 with the method used in Section 11.4.

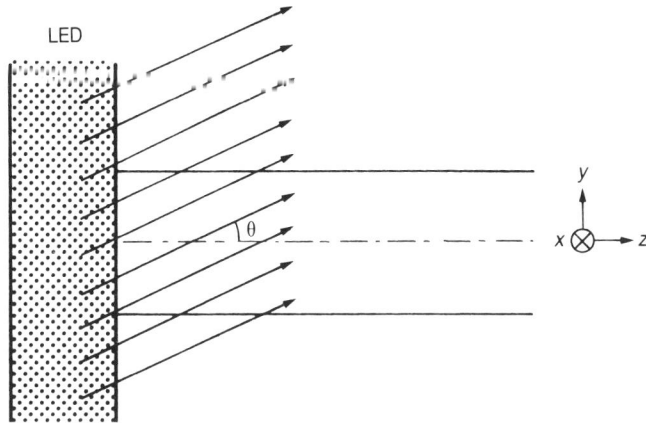

Figure 11.21 Model of the superposition of beams that make an angle θ with the fiber axis. (Source: reproduced with permission from *Journal of Optical Communications*, vol. 9, no. 3, 1988, pp. 100–1, 'Coupling of LED light into a single-mode fiber', by W. van Etten.)

In the first instance, we only consider linear transverse polarization parallel to the fiber end-face. Moreover, we consider a beam that propagates parallel to the y-axis of the coordinate system, as shown in Figure 11.21. We shall now consider the coupling of this particular beam to the fiber field; we shall subsequently remove the restricting assumptions in order to arrive at the coupling efficiency of the LED beam to the fiber. For the time being, we shall consider an incident beam with a uniform field equal to unity in the direction θ. Bearing these assumptions in mind, we can find the coupling coefficient in the x-direction by using equations (11.105) and (11.106)

$$c_x = \frac{\sqrt[4]{2}}{\sqrt[4]{\pi}} \frac{1}{\sqrt{w_f}} \int_{-\infty}^{\infty} \exp\left(-\frac{x^2}{w_f^2}\right) dx = \frac{\sqrt[4]{2}}{\sqrt[4]{\pi}} \sqrt{\pi} \sqrt{w_f} \quad (11.110)$$

In the y-direction, the phase of the incoming beam needs the extra term $ky \tan \theta$, so that the coupling in the y-direction becomes

$$c_y = \frac{\sqrt[4]{2}}{\sqrt[4]{\pi}} \frac{1}{\sqrt{w_f}} \int_{-\infty}^{\infty} \exp\left(-\frac{y^2}{w_f^2} - jky \tan \theta\right) dy$$

$$= \frac{\sqrt[4]{2}}{\sqrt[4]{\pi}} \sqrt{\pi} \sqrt{w_f} \exp\left(-\frac{1}{4} k^2 \tan^2(\theta) w_f^2\right) \quad (11.111)$$

And the power coupling coefficient for the beam becomes

$$|c_x c_y|^2 = 2\pi w_f^2 \exp\left(-\tfrac{1}{2} k^2 \tan^2(\theta) w_f^2\right) \quad (11.112)$$

We assume that the spectral width of the LED is relatively small, so that the wave number k and the field parameter w_f are considered to be constant for all the spectral components. We now introduce equation (11.5) in order to calculate the total coupled power of all beams $0 < \theta < \pi/2$. The integration over S has already been performed in equations (11.110) and (11.111). Since we consider only one direction of the polarization L has to be replaced by $L/2$, whilst integrating over w requires the factor given in equation (11.12) to be introduced. Using equation (11.2) yields the coupled power

$$P_\| = 2\pi^2 L \int_0^\infty \frac{ww_f^2}{(1 + w^2)^2} \exp\left(-\tfrac{1}{2} k^2 w^2 w_f^2\right) dw$$

$$= \pi^2 L w_f^2 \int_0^\infty \frac{\exp\left(-\tfrac{1}{2}k^2 \xi w_f^2\right)}{(1 + \xi)^2} d\zeta \qquad (11.113)$$

For most practical situations $k^2 w_f^2 \gg 1$; therefore, $\exp\left(-\tfrac{1}{2}k^2 \xi w_f^2\right)$ decays much more rapidly than $1/(1 + \xi)^2$, giving

$$P_\| \approx \frac{\lambda^2 L}{2} \qquad (11.114)$$

When considering beams that are transversely polarized perpendicular to the x-direction, a factor $\cos\theta$ has to be introduced into the coupling coefficients. However, in the approximation used to arrive at the result of equation (11.114) this does not have any significant effect on parallel polarized beams, and total coupled power becomes

$$P = \lambda^2 L \qquad (11.115)$$

It is assumed that the incident beam is randomly polarized, which means that each of the polarizations considered carries half of the total power. It is stressed that the coupled power does not depend on the fiber core radius or on the mode field diameter $2w_f$, but only on the wavelength and the radiance of the source. By means of equation (11.6), we can obtain the power coupling efficiency

$$\eta = \frac{\lambda^2}{\pi^2 b^2} \qquad (11.116)$$

It is interesting to note that the solution of equation (11.115) can also be obtained by considering the single-mode fiber to be a multimode step index fiber with a core radius of w_f and NA equal to $\lambda/(\pi w_f)$, the divergence angle of a Gaussian beam with the beam parameter as w_f (compare this with equations 11.10 and 11.11). The result of equation (11.115) has been verified experimentally [20].

Example 11.4
Consider an LED with a circular emitting area of diameter 50 μm, which is equal to the core diameter of a standard multimode fiber. The wavelength is 1300 nm. For these values the efficiency according to equation (11.116) is $\eta = 2.74 \times 10^{-4}$. When we compare this with the coupling efficiency of a

multimode step index fiber, this value is less by 21.6 dB; however, differences in the order of 20 dB are typical for the coupling efficiency of an LED to a multimode versus a single mode fiber.

11.6 Single-mode fiber–fiber coupling

In this section we will consider similar misalignments to those for multimode fiber–fiber coupling. The field of a transmitting fiber and a receiving fiber can both be described as Gaussian distributions. For the coupling of Gaussian beams we have developed the general method presented in Section 11.4, and this enables us to combine both the extrinsic and the intrinsic losses (excluding Fresnel losses) into a single expression. In this case, we have designated the parameters of the transmitting fiber with the subscript '1' and the parameters of the receiving fiber with the subscript '2'.

11.6.1 Coupling between single-mode fibers with a radial offset

According to the notation above and using equation (11.105), the field of the transmitting fiber can be written as

$$E_1 = \frac{\sqrt[4]{2}}{\sqrt[4]{\pi} \sqrt{w_1}} \exp\left(-\frac{x^2}{w_1^2}\right) \frac{\sqrt[4]{2}}{\sqrt[4]{\pi} \sqrt{w_1}} \exp\left(-\frac{y^2}{w_1^2}\right) \tag{11.117}$$

The field of the receiving fiber is the same, with the subscripts '1' being replaced by '2'. We assume that the fiber end-faces are parallel and the fiber axes are in the same direction, although displaced by x_0 in the x-direction and y_0 in the y-direction. Then, for the coupling coefficients we find that

$$
\begin{aligned}
c_x &= \frac{\sqrt{2}}{\sqrt{\pi}} \frac{1}{\sqrt{w_1 w_2}} \int_{-\infty}^{\infty} \exp\left(-\frac{x^2}{w_1^2}\right) \exp\left[-\frac{(x-x_0)^2}{w_2^2}\right] dx \\
&= \frac{\sqrt{2}\sqrt{w_1 w_2}}{\sqrt{w_1^2 + w_2^2}} \exp\left(-\frac{x_0^2}{w_1^2 + w_2^2}\right)
\end{aligned}
\tag{11.118}
$$

and

$$c_y = \frac{\sqrt{2}\sqrt{w_1 w_2}}{\sqrt{w_1^2 + w_2^2}} \exp\left(-\frac{y_0^2}{w_1^2 + w_2^2}\right) \tag{11.119}$$

From equations (11.92) and (11.107), the power coupling efficiency becomes

$$\eta = |c_x c_y|^2 = 4\left(\frac{w_1 w_2}{w_1^2 + w_2^2}\right)^2 \exp\left(\frac{-2d^2}{w_1^2 + w_2^2}\right) \tag{11.120}$$

with

$$d \triangleq \sqrt{x_0^2 + y_0^2} \tag{11.121}$$

the total radial displacement of the fiber axes.

11.6.2 Coupling between single-mode fibers with an axial offset

In this section we assume that the fiber end-faces are parallel and the fiber axes are aligned. Due to the gap between the end-faces, the Gaussian field of the transmitting fiber expands according to (see Appendix 3)

$$w_1(z) = w_1 \sqrt{1 + \frac{4z^2}{k^2 w_1^4}} \tag{11.122}$$

where z is the distance between the fiber end-faces along the axis. In this equation k is the wave number in the gap. For the power coupling efficiency we can use the result of equation (11.109) and insert $w_f = w_1(z)$ and $w_x = w_y = w_2$, so that we get

$$\eta = 4 \bigg/ \bigg\{ \bigg[\frac{w_1}{w_2} \sqrt{1 + \frac{4z^2}{k^2 w_1^4}} + \frac{w_2}{w_1 \sqrt{1 + (4z^2/k^2 w_1^4)}} \bigg]^2$$
$$+ \frac{k^2 w_1^2(z) w_2^2}{4R^2(z)} \bigg\} \tag{11.123}$$

$R(z)$ is the radius of curvature of the transmitted Gaussian beam at the end-face of the receiving fiber (see Appendix 3). In the last term of the denominator of equation (11.123) we used the parameters k and $R(z)$ for the gap, although we should have used the values of the fiber material. However, it is shown in Appendix 3 that the quotient k/R is independent of the medium, i.e. its value does not change at the gap–fiber interface. Using (see Appendix 3)

$$R(z) = z + \frac{k^2 w_1^4}{4z} \tag{11.124}$$

and introducing the normalized gap width

$$\zeta \triangleq \frac{z}{kw_1 w_2} \tag{11.125}$$

the efficiency is written as

$$\eta = \frac{4\left(4\zeta^2 + \dfrac{w_1^2}{w_2^2}\right)}{\left(4\zeta^2 + \dfrac{w_1^2 + w_2^2}{w_2^2}\right)^2 + 4\zeta^2 \dfrac{w_2^2}{w_1^2}} \tag{11.126}$$

As in the case of multimode fibers, the Fresnel reflections at the fiber end-faces can lead to an oscillating component for the loss; this is superimposed on the loss that follows from equation (11.126). Since the calculations related to these losses are rather lengthy and tedious we shall not give them here, but the method is similar to that shown in Section 11.3.2., provided that the ray propagation in the gap is replaced by the propagation of a Gaussian beam. The result looks like that for multimode fibers, especially with an LED as the light source. If the light source is a laser, the low-frequency modulation shown in

Figure 11.15 is absent. This modulation is caused by the fact that a multimode fiber has a finite numerical aperture, beyond which no guided rays can be accepted, whereas a single mode fiber has no sharp cut-off point (see equation 11.111).

11.6.3 Coupling between single-mode fibers with an angular misalignment

We now consider the situation of Figure 11.16(b), which has been redrawn in Figure 11.22 in a slightly different way. A description of the fields in the fibers has been given in the preceding sections; looking at Figure 11.22 it can be seen that all points of the exciting field, for a fixed value of x and in a plane perpendicular to the z-axis in the receiving fiber, have the same phase. Thus, the coupling coefficient (using equation 11.118 with $x_0 = 0$) becomes

$$c_x = \frac{\sqrt{2} \sqrt{w_1 w_2}}{\sqrt{w_1^2 + w_2^2}} \tag{11.127}$$

If we consider the same reference plane, but when the points have a fixed value of y, it can be concluded from Figure 11.22 that these points have a relative phase difference equal to $k2\Delta z = k2y \tan \alpha$. Consequently, c_y becomes

$$
\begin{aligned}
c_y &= \frac{\sqrt{2}}{\sqrt{\pi}} \frac{1}{\sqrt{w_1 w_2}} \int_{-\infty}^{\infty} \exp\left[-y^2\left(\frac{1}{w_1^2} + \frac{1}{w_2^2}\right) - j2ky \tan \alpha\right] dy \\
&= \frac{\sqrt{2} \sqrt{w_1 w_2}}{\sqrt{w_1^2 + w_2^2}} \exp\left[-\frac{(2\pi n(\tan \alpha)w_1 w_2)^2}{\lambda_0^2(w_1^2 + w_2^2)}\right]
\end{aligned}
\tag{11.128}
$$

and the power coupling efficiency is

$$\eta = |c_x c_y|^2 = 4\left(\frac{w_1 w_2}{w_1^2 + w_2^2}\right)^2 \exp\left[-\frac{2(2\pi n(\tan \alpha)w_1 w_2)^2}{\lambda_0^2(w_1^2 + w_2^2)}\right] \tag{11.129}$$

In this equation n is the refractive index of the fiber material.

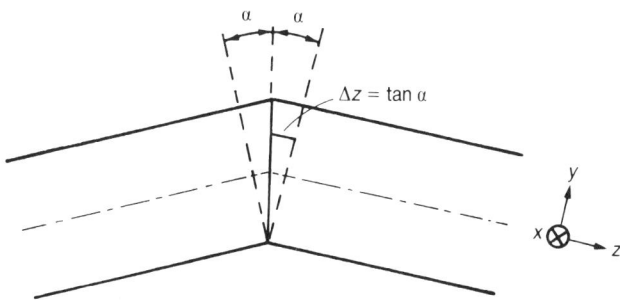

Figure 11.22 The angular misalignment of two fibers.

11.6.4 Concluding remarks

If in the results from Sections 11.6.1–3 for the misalignment parameter d, ζ and α become zero, all the corresponding power efficiency expressions are equal to the intrinsic efficiency

$$\eta_{\text{intr}} = \left(\frac{2w_1 w_2}{w_1^2 + w_2^2} \right)^2 \tag{11.130}$$

This, in turn, becomes unity when the parameter w_1 equals w_2. Unlike the multimode case (equation 11.22), joining the single-mode fibers with $w_1 < w_2$ leads to an inefficient coupling. An optimal coupling is only achieved for fibers with the same values of the w parameters.

Comparing equations (11.120), (11.126) and (11.129), it is obvious that the coupling of single-mode fibers, like that of multimode fibers, is not so sensitive to axial misalignment as for the other two misalignments. Equation (11.126) decays at a rate of approximately $1/\zeta^2$, whereas equations (11.120) and (11.129) decay exponentially. It was further concluded that the radial and angular offset of multimode fibers behave similarly. The coupling efficiency in equation (11.120) decreases to $1/e$ of its maximum value when

$$d_0 = \sqrt{\frac{w_1^2 + w_2^2}{2}} \tag{11.131}$$

From equation (11.129) the $1/e$ value is given by

$$\tan \alpha_0 = \sqrt{\frac{w_1^2 + w_2^2}{2}} \; \frac{\lambda_0}{2\pi n w_1 w_2} \tag{11.132}$$

For a connection between identical fibers, putting $w_1 = w_2$, and combining equations (11.131) and (11.132) produces

$$d_0 \tan \alpha_0 = \frac{\lambda_0}{2\pi n} \tag{11.133}$$

It is important to note that designing a fiber that is more tolerant to radial offset – by increasing the beam widths $2w_1$ and $2w_2$ (see equation 11.133) of both fibers being joined by enlarging the widths of their waists with lenses – produces a fiber that is less tolerant to angular misalignment, and vice versa. It is necessary to compromise between these two defects.

11.7 Fiber–detector coupling

Coupling light leaving a fiber to a detector should not result in any serious problems, because the photosensitive area of a detector is usually larger than the area of the fiber core, and therefore a butt joint should not incur any extrinsic loss. The photosensitive area is often protected by a glass window, rendering a butt joint unsuitable; in this case, an image of the fiber end-face can be made with a lens on the sensitive area without causing substantial loss.

References

[1] J. Gowar, *Optical Communication Systems* (London: Prentice Hall, 1984) pp. 551–2.

[2] M. C. Hudson, 'Calculation of the maximum optical coupling efficiency into multimode optical waveguides', *Applied Optics*, vol. 13, no. 5, May 1974, pp. 1029–33.

[3] M. Born and E. Wolf, *Principles of Optics*, 5th edn (Oxford: Pergamon Press, 1975).

[4] T. H. Zachos and J. E. Ripper, 'Resonant modes of GaAs junction lasers', *IEEE Journal of Quantum Electronics*, vol. QE-5, January 1969, pp. 29–37.

[5] H. Kogelnik and T. Li, 'Laser beams and resonators', *Applied Optics*, vol. 5, no. 10, October 1966, pp. 1550–67.

[6] H. R. D. Sunak and M. A. Zamprônio, 'Launching light from semiconductor lasers into plane-ended multimode optical fibers', *Applied Optics*, vol. 22, no. 15, August 1983, pp. 2337–43.

[7] Wood-Hi Cheng, 'The optimum power coupling from GaAs lasers into spherical-ended fibers', *Proceedings IEEE*, vol. 69, no. 3, March 1981, pp. 396–7.

[8] F. L. Thiel and R. M. Hawk, 'Optical waveguide cable connection', *Applied Optics*, vol. 15, no. 11, November 1976, pp. 2785–91.

[9] E. Neumann and W. Weidhaas, 'Loss due to radial offsets in dielectric optical waveguides with arbitrary index profiles', *Archiv für Elektronik und Übertragungstechnik*, vol. 30, no. 11, November 1976, pp. 448–50.

[10] W. van Etten, W. Lambo and P. Simons, 'Loss in multimode fiber connections with a gap', *Applied Optics*, vol. 24, no. 7, April 1985, pp. 970–6.

[11] R. E. Wagner and C. R. Sandahl, 'Interference effects in optical fiber connections', *Applied Optics*, vol. 21, no. 8, April 1982, pp. 1381–5.

[12] F. Budé and W. van Etten, 'Attenuation and interference in multimode fiber connections', *Applied Optics*, vol. 26, no. 7, April 1987, pp. 1158–61.

[13] I. Gradshteyn and I. Ryzhik, *Table of Integrals, Series and Products*, 4th edn (New York: Academic Press, 1980).

[14] D. Opielka and D. Rittich, 'Transmission loss caused by an angular misalignment between two multimode fibers with arbitrary profile exponents', *Applied Optics*, vol. 22, no. 7, April 1983, pp. 991–4.

[15] P. Di Vita and U. Rossi, 'Theory of power coupling between multimode optical fibres', *Optical and Quantum Electronics*, vol. 10, 1978, pp. 107–17.

[16] D. Marcuse, 'Loss analysis of single-mode fiber splices', *Bell System Technical Journal*, vol. 56, no. 5, May–June 1977, pp. 703–18.

[17] A. W. Snyder and J. D. Love, *Optical Waveguide Theory* (London: Chapman & Hall, 1983).

[18] S. Ramo, J. R. Whinnery and Th. van Duzer, *Fields and Waves in Communication Electronics*, 2nd edn (New York: Wiley, 1984).

[19] H. Kogelnik, 'Coupling and conversion coefficients for optical modes', *Microwave Research Institute Symposia Series*, vol. 14, pp. 333–47 (New York: Polytechnic Press, 1964).

[20] W. van Etten, 'Coupling of LED light into a single-mode fibre', *Journal of Optical Communications*, vol. 9, no. 3, September 1988, pp. 100–1.

Problems

11.1 Make an estimate of the ratio of the internal quantum efficiency of a surface emitter LED and its external efficiency.

11.2 The radiant intensity of an LED is written as $I = I_0 \cos^n \theta$. Calculate the efficiency when butt-coupling this LED to a multimode step index fiber with core radius a and numerical aperture NA.

11.3 Use a simple geometrical construction to show that equation (11.28) is valid for small values of d/a.

11.4 Consider the coupling of two multimode step index silica fibers. The fiber axes are aligned and the end-faces, which are perpendicular to the axes, are parallel. Show that our calculations in equations (11.57)–(11.72) approximate the exact value with an error smaller than 0.5 percent.

11.5 Light from an LED is launched into a multimode step index fiber. The multimode fiber is butt-coupled to a single-mode fiber; the fiber axes are aligned without gap, radial displacement or angular misalignment.

Calculate the efficiency of this fiber–fiber coupling.

12

Receiver principles and signal-to-noise ratio in analog receivers

12.1   Connection diagram and equivalent scheme of photodetectors
12.2   The impulse response of a PIN photodiode
12.3   Signal-to-noise ratio in analog receivers
12.4   The thermal noise in front-end amplifiers
      References
      Problems

The purpose of the receiver in an optical communication system is to recover the information signal, which has been modulated (by analog methods or digitally), on the optical carrier leaving the fiber at the receiving end. A diagram of the receiver is given in Figure 8.1. The differences between an analog receiver and a digital receiver are in the method of processing. Both receivers have a detector, and amplification of the detected (weak) signal is the same in each case. However, it is necessary to define the quality required in order to be able to judge the performance of a system. For analog systems, the mean squared error is taken, and for digital systems the bit error rate is used.

In addition to the thermal noise of both the active components and the resistors of a receiver, the shot noise is very important in optical receivers. Shot noise is produced when carriers are generated by photons in the photodiode. The photons arrive randomly, and therefore the generation of carriers occurs at random points in time with a Poisson distribution. In Appendix 4 various properties of Poisson processes are deduced. In this chapter we shall use these properties, and those of thermal noise, in order to calculate the signal-to-noise ratio of analog optical communication systems. In doing so, we shall restrict ourselves to the photoconductive mode of the photodiode and to direct detection, which means that the photodiode current is proportional to the optical power received. The phase information of the optical wave is ignored; this is contrary to coherent detection, where the frequency and phase of the optical wave can be used in the receiver to recover information that will eventually modulate these signal parameters (see Chapter 16).

12.1 Connection diagram and equivalent scheme of photodetectors

In Chapter 8 it was shown that semiconductor photodiodes can be used in two different modes, namely the photovoltaic and the photoconductive modes. In the former mode the impedance of the detector device is quite low, so that the load resistance must also be kept low. The best method of operating this mode

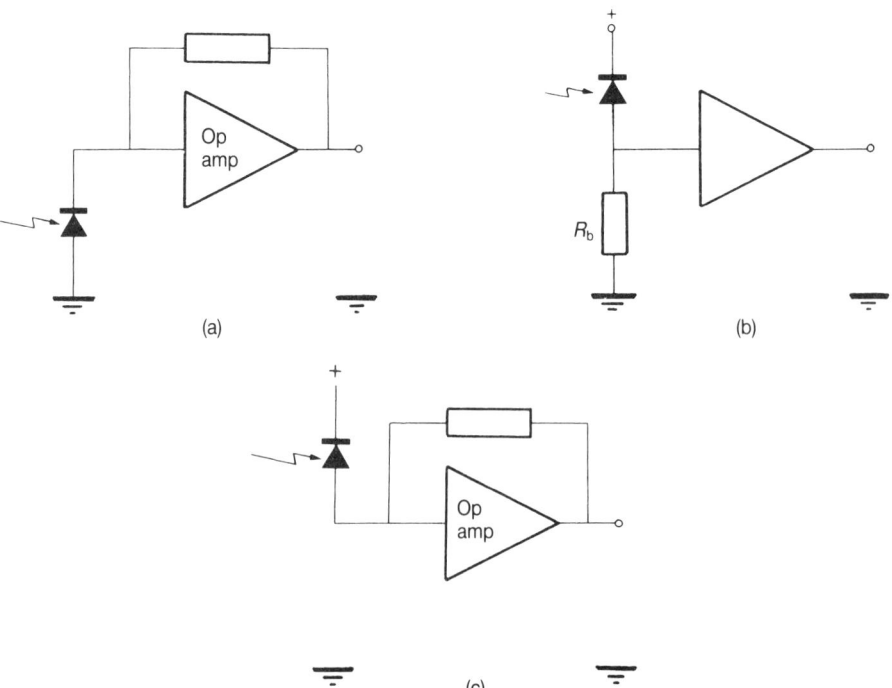

Figure 12.1 Connection diagrams for semiconductor photodiodes. (a) Photovoltaic mode with transimpedance amplifier. (b) Photoconductive mode with high input impedance. (c) Photoconductive mode with transimpedance amplifier.

is to short-circuit the diode, so that it behaves as a current source (see Figure 8.19). However, in this mode the diode is slow, so that the mode is only suitable for measuring purposes, when the modulating frequencies are low; it is not suitable for fast modulation. A possible circuit diagram is presented in Figure 12.1(a); there the diode is connected to the virtual earth, which consists of the input of a feedback operational amplifier.

Semiconductor photodiodes are used in the photoconductive mode for broadband transmission systems. As seen from Figure 8.19, the impedance of the diode is high in this mode and the diode behaves like a current source. For amplifying and processing it is more convenient to make the signal available as a voltage; this can be achieved when a resistor is placed in series with the photodiode and a high input impedance amplifier is connected in parallel to the resistor. Consequently, the circuit diagram looks like Figure 12.1(b). A third front-end circuit is shown in Figure 12.1(c), where a photodiode in the photoconductive mode is connected to a transimpedance amplifier.

For the analysis we shall use the diagram shown in Figure 12.1(b), but the other two diagrams should not give significantly different results. The front-end circuit is shown in Figure 12.2; in this circuit C_s represents the capacitance of the power supply, C_d is the parasitic capacitance of the photodiode and R_b is the load resistance. In practical situations the power supply, with a parallel capacitance C_s, may be considered to be short-circuited with regard to the information signal. The equivalent signal circuit is therefore like that depicted in Figure 12.3.

The average diode current can be found from equation (8.24)

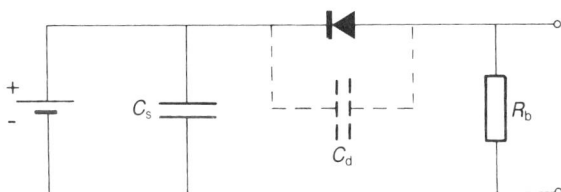

Figure 12.2 Front-end circuit of a photodiode with load resistance.

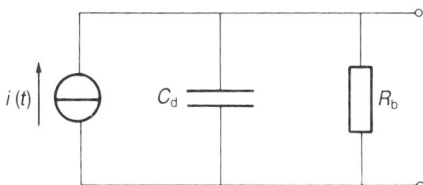

Figure 12.3 The equivalent circuit of a photodiode with a load resistance R_b.

$$\mathrm{E}[i(t)] = eG\,\lambda(t) = \frac{e\eta G}{h\nu}\,p_{\mathrm{s}}(t) + eG\lambda_{\mathrm{d}}$$

$$= R\,p_{\mathrm{s}}(t) + eG\,\lambda_{\mathrm{d}} = \mathrm{E}[i_{\mathrm{s}}(t)] + \mathrm{E}[i_{\mathrm{d}}(t)] \tag{12.1}$$

where R is the responsivity of the photodiode, $p_{\mathrm{s}}(t)$ is the incident optical power, $i_{\mathrm{s}}(t)$ is the signal current and $i_{\mathrm{d}}(t)$ is the dark current of the photodiode. As a result of the parasitic capacitance C_{d} of the photodiode and the load resistance R_{b}, the diode current $i(t)$ is filtered. The arrival of photons and the spontaneous generation of the carriers that produce the dark current follow Poisson processes (see Figure 12.4a). The expected value of these Poisson processes is $\lambda(t)$, the mean number of hole–electron pairs generated per unit of time in the photodiode material. When a PIN diode is used as an opto-electrical converter in the receiver, analysis of the signal-to-noise ratio requires a knowledge of filtered Poisson processes. For avalanche photodiodes (APDs) the internal gain is a random variable, and complicates the analysis of the signal-to-noise ratio considerably. Poisson processes with random impulse gains are called marked Poisson processes (see Figure 12.4b). The filtered version of such a process is called a marked and filtered Poisson process (see Appendix 4).

12.2 The impulse response of a PIN photodiode

Up to this point, it has been assumed that the photodiode itself is ideal, so that the generation of a hole–electron pair gives an impulse current of an amplitude e in the outer circuit. However, for explaining fast modulation of the optical

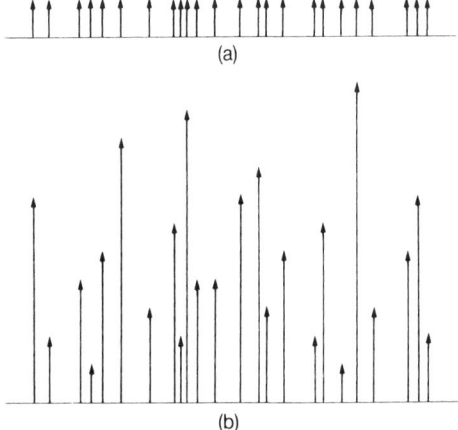

Figure 12.4 (a) The Poisson process: all impulses have the same amplitude e. (b) Marked Poisson process: the lth impulse has an amplitude of eg_l where g_l is the random gain of the APD.

signal, that assumption is too simple. The carriers show a certain transit time through the depletion layer and this gives an impulse response of non-zero width (see below). In that case, the current $i(t)$ is a filtered Poisson process which excites the outer circuit consisting of C_d and R_b. The impulse response of the photodiode and that of the outer circuit can be considered as a series connection of two linear, time-invariant filters.

As an illustration we consider a uniformly side-illuminated diode. When calculating the impulse response we assume that at $t = 0$ there is a uniform generation of carriers in the depletion layer and that no more carriers are subsequently created. The carriers can be considered as space charges that are forced by the electric field to move towards the ohmic contacts on both sides of the depletion layer. This movement induces a current in the outer circuit (see Figure 12.5a). The electrons move with a velocity of v_e, whereas the holes move

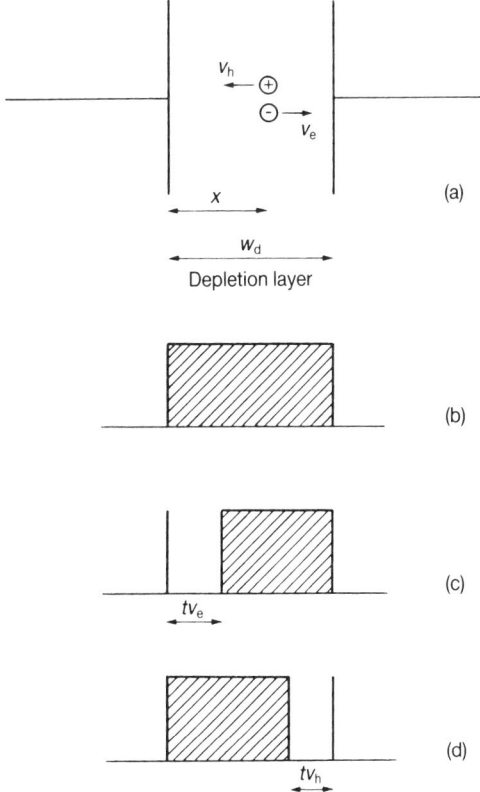

Figure 12.5 Space charge in a side-illuminated PIN photodiode. (a) The opposite movement of electrons and holes in the depletion region. (b) Electron and hole space charge at $t = 0$. (c) Electron space charge at $t \neq 0$ $(t > 0)$. (d) Hole space charge at $t \neq 0$ $(t > 0)$.

in the opposite direction with a generally smaller velocity of v_h (due to their reduced mobility). Considering a slice of the depletion layer dx thick and assuming that the space charge density in this slice is ρ, while it moves with velocity v_e, then this slice of moving charge contributes to the external current (see [1, p. 122])

$$dI_e = \rho A v_e \frac{dx}{w_d} \tag{12.2}$$

where A is the cross-section of the depletion layer and w_d is its width. The total current at time t is found by integrating equation (12.2) over the interval $tv_e \le x \le w_d$ (see Figure 12.5c). This yields

$$I_e(t) = \left[U(t) - U\left(t - \frac{w_d}{v_e} \right) \right] \int_{tv_e}^{w_d} \frac{\rho A v_e}{w_d} \, dx$$

$$= \rho A v_e \left(1 - \frac{v_e}{w_d} t \right) \left[U(t) - U\left(t - \frac{w_d}{v_e} \right) \right] \tag{12.3}$$

where $U(t)$ is the unit step function. For the hole current a similar derivation can be made (see Figure 12.5d), leading to

$$I_h(t) = \rho A v_h \left(1 - \frac{v_h}{w_d} t \right) \left[U(t) - U\left(t - \frac{w_d}{v_h} \right) \right] \tag{12.4}$$

The total current consists of the sum of the electron current and the hole current

$$I(t) = I_e(t) + I_h(t) \tag{12.5}$$

and is depicted in Figure 12.6. When deriving the total current, the velocity of the carriers is assumed to be constant. In [2] it can be seen that this is a

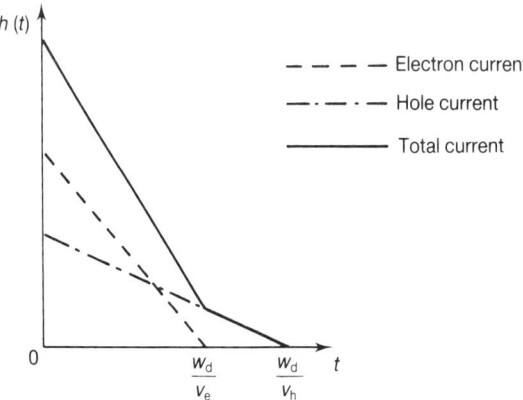

Figure 12.6 The impulse response of a uniformly illuminated side-illuminated PIN photodiode and its components.

reasonable assumption; also, it shows that the carriers created reach their saturation velocity in a much shorter time interval than the transit time. For Si the saturation velocities are, $v_e = v_h = 10^5 \, \text{m s}^{-1}$; for Ge they are: $v_e = 10^5 \, \text{m s}^{-1}$, $v_h = 0.8 \times 10^5 \, \text{m s}^{-1}$.

In the case of a front-illuminated PIN diode the reasoning is similar, but now the space charge density is a function of x (see Chapter 8)

$$\rho(x) = \rho(0) \exp(-\alpha x) \tag{12.6}$$

Figure 12.7 shows the space charges at $t = 0$ and $t > 0$. Using this picture we arrive at the following expression for the electron current, by integrating the shaded area of Figure 12.7(b)

$$I_e(t) = \left[U(t) - U\left(t - \frac{w_d}{v_e} \right) \right] \int_0^{w_d - v_e t} \rho(0) \exp(-\alpha x) \frac{A v_e}{w_d} \, dx$$

$$= \frac{A v_e \rho(0)}{\alpha w_d} [1 - \exp(-\alpha w_d + \alpha v_e t)] \left[U(t) - U\left(t - \frac{w_d}{v_e} \right) \right] \tag{12.7}$$

In a similar way we can find the hole current by integrating the shaded area of Figure 12.7(c)

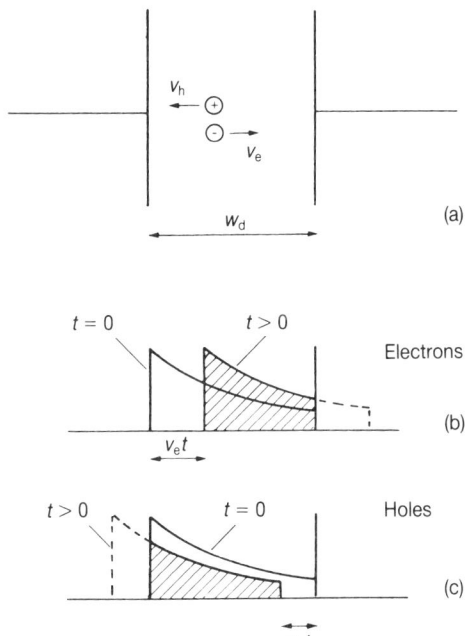

Figure 12.7 Space charge in a front-illuminated PIN photodiode. (a) Opposite movement of electrons and holes in the depletion layer. (b) Electron space charge at $t = 0$ and $t > 0$. (c) Hole space charge at $t = 0$ and $t > 0$.

$$I_h(t) = \left[U(t) - U\left(t - \frac{w_d}{v_h} \right) \right] \int_{v_h t}^{w_d} \rho(0) \exp\left(-\alpha x \right) \frac{A v_h}{w_d} \, \mathrm{d}x$$

$$= \frac{A v_h \rho(0)}{\alpha w_d} \left[\exp\left(-\alpha v_h t \right) - \exp\left(-\alpha w_d \right) \right] \left[U(t) - U\left(t - \frac{w_d}{v_h} \right) \right]$$

$$(12.8)$$

The whole impulse response function comprises the sum of equations (12.7) and (12.8) and this is shown in Figure 12.8 (solid line). When measuring this response, the theoretical curve is always convolved with the impulse response of the low-pass circuit of the measuring equipment and the parasitic components of the detector itself. Due to these effects the measured curve always looks like the broken line in Figure 12.8.

12.3 Signal-to-noise ratio in analog receivers

The general model of an analog optical receiver is shown in Figure 12.9. The process $i(t)$, which represents the photodiode current, consists of a marked Poisson process

$$i(t) = i_s(t) + i_d(t) \tag{12.9}$$

In this equation $i_s(t)$ is the signal current given by

$$i_s(t) = e \sum_l g_l \delta(t - t_l) \tag{12.10}$$

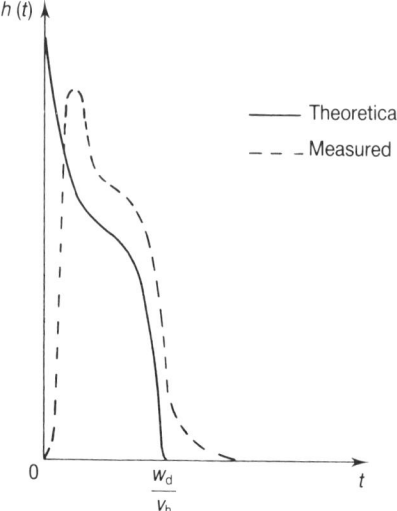

Figure 12.8 The impulse response of a front-illuminated PIN photodiode.

The number of random points of $\{t_l\}$ are Poisson distributed with an intensity of

$$\lambda_s(t) = \frac{\eta}{h\nu} P_s(t) \tag{12.11}$$

When the optical power also contains background radiation, except for the signal, $\lambda_s(t)$ has to be replaced by $\lambda_s(t) + \lambda_b$ with

$$\lambda_b = \int \frac{\eta(\nu)S_{bb}(\nu)}{h\nu}\, d\nu = \int \frac{\eta(\lambda)S_{bb}(\lambda)}{hc}\, \lambda\, d\lambda \tag{12.12}$$

In this equation $S_{bb}(\cdot)$ is the power density spectrum of the background radiation. Note the different notation: λ with a subscript refers to the intensity of the Poisson processes, whereas λ without a subscript is the optical wavelength. In addition, the intensity of carriers that are generated by the signal (equation 12.11) is written in a different way to the intensity of carriers that originate from the background radiation (equation 12.12). This difference results from the implicit assumption that the signal source is a quasi-monochromatic one and the background radiation has a broad spectral distribution.

The term $i_d(t)$ in equation (12.9) originates from the dark current, which results from the spontaneous creation of electron–hole pairs in the detector material, due to thermal excitation. This current is described by a Poisson process

$$i_d(t) = e\sum_m g_m \delta(t - t_m) \tag{12.13}$$

The intensity of this process is λ_d.

Referring to Figure 12.9, then the box containing impulse response $h(t)$, representing the filtering of the detector itself and the outer circuit, is the same as that described in Sections 12.1 and 12.2. Moreover, $h(t)$ includes the amplification and further filtering or equalization. The output from this box is described by a marked and filtered Poisson process. The model in Figure 12.9 shows that the thermal noise contributed by both the input circuit and the amplifier is represented by the additive term $v_{th}(t)$ at the output of the receiver. The output now becomes

$$v_{out}(t) = e\sum_n g_n h(t - t_n) + v_{th}(t) \tag{12.14}$$

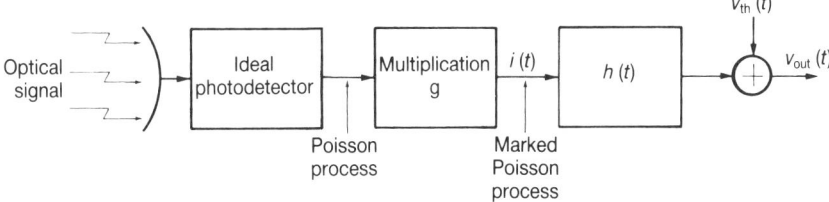

Figure 12.9 The general model of an optical analog receiver.

where the random points in time $\{t_n\} = \{t_l\} \cup \{t_m\}$ are determined by the two previously mentioned Poisson processes. The signal-to-noise ratio is defined by

$$\frac{S}{N} \triangleq \frac{\mathrm{E}^2[v_i(t)|\lambda_b = \lambda_d = v_{th} = 0]}{\sigma^2_{v_{out}(t)}} \qquad (12.15)$$

The part of $v_{out}(t)$ that carries information is denoted by $v_i(t)$ and this is the relevant signal that we want to detect. An explicit expression for this signal cannot be given in general terms, but it has to be found for each specific case.

In order to find the signal-to-noise ratio we need the expectation and variance of a marked and filtered Poisson process. From Appendix 4 it is known that these are

$$\mathrm{E}[s(t)] = \mathrm{E}[g] \int \lambda(\tau)\, h(t - \tau)\, \mathrm{d}\tau \qquad (12.16)$$

$$\sigma^2_s = \mathrm{E}[g^2] \int \lambda(\tau)\, h^2(t - \tau)\, \mathrm{d}\tau \qquad (12.17)$$

Inserting these equations into equation (12.15) and assuming that the thermal noise is independent of the shot noise gives

$$\frac{S}{N} = \frac{e^2 \mathrm{E}^2[g]\{\int_{-\infty}^{\infty} h(t - \tau)\lambda_i(\tau)\mathrm{d}\tau\}^2}{e^2 \mathrm{E}[g^2]\int_{-\infty}^{\infty} h^2(t - \tau)\{\lambda_s(\tau) + \lambda_b + \lambda_d\}\mathrm{d}\tau + \sigma^2_{th}} \qquad (12.18)$$

Here σ^2_{th} is the power of thermal noise at the output of the receiver and equals the variance of $v_{th}(t)$. For the time being, this quantity will not be elaborated, but it will be considered in detail in Section 12.4, where the thermal noise properties of various types of front-end amplifiers are examined.

We shall now consider equation (12.18) for two special cases.

12.3.1 Unmodulated optical carrier

The optical power $p_s(t)$ in this case has a constant value of P_0. In general, according to equation (12.18), the signal-to-noise ratio is a time variable, but it is constant in this special case. The intensity of the Poisson distributions reads

$$\lambda_i = \lambda_s = \frac{\eta P_0}{h v} \qquad (12.19)$$

so that

$$\frac{S}{N} = \frac{(\eta P_0/h v)^2}{2B \dfrac{\mathrm{E}[g^2]}{\mathrm{E}^2[g]} \left\{ \dfrac{\eta P_0}{h v} + \lambda_b + \lambda_d \right\} + \dfrac{\sigma^2_{th}}{e^2 \mathrm{E}^2[g]H^2(0)}}$$

$$= \frac{(RP_0)^2}{2Be \dfrac{\mathrm{E}[g^2]}{\mathrm{E}^2[g]} \mathrm{E}[g]\{RP_0 + eG(\lambda_b + \lambda_d)\} + \dfrac{\sigma^2_{th}}{H^2(0)}} \qquad (12.20)$$

where B is the equivalent noise bandwidth of the filter characteristic of $h(t)$ and is given by [3]

$$B \triangleq \frac{\int_0^\infty |H(f)|^2 \, df}{|H(0)|^2} = \frac{\int_{-\infty}^\infty h^2(t - \tau) \, d\tau}{2\{\int_{-\infty}^\infty h(t - \tau) \, d\tau\}^2} \tag{12.21}$$

The signal-to-noise ratio of equation (12.20) also can be expressed in terms of the currents as follows:

$$\frac{S}{N} = \frac{I^2_s}{2Be \dfrac{E[g^2]}{E^2[g]} (I_s + I_b + I_d) + \dfrac{\sigma^2_{th}}{E^2[g]H^2(0)}} \tag{12.22}$$

where I_s, I_b and I_d are the primary signal current, the background current and the dark current, respectively. The primary current is the current produced by the carriers that are generated directly by photons or thermal excitation. In equation (12.22) the signal and the shot noise are related to the primary currents; therefore, the thermal noise term has to be divided by the square of the total receiver amplification. For detectors without an internal gain ($g = 1$), the latter term in the denominator of equation (12.22) usually determines the ratio S/N; for photomultiplier tubes or APDs the first term in the denominator is usually the most important. As will be demonstrated below, the ratio S/N for these detectors show an optimum as a function of the mean gain. In this optimum none of the terms in the denominator may be neglected.

The factor

$$F \triangleq \frac{E[g^2]}{E^2[g]} = 1 + \frac{\sigma^2_g}{E^2[g]} > 1 \tag{12.23}$$

is called the excess noise factor, and determines the size of the factor when comparing the shot noise of a photodiode with an internal gain and a photodiode without internal gain. For the latter, this factor is unity. The excess noise factor of APDs is related to the mean gain $G \triangleq E[g]$ as follows [4]:

$$F = kG + \left(2 - \frac{1}{G}\right)(1 - k) \tag{12.24}$$

Here k is the ratio of the ionization rate of a hole and the ionization rate of an electron, whilst the ionization rate (or ionization coefficient) is defined as the average number of secondary carrier pairs created by an ionizing collision of a carrier per unit distance. In Figure 12.10 the excess noise factor F is shown as a function of G for various values of k. Good Si detectors have a k value between 0.01 and 0.03. Since the ionization rate of holes in Ge and InGaAs is larger than for electrons, we have to replace $k' = 1/k$ in equation (12.24).

The ionization rates of various important photodiode materials are given in Figure 12.11. It can be seen that these rates depend largely on the electric field in the multiplication region; however, the ratio k is nearly independent of the field. For practical photodiodes, the effective value of k is generally slightly

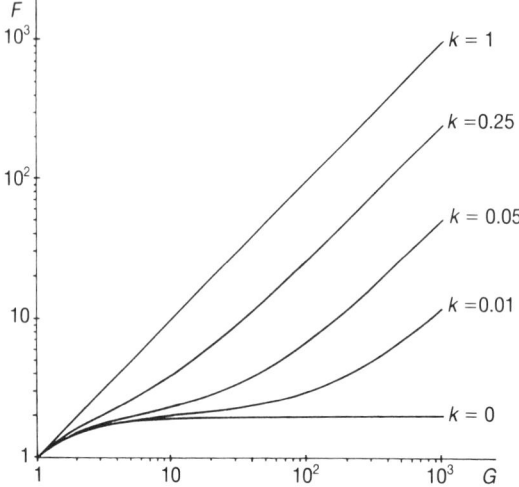

Figure 12.10 The excess noise factor of an APD as a function of the mean gain G.

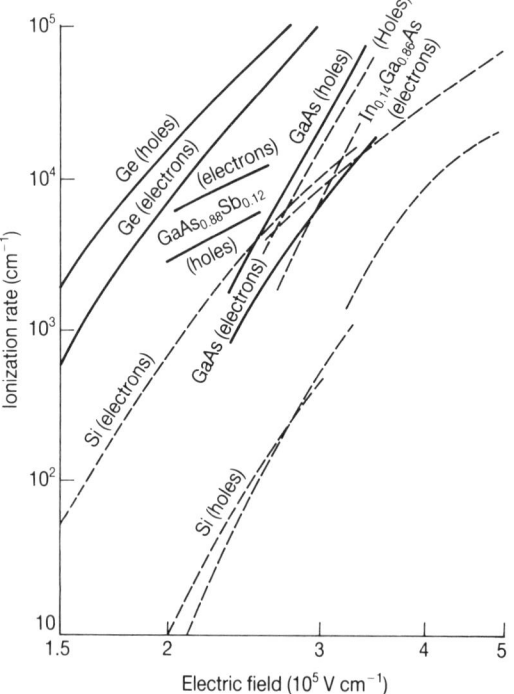

Figure 12.11 Experimental ionization rates of holes and electrons for various photodiode detector materials. (Source: reproduced with permission from *Physics Today*, November 1977, pp. 32–9, 'Detectors for lightwave communication', by H. Melchior.)

worse than that obtained from Figure 12.11. This is due to avalanche multiplication being initiated by both the carriers with a high ionization rate, and by the few carriers of the low ionization rate that are involved.

Equation (12.24) is often replaced by the simpler approximation

$$F = G^x \tag{12.25}$$

Good Si detectors have $x = 0.25$ and good Ge detectors have $x = 0.9$.

It follows from equation (12.22) that the thermal noise decreases with respect to the signal when the gain G is increased. If, however, the gain is increased to a very high value the shot noise term dominates the denominator of equation (12.22). A further increase of G decreases the signal-to-noise ratio, as F increases. This indicates that the signal-to-noise ratio can reach an optimal value. The characteristics of the signal-to-noise ratio as a function of G are rather flat in the vicinity of the optimum. As shown in Figure 8.28, the gain G is controlled by the reverse voltage over the APD. For large signal-to-noise ratios the optimal value of G tends to unity so that, in this case, an APD is ineffective.

Using equation (12.25) the receiver can be optimized analytically when the x of the APD is known

$$G_{opt} = \left[\frac{\sigma^2_{th}}{Be(I_s + I_b + I_d) H^2(0)x} \right]^{1/(2+x)} \tag{12.26}$$

and

$$\left(\frac{S}{N} \right)_{max} = \frac{I_s^2 H^2(0) \left[\dfrac{\sigma^2_{th}}{Be(I_s + I_b + I_d)H^2(0)x} \right]^{2/(2+x)}}{\sigma^2_{th}(2 + x)/x} \tag{12.27}$$

When equation (12.24) is used, optimization of the system requires a numerical evaluation of the signal-to-noise ratio; see Example 12.1 below.

12.3.2 Modulated optical carrier

In this section we assume that the optical power received is written as

$$p_s(t) = P_0[1 + \gamma \tilde{m}(t)], \qquad 0 < \gamma \le 1 \tag{12.28}$$

with

$$\tilde{m}(t) \triangleq \frac{m(t)}{|m(t)|_{max}} \tag{12.29}$$

Here $m(t)$ is the information-carrying signal and γ is the modulation index [3], so that

$$\lambda_i(t) = \frac{\eta \gamma P_0 \tilde{m}(t)}{h\nu} \tag{12.30}$$

If this is inserted into equation (12.18), then in the numerator we get the convolution of $\widetilde{m}(t)$ and $h(t)$; however, we want to separate the information signal $\gamma P_0 \widetilde{m}(t)$ from the carrier component P_0. For that purpose the receiver circuit must be provided with a DC blocking element, e.g. a series capacitor. Consequently, the receiver has a bandpass characteristic. Assuming that the signal $\widetilde{m}(t)$ is not distorted during transmission through $h(t)$, then $H(f)$ must have a flat characteristic over the spectral range of $M(f)$. The value of $H(f)$ in that range is denoted by H. The signal-to-noise ratio now becomes

$$
\frac{S}{N} = \frac{e^2 E^2[g] H^2 \gamma^2 \widetilde{m}^2(t) (\eta P_0/h\nu)^2}{e^2 E[g^2] \int_{-\infty}^{\infty} h^2(t-\tau) \left\{ \dfrac{\eta}{h\nu} P_0[1 + \gamma\widetilde{m}(\tau)] + \lambda_b + \lambda_d \right\} d\tau + \sigma_{th}^2}
$$

(12.31)

As opposed to Section 12.3.1, in this case the signal-to-noise ratio is time dependent; therefore, we define the signal-to-noise as the mean squared signal divided by the mean noise variance. The bandpass character of the signal $\widetilde{m}(t)$ requires that it has a mean value of zero. With this signal-to-noise ratio definition we get

$$
\frac{\langle S \rangle}{\langle N \rangle} = \frac{(\eta P_0/h\nu)^2 \gamma^2 \langle \widetilde{m}^2(t) \rangle}{2B \dfrac{E[g^2]}{E^2[g]} \left(\dfrac{\eta}{h\nu} P_0 + \lambda_b + \lambda_d \right) + \dfrac{\sigma_{th}^2}{e^2 E^2[g] H^2}}
$$

(12.32)

where $\langle \cdot \rangle$ denotes the time average of the factor in the angle brackets. For the present situation the equivalent noise bandwidth B is defined as [3]

$$
B \triangleq \frac{\int_{-\infty}^{\infty} h^2(t-\tau) \, d\tau}{2H^2} = \frac{\int_0^{\infty} |H(f)|^2 \, df}{H^2}
$$

(12.33)

The crest factor of a signal is defined by

$$
C \triangleq \frac{|m(t)|_{\max}}{\sqrt{\langle m^2(t) \rangle}}
$$

(12.34)

and inserting this equation into equation (12.32) yields

$$
\frac{\langle S \rangle}{\langle N \rangle} = \frac{(\eta P_0/h\nu)^2 \gamma^2}{2BC^2 F \left(\dfrac{\eta}{h\nu} P_0 + \lambda_b + \lambda_d \right) + \dfrac{\sigma_{th}^2 C^2}{e^2 G^2 H^2}}
$$

$$
= \frac{R^2 P_0^2 \gamma^2}{2BeC^2 FG[RP_0 + G(I_b + I_d)] + \dfrac{\sigma_{th}^2 C^2}{H^2}}
$$

(12.35)

Subcarrier modulation systems are systems in which the information signal $m(t)$ modulates a sine or cosine waveform whose frequency lies in the radio frequency (r.f.) range; this carrier is called a subcarrier. This modulated signal in its turn modulates the intensity of an optical carrier. For systems with subcarrier modulation, equation (12.35) represents the signal-to-noise ratio of the r.f. signal. Frequency or phase modulation of the r.f. subcarrier leads to $C = \sqrt{2}$, just like the case where $m(t)$ is a pure harmonic signal. For the relationship between the signal-to-noise ratio of the r.f. signal and the signal-to-noise ratio of the baseband signal, see [3].

Example 12.1

We consider an optical fiber system as an example of a subcarrier modulation system designed for the transmission of 16 audio channels in the FM broadcast band [5]. Due to the rather high frequencies involved, namely in the 88–108 MHz band, a laser diode is the obvious light source. The laser has been biased so that the mean optical power emitted amounts to 3 mW. The fiber coupling efficiency is 10 percent, so that the power on the input side of the fiber is 300 μW. The fiber has a length of 1100 m and an attenuation of 5 dB km^{-1}. As a consequence, the optical power that strikes the photosensitive area of the photodiode is about 30 μW. An APD has been taken as the detector and the input impedance of an FM tuner as a load resistance of 75 ohms. We shall calculate the r.f. signal-to-noise ratio of a single FM audio channel, the bandwidth of which is taken to be 200 kHz, at the input of the tuner. Since the transmitter is designed so that the laser is fully modulated when all the channel signals are in phase, the modulation index is taken as $\gamma = 1/16$. The following data are still needed in order to be able to perform the calculations:

$$R = 6.5 \text{ A W}^{-1} \text{ (responsivity of the photodiode)}$$

$$\left. \begin{array}{l} k = 0.019 \\ G = 10 \end{array} \right\} F = 2.05$$

The thermal noise power becomes $2k\theta R_b 2B$, which is the mean squared voltage over the load resistance R_b. Since the signal-to-noise ratio over this load resistance is required, it follows that $H = R_b$. When all these data are inserted into equation (12.35), as well as $\lambda_d = \lambda_b = 0$, we get

$$\frac{\langle S \rangle}{\langle N \rangle} = 53.5 \text{ dB}$$

In Figure 12.12 $\langle S \rangle / \langle N \rangle$ is shown as a function of the mean avalanche gain G, which gives a maximum signal-to-noise ratio of 53.6 dB at $G = 17$ (broken line). Moreover, it can be seen that this function is rather flat in the vicinity of the maximum. From the thermal noise term it follows that the signal-to-noise ratio increases when a larger value is chosen for the load resistance R_b. This influence can be examined with the aid of the signal-to-noise ratio depicted by the solid line in Figure 12.12 for a load resistance value of $R_b = 750$ ohms. In this case, the maximum signal-to-noise ratio amounts to 54.4 dB at $G = 5$. Two

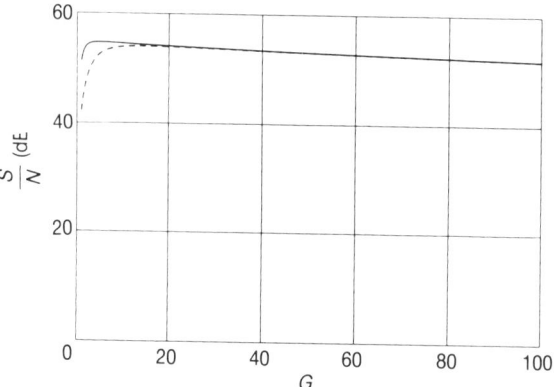

Figure 12.12 The r.f. signal-to-noise ratio of a single FM audio signal out of a 16-channel multiplexed subcarrier modulation system with a glass fiber. The signal-to-noise ratio is given as a function of the APD avalanche gain with the load resistance R_b of the APD as a parameter. Broken line: $R_b = 75\ \Omega$, solid line: $R_b = 750\ \Omega$.

things are significant in this figure. Firstly, it appears that the optimal value of the signal-to-noise ratio is not very sensitive to changes of R_b. Enlarging this resistance by a factor of 10 only improves the signal-to-noise ratio by 0.8 dB. Secondly, the derivative of the characteristic is relatively large for G values smaller than the optimum value, but it is relatively small for G values above the optimum value. It is therefore advisable to make the APD a gain value slightly greater than the optimum. Then the signal-to-noise ratio is slightly smaller than the optimal value and changes little if G changes due to temperature variations.

With respect to the signal-to-noise ratio we now consider two asymptotic situations. If the optical power to be measured is large, the dark current, the background radiation current and the thermal noise become negligible compared with the signal shot noise. In this case it does not make sense to use an APD, thus $F = 1$. We get an absolute limit for $\eta = 1$ and for this limit we find that the signal-to-noise ratio becomes

$$\frac{S}{N} = \frac{\gamma^2 P_0}{2BC^2 h\nu} \tag{12.36}$$

This ratio is called the quantum limit, since only noise due to the quantum effect of the optical power has been considered. It means that equation (12.36) describes an absolute upper limit of the signal-to-noise ratio.

If, on the other hand, the shot noise is negligible with respect to the thermal noise, then the thermal noise limit is

$$\frac{S}{N} = \frac{\gamma^2 P^2{}_0}{C^2 h^2 v^2} \frac{e^2 G^2 H^2}{\sigma^2{}_{th}}$$

(12.37)

In equation (12.37) the left-hand part of the product represents the square of the number of photons which comprises the information carrying part of the signal. The factor $e^2 G^2 H^2 / \sigma^2{}_{th}$ is the signal-to-thermal-noise ratio when the receiver is excited by a single photon which creates a hole–electron pair. This parameter is a figure of merit for the preamplifiers used for optical fiber systems [6]. The classical figures of merit for amplifiers, such as the noise figure, are not important here, because they are based on a matched load condition [7], which is certainly not the case here as the photodiode behaves like a current source. Note that the signal-to-noise ratio in the quantum limit varies with P_0, whereas in the thermal noise limit it varies with P_0^2.

The performance of a receiver is sometimes given in terms of its sensitivity, namely the amount of optical power that must be supplied to the receiver in order to produce a specific signal-to-noise ratio. Usually, sensitivity is expressed in dBm, the logarithmic power measure with respect to a level of 1 mW. On the strength of equations (12.34)–(12.36), sensitivity can be determined easily [6, 8].

Two problems arise from the use of analog modulation for optical communication systems. Firstly, it is difficult to fulfil the linearity requirements of the components (light sources and detectors). Consequently, only small values for the optical power P_0 and modulation index γ can be obtained. Secondly, the signal-to-noise ratio requirements for analog systems are high, some 50 or 60 dB; therefore, only small losses along the optical transmission path between the transmitter and receiver can be tolerated.

12.4 The thermal noise in front-end amplifiers

The equivalent circuit for the photodiode and load resistance is presented in Figure 12.3. The detected signal is supplied to the input of the front-end amplifier (see Figure 12.2). In Figure 12.13 the combination of photodiode, load resistance and amplifier is depicted in the form of an equivalent circuit. The amplifier is represented by a voltage-controlled current source with a transconductance of g. The input impedance of the amplifier comprises a parallel connection of the input resistance R_a and the input capacitance C_a. Furthermore, two thermal noise sources, $i_1(t)$ and $i_2(t)$, are shown that represent the noise of the input circuit and the noise of the output circuit of the amplifier, respectively. These noise sources are considered to be statistically independent, showing Gaussian probability density functions and flat spectra with densities of S_1 and S_2, respectively. The feedback impedance Z_f consists of parasitic elements, for instance the Miller capacitance [9, p. 137] or a feedback resistance

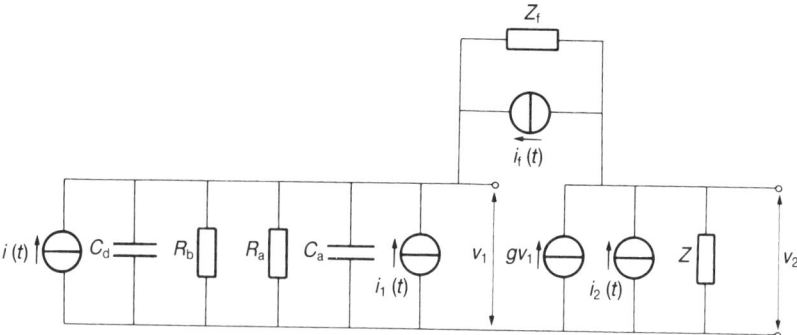

Figure 12.13 Equivalent circuit of detector amplifier combination.

like that seen in a transimpedance amplifier (see Section 12.3.3). For the time being, we ignore Z_f, which means that $|Z_f| \to \infty$.

The photodiode current $i(t)$ and the noise current $i_1(t)$ provide a voltage of v_1 across the total impedance of the input circuit. This impedance is written as

$$Z_i = \left[\frac{1}{R_b} + \frac{1}{R_a} + j2\pi f(C_d + C_a)\right]^{-1} = \left(\frac{1}{R_t} + j2\pi f C_t\right)^{-1} \qquad (12.38)$$

where

$$R_t \triangleq \frac{R_a R_b}{R_a + R_b} \qquad (12.39)$$

$$C_t \triangleq C_d + C_a \qquad (12.40)$$

In Section 12.3.2 it has been assumed that the information signal is not distorted during transmission through the detector and the amplifier. For this purpose the input circuit must be a broadband circuit; however, for a large signal-to-noise ratio the resistance R_t should be as large as possible (although its effect if combined with an APD is small). A large value of R_t, together with the capacitance C_t, however, gives a narrow-band input circuit. An escape from this dilemma can be found by including an equalizer after the amplifier. Such an equalizer must have a differentiating action in order to cancel out the integrating action of the input circuit (see Figure 12.14).

In Figure 12.13 only the input stage of the amplifier has been drawn. Let A_1 be the amplification of the later stages and suppose that the output impedance of the first stage Z is real; then the overall amplification becomes

$$\frac{V_{out}(f)}{I(f)} = Z_i g Z A_1 \frac{1}{Z_i} = \begin{cases} A, & \text{for } |f| < B \\ \\ 0, & \text{for } |f| > B \end{cases} \qquad (12.41)$$

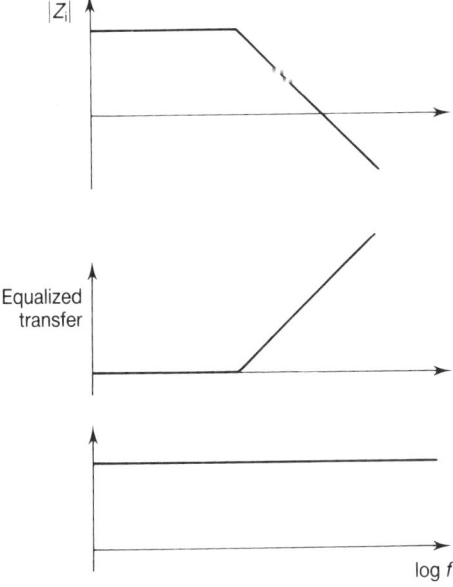

Figure 12.14 The frequency transfer characteristic of the input impedance Z_i and the equalizer.

When arriving at this equation it has been assumed that the equalizer has a transfer equal to $1/Z_i$ and that an ideal low-pass filter of bandwidth B is inserted somewhere in the amplifier; moreover, it has been defined

$$A \overset{\triangle}{=} gZA_1 \tag{12.42}$$

In the subsequent noise calculations it is assumed that the thermal noise of the amplifier is determined by the noise behaviour of the input stage; consequently, the amplifier A_1 and the equalizer are noise free.

Once all these things have been considered, the noise at the amplifier input can be described in terms of its power spectral density as follows

$$S_v(f) = S_1 \left| Z_i g Z A_1 \frac{1}{Z_i} \right|^2 + S_2 \left| \frac{ZA_1}{Z_i} \right|^2$$

$$= S_1 A^2 + S_2 \frac{A^2}{g^2 |Z_i|^2} \tag{12.43}$$

The thermal noise variance is found by integrating this spectral density over the bandwidth B, in order to produce

$$\sigma^2_{th} = \int_{-B}^{B} S_v(f) \, df = S_1 2A^2 B + \frac{S_2 2A^2 B}{g^2 R_t^2} + \frac{S_2 2A^2 B^3 (2\pi C_t)^2}{3g^2} \tag{12.44}$$

Thus, the thermal noise term has been established, and can be inserted in equations (12.22) and (12.35) with $H(0) = H = A$ in order to obtain the signal-to-noise ratio. It can be seen that the thermal noise contains a term that is proportional to B and a term proportional to B^3; however, the value of the amplification A is not important as far as the signal-to-noise ratio is concerned, because this factor vanishes when included in equations (12.22) and (12.35).

On the basis of equation (12.44) some tentative conclusions can be drawn:

- The load resistance should be as large as possible.
- The input capacitance should be as small as possible.
- The gain g of the first transistor stage must be large.
- The spectral densities S_1 and S_2 should be as small as possible.

In Sections 12.4.1–3 below, equation (12.44) will be elaborated for various amplifier types.

12.4.1 The FET as input stage

As an equivalent circuit we can use Figure 12.13, in which we have, for the time being, ignored Z_f. The gate bias resistance of an FET can be very high, so that $1/R_a$ becomes negligible with respect to R_b. The noise source $i_1(t)$ consists of the thermal noise of R_t and, as a consequence, its spectral density becomes $S_1 = 2k\Theta/R_t$. The noise source $i_2(t)$ originates from the thermal noise of the FET channel between the source and the drain. Its power spectral density is written as $S_2 = \alpha 2k\Theta g$ [10]. Here, α is a factor that depends on the type of FET and may theoretically have numerical values in the range $1/2$–$2/3$ [10]. Practical values are somewhat larger, 0.75–1, due to the so-called gate-induced noise and its correlation with the channel noise [11]. Let us take the worst case with a value of unity, so that $S_2 = 2k\Theta g$, where g is the transconductance of the FET, which may have values of the order of $5 \, \text{mA V}^{-1}$. When the given expressions of S_1 are inserted in equation (12.44), we get

$$\sigma^2_{\text{th}} = \frac{4k\Theta}{R_t} A^2 B + 4 \frac{k\Theta A^2 B}{gR_t^2} + \frac{4}{3} \frac{k\Theta A^2 B^3 (2\pi C_t)^2}{g} \tag{12.45}$$

As has been mentioned previously and following from this equation, the load resistance R_t should be as large as possible. With an FET as the input stage the first two terms in equation (12.45) can often be neglected. The third term depends on C_t^2/g. For a given material and a certain FET geometry, C_a/g has a fixed value. At a given detector capacitance C_d, the function $(C_d + C_a)^2/g$, under the condition of a fixed value C_a/g, can be minimized using the Lagrange multiplier. The optimal value appears to be $C_a = C_d$. Optimally, the thermal noise yields

$$\sigma^2_{\text{th}} = \frac{4k\Theta}{R_t} A^2 B + 4 \frac{k\Theta A^2 B}{gR_t^2} + \frac{64}{3} k\Theta A^2 B^3 \pi^2 C_d \frac{C_a}{g} \tag{12.46}$$

The value of C_a/g for a GaAs FET is smaller than the value for a Si FET, which makes the GaAs FET, together with its faster response, more suitable for wideband receivers.

12.4.2 Bipolar input stage

We now consider the bipolar common emitter stage and use the equivalent circuit in Figure 12.13, with $Z_f \to \infty$. The transconductance g equals the quotient of the current gain factor β and the input resistance of the transistor $R_i = k\Theta/eI_b$, where I_b is the base bias current. The base bias resistors have a resistance that is usually much larger than R_i and can be ignored as far as the noise calculations are concerned. The noise term $i_1(t)$ now has two sources: the thermal noise of the load resistance R_b and the shot noise of the base bias current, which is also a thermal noise source. The spectral density of $i_1(t)$ becomes

$$S_1 = \frac{2k\Theta}{R_b} + eI_b = \frac{2k\Theta}{R_b} + \frac{k\Theta}{R_i} \qquad (12.47)$$

However, the noise term $i_2(t)$ stems from the shot noise of the collector bias current and its spectral density reads

$$S_2 = eI_c = e\beta I_b = \frac{\beta k\Theta}{R_i} \qquad (12.48)$$

When these spectral densities are inserted in equation (12.44), together with $g = \beta/R_i$, it is found that

$$\sigma^2_{th} = 2k\Theta A^2 B \left(\frac{2}{R_b} + \frac{1}{R_i} \right) + \frac{2k\Theta A^2 B R_i}{\beta R_t^2} + \frac{2k\Theta A^2 B^3 (2\pi C_t)^2 R_i}{3\beta} \qquad (12.49)$$

For a common emitter circuit, where $R_b \gg R_i$, we get

$$\sigma^2_{th} \approx \frac{2k\Theta A^2 B}{R_i} + \frac{2k\Theta A^2 B}{\beta R_i} + \frac{2k\Theta R_i A^2 B^3 (2\pi C_t)^2}{3\beta} \qquad (12.50)$$

In this equation the term containing B is inversely proportional to R_i, whereas the term involving B^3 is proportional to R_i. In bipolar transistors, however, R_i is inversely proportional to the base bias current; this was stressed at the beginning of this section. As a consequence, the base bias current can be used to control the noise behaviour of the bipolar input stage. It can be confirmed easily that equation (12.50) has a minimum value for

$$R_{i,opt} = \frac{\sqrt{3(\beta + 1)}}{2\pi BC_t} \approx \frac{\sqrt{3\beta}}{2\pi BC_t} \qquad (12.51)$$

and, associated with it, the optimum value of the base bias current is

$$I_{b,opt} = \frac{k\Theta}{e} \frac{2\pi BC_t}{\sqrt{3\beta}}$$ (12.52)

This leads to the minimum value of the thermal noise variance

$$\sigma^2_{th,min} = \frac{4k\Theta BA^2}{R_{i,op}} = \frac{4k\Theta A^2 B^2 2\pi C_t}{\sqrt{3\beta}}$$ (12.53)

For an optimized bipolar input stage the thermal noise power increases with the square of the bandwidth, but for an FET front-end this noise power increases with the third power of the bandwidth. Due to the high impedance level of an FET, such an input stage performs better for small bandwidths than the bipolar input. With an increasing bandwidth, beyond a certain point the bipolar input gets better. For a Si FET, this crossover point is situated between 25 and 50 MHz. The GaAs FET has a larger bandwidth than a Si FET; moreover, the GaAs FET can be integrated with a GaAs photodetector on the same chip rather easily. In this way, the capacitance C_t of such GaAs input devices can become very small, which is a very attractive aspect of these devices and enables them to perform better than the bipolar input for all the bandwidths in use today, especially in the long-wavelength region.

12.4.3 The transimpedance amplifier

A transimpedance amplifier is obtained if the feedback impedance Z_f is given a finite value in Figure 12.13, in contrast to what has been assumed so far. We shall analyze the case where the feedback impedance is placed between the amplifier input and the output of the entire amplifier circuit. The equalizer, however, remains outside the feedback loop (see Figure 12.15). The signal transfer from the amplifier input to the equalizer input reads

$$\frac{V_3(f)}{I(f)} = -Z_f$$ (12.54)

This result is achieved under the assumption that

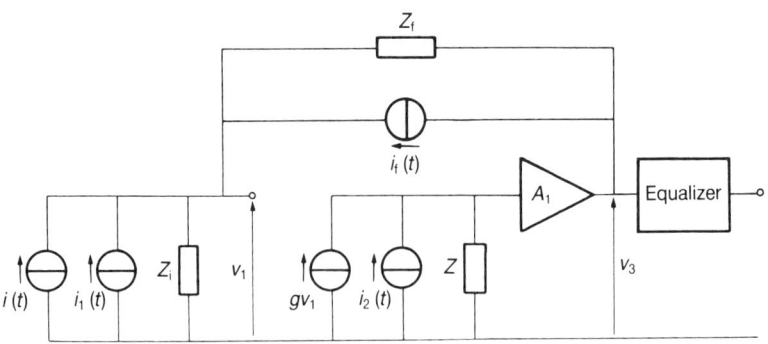

Figure 12.15 Equivalent circuit of a transimpedance amplifier.

$$\left| gZA_1 \frac{Z_i}{Z_i + Z_f} \right| \gg 1 \tag{12.55}$$

The transfer from the photodiode current to amplifier output is taken to be

$$\frac{V_{out}(f)}{I(f)} = \begin{cases} A, & \text{for } |f| < B \\ \\ 0, & \text{for } |f| > B \end{cases} \tag{12.56}$$

As a consequence, the equalizer includes the transfer $V_{out}(f)/V_3(f) = -A/Z_f$. For the power spectral density of the thermal noise at the amplifier output it can be deduced that

$$S_v(f) = S_1 A^2 + S_f A^2 + S_2 \frac{A^2}{g^2} \left| \frac{Z_i + Z_f}{Z_i Z_f} \right|^2 \tag{12.57}$$

where S_f is the spectral density of the noise current caused by Z_f.

We now assume that

$$|Z_f| \gg |Z_i| \tag{12.58}$$

Together with equation (12.55) this requires

$$|gZA_1| \gg 1 \tag{12.59}$$

The spectral density of the output noise becomes

$$S_v(f) = S_1 A^2 + S_f A^2 + S_2 \frac{A^2}{g^2 |Z_i|^2} \tag{12.60}$$

When this equation is compared to equation (12.43), it is seen that the noise of the transimpedance amplifier comprises the noise without feedback plus the noise of the feedback impedance. Referring to [12], it follows that

$$S_f = 2k\Theta \operatorname{Re} \left\{ \frac{1}{Z_f} \right\} \tag{12.61}$$

Thus, it can be seen from equation (12.58) that if Z_f is almost ohmic, then the excess noise contributed by the feedback impedance is negligible. Nevertheless, the noise increases due to the feedback. At this point, it is difficult to understand why transimpedance amplifiers are used at all; however, they do have a larger dynamic range than amplifiers without a feedback loop. In the latter type, amplification of the higher frequencies must be considerable, because these frequencies are greatly attenuated by integration by the input circuit. This great amplification can cause saturation in various parts of the amplifier, and can thus lead to distortion. Distortion can be prevented by reducing the dynamic range of the amplifier. The transimpedance amplifier has a larger dynamic range, since the time constant associated with the feedback impedance is usually much smaller than the time constant of the input circuit. This means that the equalizer can often be omitted.

Let us further consider the situation that

$$|Z_f| \ll |Z_i|$$ (12.62)

Then the spectral density of the noise at the output becomes

$$S_v = S_1 A^2 + S_f A^2 + S_2 \frac{A^2}{g^2 |Z_f|^2}$$ (12.63)

When this equation is compared to equation (12.43) it can be seen that the noise is greatly increased under the present conditions. Therefore the conditions that lead to equation (12.63) must be avoided, as far as the signal-to-noise ratio of the receiver amplifier is concerned.

References

[1] S. Ramo, J. R. Whinnery and Th. van Duzer, *Fields and Waves in Communication Electronics,* 2nd edn (New York: Wiley, 1984).

[2] J. Gowar, *Optical Communication Systems* (London: Prentice Hall, 1984).

[3] A. B. Carlson, *Communication Systems,* 3rd edn (New York: McGraw-Hill, 1986).

[4] R. J. McIntyre, 'Multiplication noise in uniform avalanche diodes', *IEEE Transactions on Electronic Devices,* vol. ED-13, January 1966, pp. 164–8.

[5] W. van Etten and T. Lammers, 'Transmission of FM-modulated audiosignals in the 87.5–108 MHz broadcast band over a fiber optic system', Eindhoven University of Technology, TH-report 80-E-108, April 1980.

[6] S. D. Personick, *Optical Fiber Transmission Systems* (New York: Plenum, 1981).

[7] W. B. Davenport, Jr and W. L. Root, *An Introduction to the Theory of Random Signals and Noise* (New York: IEEE Press, 1987; originally published by McGraw-Hill, 1958).

[8] S. E. Miller and A. G. Chynoweth, *Optical Fiber Telecommunications* (New York: Academic Press, 1979).

[9] R. Gregorian and G. C. Temes, *Analog Mos Integrated Circuits for Signal Processing* (New York: Wiley, 1986).

[10] A. van der Ziel, *Noise in Solid State Devices and Circuits* (New York: Wiley, 1986).

[11] B. L. Kasper and J. C. Campbell, 'Multigigabit-per-second avalanche photodiode lightwave receivers', *IEEE/OSA Journal of Lightwave Technology,* vol. LT-5, no. 10, October 1987, pp. 1351–64.

[12] A. Papoulis, *Probability, Random Variables, and Stochastic Processes,* 2nd edn (New York: McGraw-Hill, 1984).

Problems

12.1 The drift velocities in Si are: $v_e = 0.75 \times 10^5 \, \mathrm{m\,s^{-1}}$, $v_h = 0.5 \times 10^5 \, \mathrm{m\,s^{-1}}$. The thickness of the depletion layer of a Si PIN photodiode is 13.3 μm. It is a front-illuminated photodiode; the light enters the diode at the p side and all photons are absorbed at the p–i interface.

(a) Draw a picture of the impulse response of the photodiode.
(b) What is the maximum bit rate when this photodiode is used in a digital communication link?

12.2 A PIN photodiode has the following responsivity as a function of the wavelength:

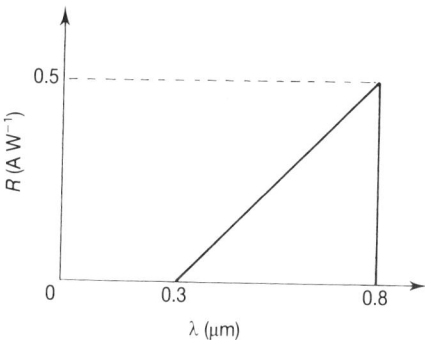

The photodiode, which is at room temperature (290 K), is loaded with a resistance of 50 ohms. This photodiode is used to measure the optical power of a laser ($\lambda_0 = 700$ nm); the minimum value of this power is 10 nW. The dark current of the photodiode can be neglected. A signal-to-noise ratio of 30 dB is required, at an equivalent noise bandwidth of 10 Hz. The spectral density of the background radiation as a function of the wavelength is constant.

What is the maximum value of this spectral density in order to achieve the required signal-to-noise ratio?

12.3 Twelve audio signals are transmitted via an optical fiber. These signals are frequency multiplexed in the FM broadcast band and this multiplexed signal is used to modulate an LED ($\lambda_0 = 1300$ nm). A single-frequency modulated audio signal has a bandwidth of 200 kHz. The derivative of the $P-I$ curve of the LED amounts to 0.04 W A^{-1}. The coupling efficiency from light source to fiber is 1 percent and the fiber has a length of 10 km and attenuation of 0.3 dB km^{-1}. No loss occurs at the coupling from fiber to detector. This detector consists of an APD with: $k = 0.1$, $G = 10$, $R = 8$ A W^{-1}, dark current $I_d = 10$ μA.

The APD is loaded with the input impedance of an FM receiver (75 ohms). The signal-to-noise ratio at the receiver input is required to be 50 dB for a single FM signal.

(a) How should the LED be biased in order to achieve this signal-to-noise ratio?
(b) What is the difference (in dB) between this signal-to-noise ratio and the quantum limit?

12.4 Two television signals, each with a bandwidth of 5 MHz, are frequency multiplexed and transmitted over a fiber optic link. A laser is used as the light source and an APD as the detector ($k = 0.02$, $G = 10$, $R = 6.5$ A W^{-1}). The fiber has a length of 1 km and its attenuation is 3 dB km^{-1}. The coupling efficiency from laser to fiber is 25 percent. The APD is connected to the input of a television receiver set, which has an input impedance of 75 ohms. A signal-to-noise ratio of 50 dB is required for a single television signal at the input of the television set. The laser has the following *P–I* characteristic and emits light of a wavelength $\lambda_0 = 850$ nm. A crest factor of 4 can be inserted for a television signal.

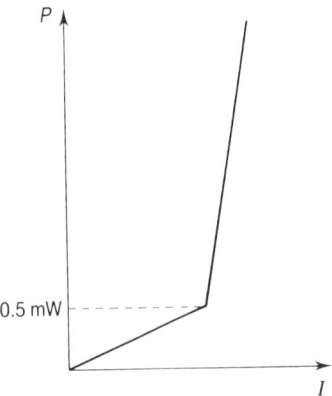

(a) How should the laser be biased in order to achieve the required signal-to-noise ratio? (Dark current and background radiation can be neglected.)

(b) Calculate the difference (in dB) between the signal-to-noise ratio and the quantum limit.

12.5 Consider the photodiode of Problem 8.5. This diode has an area of 1 mm^2 and is connected to an amplifier with an input resistance of 1 kohm.

What is the equivalent noise bandwidth of the input circuit?

13

Receivers for digital optical fiber communication systems

13.1 Introduction
13.2 Analysis of the simplified receiver model
13.3 The quantum limit
13.4 The general receiver model
       References
       Problems

13.1 Introduction

In the previous chapter some important aspects of optical detection were considered, such as shot noise (or Poisson noise) and the signal-to-noise ratio in analog receivers. Filtering of shot noise, together with the filtering of randomly multiplied Poisson noise, is analyzed in Appendix 4. In this chapter we shall study these phenomena with respect to receivers for optical fiber communication systems that are designed for transmitting digital signals [1, 2]. For this purpose a new aspect has to be taken into account, namely the intersymbol interference and its equalization in order to combat this effect [3]. The non-stationary Poisson noise does not give rise to severe problems in analog receivers, thanks to the performance criterion for this kind of receiver. In digital receivers, decisions are made at fixed, equally spaced intervals; therefore, only the signal-to-noise ratio at these moments is important, so this ratio has to be analyzed as a function of time.

Compared to the classical communication model (see [3–5]), where the noise is assumed to be additive, stationary and Gaussian, we now have a complication because shot noise does not have these properties. Analyzing the complete receiver will help us to discover the parameters that influence the signal-to-noise ratio at the moment of sampling; thus, the error probability of the system becomes the ultimate quality criterion for a digital communication system. In this chapter we shall again restrict ourselves to direct detection and the photoconductive mode of the photodiode.

However, before this complicated problem is considered, we shall first simplify the general model. In this way, we will get a better insight into the subject and an idea of the meaning of the different receiver parameters.

Let us consider a single repeater section of an optical fiber communication system by looking at the diagram shown in Figure 13.1. We then analyze the general case where an APD is used as a photodetector. The input data sequence for the communication system is denoted by $\{b_k\}$, with $b_k \in \{b_{min}, b_{max}\}$, where $b_{min} < b_{max}$. The optical power impinging on the photodiode is written as

$$p_s(t) = \sum_k b_k s_p(t - kT) \tag{13.1}$$

where k is an integer, T the bit time interval and $s_p(t)$ a pulse that is positive for all values of t. Furthermore, it is assumed that

$$\int_{-\infty}^{\infty} s_p(t)\,\mathrm{d}t = 1 \tag{13.2}$$

so that b_k represents the energy of the kth pulse received. Equation (13.1) is based on the assumption that the series connection of the transmitter and optical fiber behaves like a linear, time-invariant system.

13.2 Analysis of the simplified receiver model

In this section we shall introduce a number of simplifications related to the signal received, the noise and the way that the receiver processes this signal and noise. Based on these assumptions we can derive quality limits, in the form of

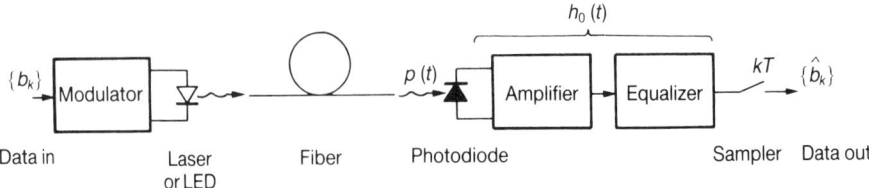

Figure 13.1 Diagram of a digital optical fiber communication system.

the error probability for digital optical fiber communication systems.
The assumptions are as follows:

- In addition to the thermal noise, the shot noise at the input to the sampler in Figure 13.1 is assumed to have a Gaussian probability density function.
- No intersymbol interference occurs in the data signal received.
- The dark current and background radiation are negligible.
- When sampling, the signal values are proportional to the number of electron–hole pairs created by a single pulse $s_p(t)$: a receiver acting in this way is called an integrate and dump receiver; it can be produced by a zero-order hold circuit [5, p. 79].

The impulse response of an integrate and dump receiving filter is shown in Figure 13.2(a) whilst Figure 13.2(b) shows a possible realization of such a filter.

Let us now analyze this simple receiver model. Using equation (A4.57a) the expected value of the voltage at the input to the sampler at the sampling times kT is

$$E[v_{out}(kT)] = eG \int_{-\infty}^{\infty} h_0(\tau)\, \lambda(kT - \tau)\, d\tau$$

$$= \frac{Ae\eta G}{h v} \int_0^T p_s(kT - \tau)\, d\tau = \frac{Ae\eta G}{h v} \int_{(k-1)T}^{kT} p_s(\tau)\, d\tau$$

$$= AeKGb_k = ARb_k \tag{13.3}$$

with

$$K \triangleq \frac{\eta}{h v} \tag{13.4}$$

The noise variance at the sampler is found by means of equation (A4.57b)

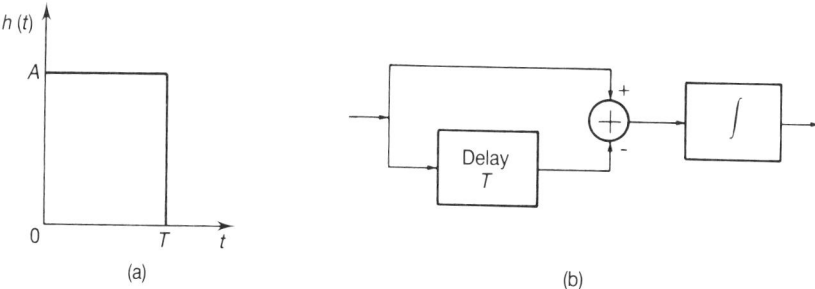

(a)   (b)

Figure 13.2 Integrate and dump receiving filter. (a) Impulse response of the receiving filter. (b) The construction of the filter consisting of a zero-order hold circuit.

$$\sigma^2(kT) = e^2 \mathrm{E}[g^2] \int_{-\infty}^{\infty} \lambda(kT - \tau)\, h_0^2(\tau)\, \mathrm{d}\tau + \sigma_{\mathrm{th}}^2$$

$$= \frac{e^2 \mathrm{E}[g^2] A^2 \eta}{h\upsilon} \int_0^T p_{\mathrm{s}}(kT - \tau)\, \mathrm{d}\tau + \sigma_{\mathrm{th}}^2$$

$$= e^2 A^2 K F G^2 b_k + \sigma_{\mathrm{th}}^2 = A^2 R^2 \frac{h\upsilon}{\eta} F b_k + \sigma_{\mathrm{th}}^2 \qquad (13.5)$$

where the thermal noise is assumed to be independent of the shot noise. The error probability of a binary communication system comprises two conditional error probabilities, namely the probability that a transmitted logical '0' is detected at a logical '1' and the probability that a transmitted logical '1' is detected as a logical '0'. The total error probability reads [5]

$$P_{\mathrm{e}} = \Pr(\hat{b}_k = b_{\max} | b_k = b_{\min})\, \Pr(b_k = b_{\min})$$

$$+ \Pr(\hat{b}_k = b_{\min} | b_k = b_{\max})\, \Pr(b_k = b_{\max}) \qquad (13.6)$$

It has been assumed that the sequence of bits can be represented unambiguously by the physical level sequence $\{b_k\}$. The probabilities $\Pr(b_k = b_{\min})$ and $\Pr(b_k = b_{\max})$ are determined by the information source, whereas the conditional probabilities $\Pr(\hat{b}_k = b_{\max} | b_k = b_{\min})$ and $\Pr(\hat{b}_k = b_{\min} | b_k = b_{\max})$ follow from the conditional probability density functions $p[\upsilon_{\mathrm{out}}(kT) | b_k = b_{\min}]$ and $p[\upsilon_{\mathrm{out}}(kT) | b_k = b_{\max}]$. In Figure 13.3 an example is given to show what these functions look like.

In accordance with one of the assumptions made at the beginning of this section, the density functions have a Gaussian shape, their mean values being given by equation (13.3) and their variances by equation (13.5). With respect to these variances we can see the most important difference from the classical communication model, namely the fact that the variance of the shot noise is proportional to the energy b_k of the received bit; this proportionality follows

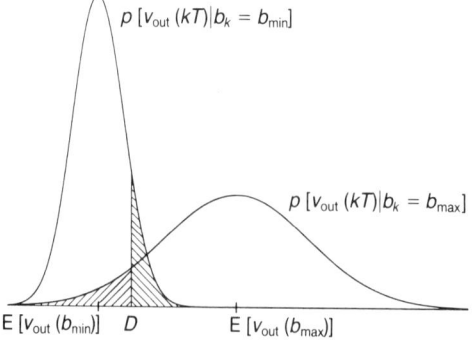

Figure 13.3 The conditional probability density functions and the threshold D which determine the conditional error probabilities.

from equation (13.5). It implies that $p[v_{out}(kT)|b_k = b_{min}]$ has a smaller variance than $p[v_{out}(kT)|b_k = b_{max}]$, whose shape is wider. As the area under the two curves has to be equal to unity, the maximum value of $p[v_{out}(kT)|b_k = b_{min}]$ must be larger than the maximum of the second curve. The bits are detected by comparing the sampled value $v_{out}(kT)$ with a later threshold D. If $v_{out}(kT) > D$, then the detector decides that $\hat{b}_k = b_{max}$, whereas if $v_{out}(kT) < D$ it is decided that $\hat{b}_k = b_{min}$. The conditional error probabilities $\Pr(\hat{b}_k = b_{max}|b_k = b_{min})$ and $\Pr(\hat{b}_k = b_{min}|b_k = b_{max})$ are given by the shaded areas in Figure 13.3. Assuming that a certain total error probability P_e is required, independent of the transmitted bit sequence, then the latter condition is met if the conditional error probabilities $\Pr(\hat{b}_k \neq b_k|b_k = b_{max})$ and $\Pr(\hat{b}_k \neq b_k|b_k = b_{min})$ are equal. This can be easily derived using equation (13.6). The probabilities can be made equal with a specific choice of threshold level D. If $b_k = b_{min}$ the probability that the noise exceeds the value of $D - E[v_{out}(b_{min})]$ must be equal to or smaller than P_e, or

$$Q\left[\frac{D - E[v_{out}(b_{min})]}{\sigma(b_{min})}\right] \leq P_e \tag{13.7}$$

where $Q(\cdot)$ is the Gaussian probability integral [3–5]. If $b_k = b_{max}$ the probability that the noise exceeds $E[v_{out}(b_{max})] - D$ has to be equal to or smaller than P_e, or mathematically

$$Q\left[\frac{E[v_{out}(b_{max})] - D}{\sigma(b_{max})}\right] \leq P_e \tag{13.8}$$

The threshold value corresponds to the conditional probabilities being equal and is found by making equations (13.7) and (13.8) have the same value and solving the resulting equation for D. This yields

$$D = \frac{E[v_{out}(b_{max})]\,\sigma(b_{min}) + E[v_{out}(b_{min})]\,\sigma(b_{max})}{\sigma(b_{max}) + \sigma(b_{min})} \tag{13.9}$$

If this equation is inserted in place of D in equation (13.7) or equation (13.8), and making it equal to P_e, we get the total error probability

$$P_e = Q(q) \tag{13.10}$$

where q is interpreted as the signal-to-noise ratio and is written as

$$q = \frac{S}{N} = \frac{E[v_{out}(b_{max})] - E[v_{out}(b_{min})]}{\sigma(b_{max}) + \sigma(b_{min})} \tag{13.11}$$

For a certain value of q, which is the parameter of the Q function, the corresponding value of P_e can be found from either a table, a characteristic or an approximating formula (see [4] or [5]). Substituting equations (13.3) and (13.5) into equation (13.11), when $P_e = 10^{-9}(q = 6)$, we get

$$eKG(E_{\max} - E_{\min}) = 6\sqrt{Ke^2FG^2E_{\min} + \sigma_{th}^2/A^2}$$

$$+ 6\sqrt{Ke^2FG^2E_{\max} + \sigma_{th}^2/A^2} \tag{13.12}$$

To emphasize that b_k can be interpreted as the energy content of the kth pulse, $b_{\max}$ has been replaced by $E_{\max}$ and $b_{\min}$ by $E_{\min}$. For the case where the light source is switched off for $b_{\min}(E_{\min} = 0)$, equation (13.12) can be solved easily

$$E_{\max} = \frac{1}{K}\left(36F + 12\frac{\sigma_{th}}{eGA}\right) \tag{13.13}$$

Thus for the special situation $E_{\min} = 0$, this equation gives the energy of the bits represented by the equivalent high level in order to get a bit error rate of 10^{-9}. On the basis of this equation some comments can be made. When thermal noise is absent we can take $F = 1$ (using an APD does not make sense in this case) and it follows from equation (13.13) that the required number of photons for $b_{\max}$ is 36. In the case where the thermal noise is not negligible for small values of G, the required number of photons will vary as $1/G$, up to the point where F becomes dominant (see Chapter 12). Figure 13.4 shows the number of photons required per high-level bit as a function of the mean APD gain G for $\sigma_{th}/eA = 1000$ and various values of the ionization rate ratio k (see Chapter 12). At gain values between 50 and 100 an optimum sensitivity is obtained. For practical receivers this sensitivity amounts to some 400 or 800 photons per pulse, which is about 13–16 dB above the shot noise limit or quantum limit. This will be dealt with in the next section.

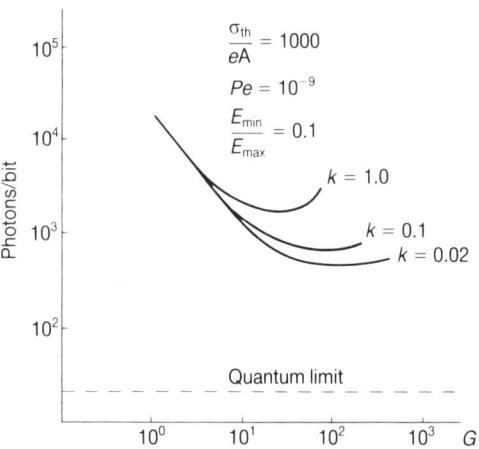

Figure 13.4 The receiver sensitivity in photons per '1' bit as a function of the mean APD gain.

13.3 The quantum limit

A lower limit for the error probability is found by assuming $E_{min} = 0$ and neglecting the thermal noise. This lower limit follows from the assumption that in a specific bit interval, for instance $(0, T)$, no electron–hole pairs are generated, whereas pulses with a mean energy E_{max} are received.

In [6] the probability of k events in the time interval $(0, T)$ is given for a Poisson distribution with a time-varying expected value $\lambda(t)$ (see also Sections A4.2.1 and A4.2.2). This probability reads

$$\Pr[k, (0, T)] = \left\{ \exp\left[-\int_0^T \lambda(t)\, dt \right] \right\} \left\{ \int_0^T \lambda(t)\, dt \right\}^k / k! \tag{13.14}$$

The probability that no electron–hole pairs are generated becomes

$$\Pr[0, (0, T)] = \exp\left[-\int_0^T \lambda(t)\, dt \right] \tag{13.15}$$

In this equation $\lambda(t)\, dt$ gives the expected number of carrier pairs that is generated in the time interval $(t, t + dt)$. The mean number of photons received in the same interval will be $p_s(t)dt/h\nu$. This gives the probability of zero generated carried pairs

$$\Pr[0, (0, T)] = \exp\left[\frac{-\eta}{h\nu} \int_0^T p(t)\, dt \right] = \exp\left(\frac{-\eta E_{max}}{h\nu} \right) \tag{13.16}$$

which equals the probability $\Pr(\hat{b}_k = b_{min} | b_k = b_{max})$. In the present case $\Pr(\hat{b}_k = b_{max} | b_k = b_{min}) = 0$, since we have assumed that $E_{min} = 0$, which implies that for $b_k = b_{min}$ no photons are transmitted and therefore no photons can be received. From equation (13.6) it follows that

$$P_e = \Pr(\hat{b}_k = b_{min} | b_k = b_{max}) \Pr(b_k = b_{max})$$

$$= \exp\left(-\frac{\eta E_{max}}{h\nu} \right) \Pr(b_k = b_{max}) \tag{13.17}$$

As a figure of merit for digital transmission systems a maximum error probability of 10^{-9} is often used. When this is substituted into equation (13.17) we get $E_{max} \geq 21 h\nu/\eta$. The maximum quantum efficiency equals unity, which means that the pulses received must consist of at least 21 photons on average in order to arrive at an error probability less than 10^{-9}. Since no photons are received during the bit intervals with $b_k = b_{min}$, it means that a bit must contain an average of 10.5 photons, assuming that the probabilities $\Pr(b_k = b_{min}) = \Pr(b_k = b_{max}) = 1/2$. The number of 21 photons per pulse, or 10.5 photons per bit is called the quantum limit. This is the lower absolute limit for the number of photons of isolated pulses found by direct detection, because such disturbances as background radiation, dark current and thermal noise have been neglected and the detector is assumed to be ideal ($\eta = 1$). In the preceding section we obtained a minimum of 36 photons per pulse. The difference between the two calculations stems from the fact that in the preceding section the shot noise was approximated by a Gaussian distribution, which is not good for noise based on so few photons.

Both in this section and the preceding one, the number of photons required per bit to give a bit error rate of 10^{-9} has been calculated. A fixed number of photons per bit means that the mean amount of received optical power depends on the bit rate. The mean optical power equals the mean energy per bit divided by the bit time. For the Poisson statistic derived in this section, this mean power becomes

$$P_p = 10.5 \frac{h\nu}{\eta T} = 10.5 \frac{hc}{\eta \lambda T} \tag{13.18}$$

whereas the Gaussian approximation of equation (13.13) makes the mean power

$$P_g = 18 \frac{hc}{\eta \lambda T} \tag{13.19}$$

When the thermal noise dominates with respect to the shot noise, it follows from equation (13.13) that the mean power thermal noise limit is

$$P_{th} = 6 \frac{h\nu}{\eta T} \frac{\sigma_{th}}{eGA} = 6 \frac{hc}{\eta \lambda eG} \sqrt{E[i_{th}^2]} \tag{13.20}$$

where i_{th} is the equivalent thermal noise current at the receiver input. The relationship between σ_{th}^2 and $E[i_{th}^2]$ is given by $E[i_{th}^2] = \sigma_{th}^2/H^2(0) = \sigma_{th}^2/(A^2 T^2)$. The quantum limits of equations (13.18) and (13.19) are inversely proportional to the bit time, which means that a larger bit rate requires more mean power. Moreover, a larger wavelength reduces the received optical power required. The thermal noise limit behaves in roughly the same way, except that the received power required increases more than proportional to the bit rate $1/T$, since an increase in the bit rate requires a larger receiver bandwidth, with the consquence that σ_{th} also increases (see Chapter 12).

13.4 The general receiver model

From now on, we drop the limiting assumptions made in Section 13.2, with one exception, namely the assumption that the filtered shot noise has a Gaussian distribution. Firstly, it is very difficult to determine the exact distribution of a marked and filtered Poisson process (see Appendix 4). Secondly, the average number of photons per pulse in practical situations is rather large, much larger than follows from the quantum limit, so that the use of a Gaussian distribution is acceptable. The Gaussian distribution has the advantage that it can be described uniquely with two parameters, namely its mean value and its variance.

In this section we shall use and combine the results of Poisson processes

(Appendix 4), thermal noise in amplifiers (Chapter 12) and the general sch of Figure 13.1 in order to describe the general receiver model with respect t signal to noise ratio at the time of sampling. The resulting model is depicte Figure 13.5, where the input circuit of the amplifier is the same as the one in Figure 12.13, except for the feedback

The input impedance of the amplifier is represented by the resistance R_a parallel to the capacitance C_a, whilst $i_a(t)$ and $e_a(t)$ are the current and voltage noise of the amplifier, respectively. In this figure $i_a(t)$ corresponds to one term of $i_1(t)$ of Figure 12.13 and $e_a(t)$ corresponds to $i_2(t)$ of the same figure, when $i_2(t)$ is transferred to the input. The thermal noise generated by R_b is given by $i_r(t)$ and is included in $i_1(t)$ of Figure 12.13. Now the expected value of the detector current $i(t)$ is given by (see Chapter 12, equation 12.1)

$$\mathrm{E}[i(t)] = \frac{e\eta G}{h\nu} p_s(t) + Ge\lambda_d \tag{13.21}$$

An equalizer is put after the amplifier in the model and this processes the signal received so that the input to the sampler is free of intersymbol interference (see [3, 5]). The output signal of the equalizer becomes (see also Figure 13.1)

$$\mathrm{E}[v_{out}(t)] = \left\{ eG\left[\frac{\eta}{h\nu} p_s(t) + \lambda_d\right]\right\} * h_i(t) * Ah_{eq}(t)$$
$$\triangleq \{eG[Kp_s(t) + \lambda_d]\} * h_o(t) \tag{13.22}$$

Here, the symbol $*$ denotes the convolution and $h_i(t)$ is the impulse response of the input circuit that converts the current $i(t)$ into the input voltage of an ideal

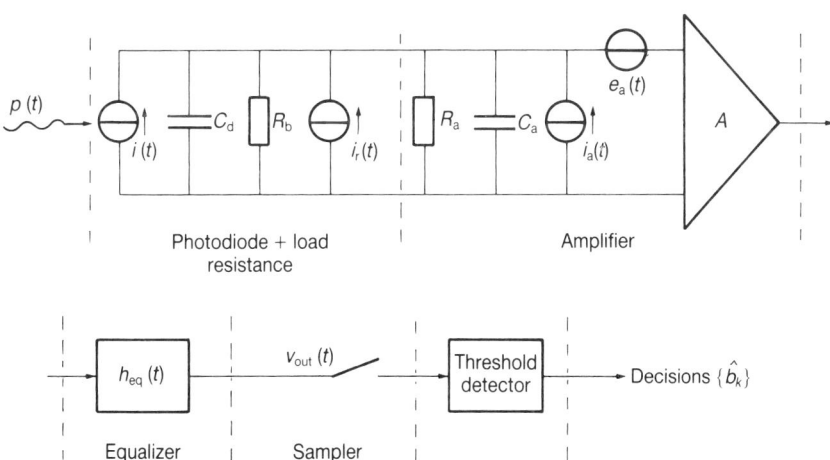

Figure 13.5 General model of an optical digital receiver, with details of its front end.

amplifier. This circuit consists of the parallel connection C_d (the detector capacitance), R_b, R_a and C_a, so that

$$h_i(t) \overset{\triangle}{=} \mathfrak{F}^{-1} \left\{ \frac{1}{\dfrac{1}{R_t} + j2\pi f(C_d + C_a)} \right\} \overset{\triangle}{=} \mathfrak{F}^{-1}\{H_i(f)\} \tag{13.23}$$

with $\mathfrak{F}^{-1}\{\cdot\}$ the symbolic notation for the inverse Fourier transform and

$$R_t = \frac{R_a R_b}{R_a + R_b} \tag{13.24}$$

The electronic amplification A is included in the impulse response for the entire receiver

$$h_o(t) \overset{\triangle}{=} h_i(t) * A h_{eq}(t) \tag{13.25}$$

As far as the expected value of the signal is concerned, an equivalent model for the entire optical fiber system can be deduced from equation (13.22). This model has been presented in Figure 13.6, from which it becomes clear that apart from a constant contribution of the dark current, $E[v_{out}(t)]$ has the following form:

$$E[v_{out}(t)] = \sum_k b_k s_{out}(t - kT) \tag{13.26}$$

This output signal is disturbed

$$v_{out}(t) = E[v_{out}(t)] + n(t) \tag{13.27}$$

where $n(t)$ represents all random deviations of $v_{out}(t)$ from its expected value $E[v_{out}(t)]$. The most important degrees of freedom in the system are those for R_b (the load resistance of the photodiode) and the impulse response $h_{eq}(t)$ of the equalizer. These degrees of freedom can be used to optimize the system in such a way that, based on comparison of the sample values $v_{out}(kT)$ with the threshold, optimal decisions with respect to the data sequence $\{\hat{b}_k\}$ can be made. As a criterion of optimality, the least probability of error is used.

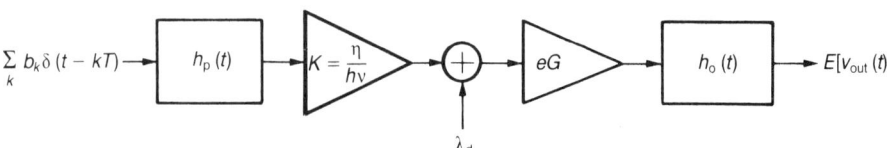

Figure 13.6 Equivalent model of the complete optical fiber communication system as far as the signal transfer is concerned.

13.4.1 The signal-to-noise ratio

We shall assume that the equalized pulses satisfy Nyquist's theorem [3], so that

$$s_{out}(lT) = \delta_l, \qquad (13.28)$$

where δ_l is the Kronecker delta function, resulting in

$$v_{out}(lT) = b_l + n(lT) \qquad (13.29)$$

The noise samples $n(lT)$ contain components that consist of randomly multiplied Poisson noise which originated from $i(t)$ (see Figure 13.5). Therefore, the samples $n(lT)$ depend, in principle on the entire sequence $\{b_k\}$, and it is this component that is responsible for the fact that an optical fiber communication system does not conform to the classical communication model, where the noise is assumed to be additive (and thus independent of the signal) and stationary. The Poisson noise or shot noise is filtered by the input circuit with an impulse response $h_i(t)$ and an equalizer with the impulse response $h_{eq}(t)$. If this part of the output noise is denoted by $n_s(t)$, then we can find the variance of this noise, in accordance with Appendix 4, equation (A4.57b),

$$\sigma_s^2(t) = E[g^2]e^2 \int_{-\infty}^{\infty} \left[K \sum_k b_k s_p(\tau - kT) + \lambda_d \right] h_0^2(t - \tau) \, d\tau \qquad (13.30)$$

As far as the signal-dependent shot noise variance is concerned, the optical fiber communication system can be represented by the equivalent model of Figure 13.7. Compare this figure with Figure 13.6 and note the differences and the similarities.

The total output noise variance also contains a signal-independent contribution, which is due to the thermal noise of the load resistance R_b and the equivalent current and voltage noise of the amplifier (see Figure 13.5). These thermal contributions are assumed to have the following properties: Gaussian probability density function, zero mean, and both an additive and a white power spectrum. In this way, we arrive at the following total output noise variance

$$\sigma_n^2(t) = \sigma_s^2(t) + \left(\frac{2k\Theta}{R_b} + S_i \right) \int_{-\infty}^{\infty} |H_0(f)|^2 \, df + S_e A^2 \int_{-\infty}^{\infty} |H_{eq}(f)|^2 \, df \qquad (13.31)$$

where

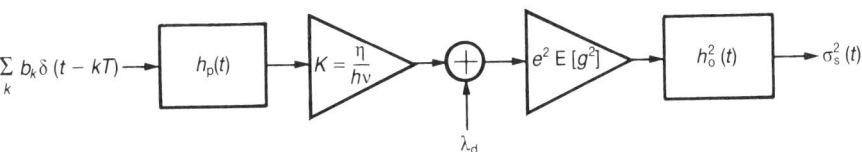

Figure 13.7 Equivalent model of the optical fiber communication system, as far as the shot noise variance is concerned.

k = Boltzmann's constant

Θ = absolute temperature in Kelvin

S_i = spectral density of $i_a(t)$

S_e = spectral density of $e_a(t)$

In order to obtain the expression in equation (13.31), it has been assumed that all the noise sources are mutually independent; this assumption is reasonable.

In equation (13.31) the terms independent of the signal are written in the frequency domain notation, whereas the signal-dependent term is described in the time domain. In order to achieve a uniform notation, equation (13.30) has to be transformed into the frequency domain. As far as the dark current term is concerned, this can be realized when using the formula of Parseval [7]. Apart from a constant factor, the first term of equation (13.30) is written as

$$\int_{-\infty}^{\infty} \sum_k b_k s_p(\tau - kT) h_0^2(t - \tau)\, d\tau$$

$$= \int_{-\infty}^{\infty} \sum_k b_k \left\{ \int_{-\infty}^{\infty} S_p(f) \exp\left[j2\pi f(\tau - kT) \right] df \right\} h_0^2(t - \tau)\, d\tau \qquad (13.32)$$

Since the sampling instance has not yet been defined, it can be taken as $t = 0$, without a loss of generality. As a consequence, it is sufficient to consider

$$\sum_k b_k \int_{-\infty}^{\infty} S_p(f) \exp\left(-j2\pi fkT \right) \left[\int_{-\infty}^{\infty} h_0^2(\tau) \exp\left(-j2\pi f\tau \right) d\tau \right] df$$

$$= \sum_k b_k \int_{-\infty}^{\infty} S_p(f) \exp\left(-j2\pi fkT \right) \left[H_0(f) * H_0(f) \right] df \qquad (13.33)$$

In order to satisfy equation (13.28) the signal spectrum $S_{out}(f)$ must have a particular shape, for instance a raised cosine [3, 5]. The input pulse shape $s_p(t)$ is determined by the transmitter and the fiber, so that $S_p(f)$ has a fixed shape. Therefore, in equations (13.31) and (13.33) we shall replace the function $H_0(f)$ by (see Figure 13.6)

$$H_0(f) = \frac{S_{out}(f)}{KeGS_p(f)} \qquad (13.34)$$

Finally, the total output noise variance at $t = 0$ reads

$$\sigma_n^2(0) = \frac{F}{K} \int_{-\infty}^{\infty} S_p(f) \sum_k b_k \exp\left(-j2\pi fkT \right) \left\{ \frac{S_{out}(f)}{S_p(f)} * \frac{S_{out}(f)}{S_p(f)} \right\} df$$

$$+ \frac{1}{K^2 e^2 G^2} \left(\frac{2k\Theta}{R_b} + S_i + e^2 E[g^2]\lambda_d \right) \int_{-\infty}^{\infty} \left| \frac{S_{out}(f)}{S_p(f)} \right|^2 df$$

$$+ \frac{1}{K^2 e^2 G^2} S_e \int_{-\infty}^{\infty} \left| \frac{S_{out}(f)}{S_p(f)} \left\{ \frac{1}{R_a} + \frac{1}{R_b} + j2\pi f(C_d + C_a) \right\} \right|^2 df \qquad (13.35)$$

Here F is the excess noise factor and G the average multiplication gain of the APD (see Chapter 12). Moreover, $H_{eq}(f)$ has been replaced by

$$H_{eq}(f) = \frac{H_0(f)}{AH_i(f)} = \frac{S_{out}(f)}{KeGS_p(f)} \frac{1}{AH_i(f)} \tag{13.36}$$

The error probability becomes minimal if $\sigma_n^2(0)$ has a minimum value. This minimum can be obtained by a correct choice of R_b and $H_{eq}(f)$. Furthermore, it should be remembered that $s_{out}(0)$ has been normalized to unity.

Based on equation (13.35), the following tentative conclusions can be drawn:

1. The load resistance R_b should be made as large as possible, independently of $S_{out}(f)$. Of course, there are practical limits to the choice of R_b.
2. The input resistance R_a of the amplifier should be as large as possible, whereas the amplifier input capacitance C_a should be as small as possible.
3. The parasitic capacitance of the photodiode C_d should be as small as possible.
4. The noise spectral densities of the amplifier S_e and S_i must be as small as possible.

For computing the signal-to-noise ratio, it must be kept in mind that the signal value at the output at time $t = lT$ equals b_l according to equation (13.29).

13.4.2 The equalized pulse shape

Starting from equation (13.35) for each $\{b_k\}$ an output pulse shape can be calculated which minimizes $\sigma_n^2(0)$, when the other system parameters are known. This approach has two disadvantages. Firstly, many different sequences $\{b_k\}$ are possible and each sequence leads to a different pulse shape. Since it is not known in advance what sequence will be received, it is impossible to choose the optimal shape. Secondly, there are other aspects than the noise and the absence of intersymbol interference (see equation 13.28). In a practical system sampling hardly ever takes place at the nominal times lT, but is subject to random fluctuations in time due to clock jitter. Therefore, at the sampling times which differ from the nominal times lT, the intersymbol interference must be kept within certain limits. In other words, the width of the eye opening is also important [3, 5]. Therefore, we shall not only look at equalized pulse shapes in order to minimize the noise, but at pulse shapes that lead to a guaranteed width of the eye opening and at the same time satisfy the Nyquist criterion given in equation (13.28).

In order to reduce the number of parameters involved and to make the result independent of the bit rate $1/T$, we shall normalize some of these parameters. For this purpose we define

$$\tilde{S}_p(f') \triangleq S_p(fT) \tag{13.37a}$$

$$\tilde{S}_{\text{out}}(f') \stackrel{\triangle}{=} \frac{1}{T} S_{\text{out}}(fT) \tag{13.37b}$$

with

$$f' \stackrel{\triangle}{=} fT \tag{13.38}$$

By the given choice of the normalized functions $\tilde{S}(f')$ and $\tilde{S}_{\text{out}}(f')$ we have

$$\int_{-\infty}^{\infty} \tilde{s}_p(t)\,dt = 1 \tag{13.39}$$

and

$$\tilde{s}_{\text{out}}(0) = 1 \tag{13.40}$$

similar to the unnormalized equivalents, equations (13.2) and (13.28). Because of these normalized parameters, we shall be able to compare systems for which $\tilde{S}_p(f')$ and $\tilde{S}_{\text{out}}(f')$ have the same shape, although with different bit rates. Note that the normalization corresponds to twice the value of the Nyquist rate [3, 5].

Well-known output pulse shapes that have a known width of the eye opening possess the so-called raised cosine characteristic [3, 5]. For these pulses the pulse shape in the time domain reads

$$s_{\text{out}}(t) = \frac{\sin\dfrac{\pi t}{T}\cos\beta\,\dfrac{\pi t}{T}}{\dfrac{\pi t}{T}\left[1 - \left(\dfrac{2\beta t}{T}\right)^2\right]} \tag{13.41}$$

and in the frequency domain

$$\tilde{S}_{\text{out}}(f') = \begin{cases} 1, & 0 \le |f'| < (1-\beta)/2 \\[2mm] \dfrac{1}{2}\left\{1 - \sin\left[\dfrac{\pi}{\beta}\left(|f'| - \dfrac{1}{2}\right)\right]\right\}, & (1-\beta)/2 \le |f'| \le (1+\beta)/2 \\[2mm] 0, & |f'| > (1+\beta)/2 \end{cases} \tag{13.42}$$

The normalized bandwidth of these pulses amounts to $(1+\beta)/2$ and both equations (13.41) and (13.42) have been drawn in Figure 13.8 for various values of β. In Figure 13.9 the eye openings [3] are shown, again for various values of β. The pulse shapes given by equations (13.41) and (13.42) combine the following properties:

1. They satisfy the Nyquist criterion for no intersymbol interference at the nominal sampling times.
2. They have a finite and well-defined bandwidth.
3. The width of the eye opening can be controlled by means of the roll-off factor β. Although values $\beta > 1$ may have some interesting properties [8], we shall restrict ourselves to $\beta \le 1$.

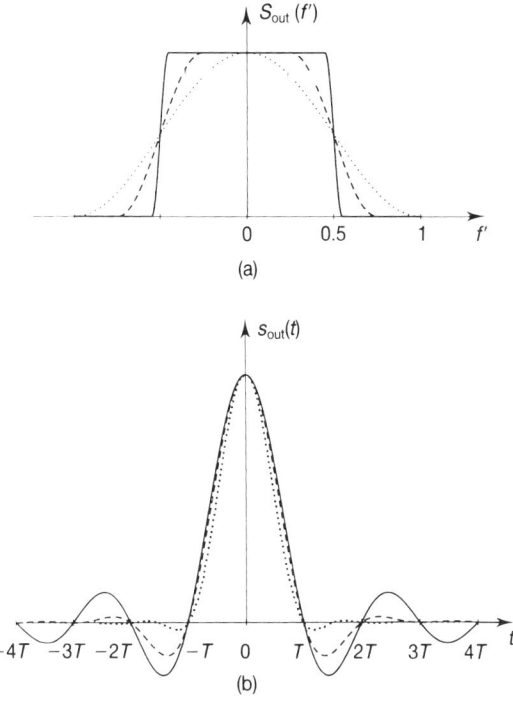

(a)

(b)

Figure 13.8 The raised cosine pulse for $\beta = 0.1$ (solid line), $\beta = 0.5$ (broken line) and $\beta = 1$ (dotted line) in: (a) the frequency domain; (b) the time domain.

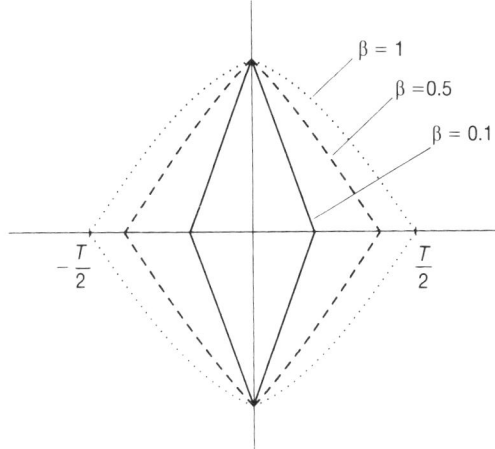

Figure 13.9 The eye opening of the raised cosine output pulse for the values of the roll-off factor β, as given in Figure 13.8.

As the optical input pulse of the receiver $s_p(t)$, we shall take a Gaussian pulse

$$s_p(t) = \frac{1}{\alpha T \sqrt{2\pi}} \exp\left(-\frac{t^2}{2\alpha^2 T^2}\right) \tag{13.43}$$

For such an input pulse and the raised cosine output pulse shape, we shall calculate the normalized shot noise variance function

$$z(t) \triangleq \tilde{s}_p(t) * \tilde{h}_0^2(t) \tag{13.44}$$

This function is vital as far as the shot noise contribution to the output is concerned (see equation 13.30). Figure 13.10 shows $z(t)/z(0)$ for various values of αT. Although $z(kT)$ for $k \neq 0$ is relatively small with respect to $z(0)$, the noise contribution consists of $z(kT)$ multiplied by b_k, so that it cannot be concluded that the shot noise originating from adjacent pulses is negligible. By means of Figure 13.10 the shot noise of each sequence $\{b_k\}$ can be computed. If the remaining noise terms, which are independent of $\{b_k\}$, are added to this outcome, then the probability that b_0 is detected in error can be calculated for each $\{b_k\}$ and a given threshold of the detector. Multiplying each probability by the probability that the given sequence $\{b_k\}$ occurs and summing these products for all the possible sequences, we arrive at the total error probability. We must then repeat the process for various values of the threshold, in order to optimize a receiver for this parameter; and similarly for the avalanche gain G. During these calculations we approximate the probability density function of the shot noise with a Gaussian function in which the variance is given by the shot noise variance function $z(t)$.

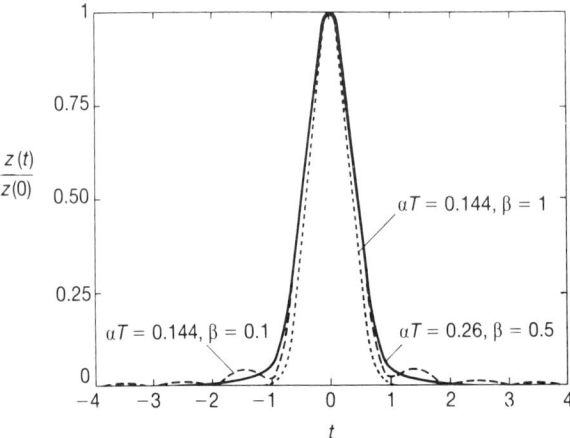

Figure 13.10 The normalized shot noise variance function for a raised cosine output pulse shape and a Gaussian input pulse.

13.4.3 The worst case noise

In order to compute the error probability exactly according to the method described in the preceding section, the statistics of the data source are needed; however, they are often not known when designing an optical receiver, or they may vary with time. Moreover, in practical situations, it is more important that the maximum acceptable error probability can be guaranteed under the most unfavourable conditions. The worst case noise follows from equation (13.35)

$$\sigma_w^2(b_0) \triangleq \max_{\{b_k, k \neq 0\}} \{\sigma_n^2(0)\}$$

$$= \frac{F}{K} \left\{ (b_0 - b_{max}) \int_{-\infty}^{\infty} S_p(f) \left[\frac{S_{out}(f)}{S_p(f)} * \frac{S_{out}(f)}{S_p(f)} \right] df \right.$$

$$\left. + b_{max} \int_{-\infty}^{\infty} \sum_{k=-\infty}^{\infty} \exp(-j2\pi fkT) S_p(f) \left[\frac{S_{out}(f)}{S_p(f)} * \frac{S_{out}(f)}{S_p(f)} \right] df \right\}$$

$$+ \frac{1}{K^2 e^2 G^2} \left(\frac{2k\Theta}{R_b} + S_i + e^2 E[g^2]\lambda_d \right) \int_{-\infty}^{\infty} \left| \frac{S_{out}(f)}{S_p(f)} \right|^2 df$$

$$+ \frac{1}{K^2 e^2 G^2} S_e \int_{-\infty}^{\infty} \left| \frac{S_{out}(f)}{S_p(f)} \left(\frac{1}{R_t} + j2\pi fC_t \right) \right|^2 df \qquad (13.45)$$

with

$$C_t \triangleq C_a + C_d \qquad (13.46)$$

If it is kept in mind that [7]

$$\sum_{k=-\infty}^{\infty} \exp(-j2\pi fkT) = \frac{1}{T} \sum_{k=-\infty}^{\infty} \delta\left(f - \frac{k}{T} \right) \qquad (13.47)$$

and using the normalized function of equation (13.37)

$$\sigma_w^2(b_0) = \frac{F}{K} \left[(b_0 - b_{max})I_1 + b_{max}\Sigma_1 \right]$$

$$+ \frac{T}{K^2 e^2 G^2} \left(\frac{2k\Theta}{R_b} + S_i + e^2 E[g^2]\lambda_d + \frac{S_e}{R_t^2} \right) I_2$$

$$+ \frac{1}{TK^2 e^2 G^2} (2\pi C_t)^2 S_e I_3 \qquad (13.48)$$

In this equation the factors I_1, Σ_1, I_2 and I_3 are defined as

$$I_1 \triangleq \int_{-\infty}^{\infty} \tilde{S}_p(f') \left[\frac{\tilde{S}_{out}(f')}{\tilde{S}_p(f')} * \frac{\tilde{S}_{out}(f')}{\tilde{S}_p(f')} \right] df' \qquad (13.49)$$

$$\Sigma_1 \triangleq \sum_{k=-\infty}^{\infty} \tilde{S}_p(k) \left[\frac{\tilde{S}_{out}(k)}{\tilde{S}_p(k)} * \frac{\tilde{S}_{out}(k)}{\tilde{S}_p(k)} \right] \qquad (13.50)$$

$$I_2 \overset{\triangle}{=} \int_{-\infty}^{\infty} \left| \frac{\tilde{S}_{\text{out}}(f')}{\tilde{S}_{\text{p}}(f')} \right|^2 df' \tag{13.51}$$

$$I_3 \overset{\triangle}{=} \int_{-\infty}^{\infty} \left| \frac{\tilde{S}_{\text{out}}(f')}{\tilde{S}_{\text{p}}(f')} \right|^2 f'^2 df' \tag{13.52}$$

In equation (13.48) the term containing $b_0 I_1$ represents the shot noise due to the pulse which is to be detected, whereas the term involving $b_{\max}(\Sigma_1 - I_1)$ represents the shot noise due to adjacent pulses. In [1] the curves shown refer to I_1, I_2, I_3 and Σ_1 for various combinations of the input pulse shape $\tilde{S}_{\text{p}}(f')$ and the output pulse $\tilde{S}_{\text{out}}(f')$. Apart from the term containing λ_{d}, the remaining terms belong to the thermal noise.

From equation (13.48) it follows that the noise variance at first decreases with an increased bit rate $1/T$ for fixed shapes of $s_{\text{out}}(t)$ and $s_{\text{p}}(t)$ and no change of the other parameters, until the point where the term involving I_3 becomes dominant. Beyond that point the noise variance increases with a rising bit rate, due to the capacitance C_{t}.

In order to determine the error probability we shall assume that the output voltage at the sampling times is a Gaussian random variable. Similar to what has been argued in Section 13.2, we will calculate the conditional probabilities $\Pr(\hat{b}_0 \neq b_0 | b_0 = b_{\min})$ and $\Pr(\hat{b}_0 \neq b_0 | b_0 = b_{\max})$. Eliminating the threshold (see Equation 13.9) produces the same equation as in equation (13.11)

$$q = \left(\frac{S}{N} \right)_w = \frac{b_{\max} - b_{\min}}{\sigma_{\text{w}}(b_{\max}) + \sigma_{\text{w}}(b_{\min})} \tag{13.53}$$

When the light source consists of an LED usually $b_{\min} = 0$, whereas for an LD the value of $b_{\min}$ depends on the threshold of the laser (see Chapters 8 and 9); for a modern laser this can lead to very small values of $b_{\min}$. For any given total error probability and $b_{\min}$ value, the value of $b_{\max}$ is determined with equation (13.53). Since this equation contains the worst-case values of the noise, the resulting error probability is smaller than the given one.

Let us now suppose that the dark current is negligible and that $b_{\min} = 0$. Then equation (13.48) changes into

$$\sigma_{\text{w}}^2(b_0) = \frac{F}{K} \left[(b_0 - b_{\max}) I_1 + b_{\max} \Sigma_1 \right] + \frac{1}{G^2} Z \tag{13.54}$$

with

$$Z \overset{\triangle}{=} \frac{T}{K^2 e^2} \left(\frac{2k\Theta}{R_{\text{b}}} + S_{\text{i}} + \frac{S_{\text{e}}}{R_{\text{t}}^2} \right) I_2 + \frac{1}{TK^2 e^2} (2\pi C_{\text{t}})^2 S_{\text{e}} I_3 \tag{13.55}$$

In equation (13.54) the term Z/G^2 represents all the thermal noise contributions. Let us consider the case where the thermal noise is dominant. Then the avalanche gain is small and we can take $F \approx 1$, so that

$$\sigma_w^2(b_0) = \frac{1}{G^2} Z$$

And by means of equations (13.53) and (13.34) it follows that

$$b_{max} = \frac{2q}{G} \sqrt{Z} \qquad (13.57)$$

The last equation is similar to the second term of equation (13.13) for the simplified model. Moreover, the mean optical power gives a similar equation to equation (13.20), namely

$$P_{th} = \frac{6}{GT} \sqrt{Z} \qquad (13.58)$$

for a bit error rate of 10^{-9}. When the shot noise cannot be neglected, the situation becomes very complicated. It appears that an optimal value of G exists, which minimizes b_{max} at a certain prescribed error probability. This optimum value of G is found by solving an eight-order equation numerically in terms of G [9].

13.4.4 Conclusions

In order to obtain an error probability as small as possible at a given optical power, or to keep the received optical power as small as possible at a prescribed bit error rate, the impedance of the input circuit should be as high as possible. Saying the same thing in another way: the load resistance of the photodiode and the input resistance of the amplifier must be as large as possible and the parasitic capacitance of the photodiode and the input capacitance of the amplifier must be as small as possible. Using a PIN diode the thermal noise usually gives the greatest noise contribution, whereas using an APD both the shot noise and the thermal noise have to be taken into account and an optimal value for the mean avalanche gain exists. The signal-dependent shot noise does not depend on the bit rate $(1/T)$, but the shot noise stemming from the dark current decreases inverse proportionally to the bit rate. When the voltage noise of the amplifier dominates the thermal noise, then the thermal noise increases proportionally to the bit rate. As a consequence, considerable equalization (large Σ_1) causes the shot noise from neighbouring pulses to be large, so that the energy per pulse has to be increased for the same error probability. Such an equalization, which is required for higher bit rates, pays the penalty of reduced performance compared to systems not requiring equalization.

It is emphasized that the raised cosine pulse is not an optimal pulse shape as far as the error probability is concerned. An optimal pulse shape, with respect to the criterion minimal bit error rate, does not, in general, satisfy Nyquist's criterion exactly for zero intersymbol interference. In practice, however, the deviations are always very small, so that the raised cosine is a good compromise, especially when the width of the eye opening is also taken into account.

References

[1] S. D. Personick, 'Receiver design for digital fiber optic communication systems, I', *Bell System Technical Journal*, vol. 52, no. 6, July–August 1973, pp. 843–74.

[2] G. L. Cariolaro, 'Error probability in digital fiber optic communication systems', *IEEE Transactions on Information Theory*, vol. IT-24, no. 2, March 1978, pp. 213–21.

[3] R. W. Lucky, J. Salz and E. J. Weldon, Jr, *Principles of Data Communication* (New York: McGraw-Hill, 1968).

[4] J. W. Wozencraft and I. M. Jacobs, *Principles of Communication Engineering* (New York: Wiley, 1965).

[5] A. B. Carlson, *Communication Systems*, 3rd edn (New York: McGraw-Hill, 1986).

[6] A. Papoulis, *Probability, Random Variables and Stochastic Processes*, 2nd edn (New York: McGraw-Hill, 1984).

[7] A. Papoulis, *The Fourier Integral and its Applications* (New York: McGraw-Hill, 1962).

[8] A. M. J. Koonen, 'Error probability in digital fiber optic communication systems', Eindhoven: University of Technology, TH-Report 79-E-99 1979.

[9] Y. Takasaki and M. Maeda, 'Receiver design for fiber optic communications optimization in terms of excess noise factors that depend on avalanche gains', *IEEE Transactions on Communications*, vol. COM-24, no. 12, December 1976, pp. 1343–6.

Problems

13.1 A digital binary optical transmission system operates at a bit rate of 1 Gb/s and at a wavelength of 1300 nm. An APD is used as the detector ($k = 0.1$, $G = 10$, $R = 6.5 \, \mathrm{AW}^{-1}$).

The length of the fiber is 25 km and the fiber has an attenuation of $0.3 \, \mathrm{dB \, km}^{-1}$. The coupling efficiency from laser source to fiber is 1 percent, whereas the coupling from fiber to detector is 100 percent. The output power from the laser is 0.5 mW for a '0' and 1 mW for a '1'. The prior probabilities for sending a '0' or a '1' are equal and independent. The thermal noise of the receiver is given by $\sigma_{\mathrm{th}}/(eA) = 3500$.

(a) Calculate the mean received optical power.
(b) Calculate the bit error of the system, when the detection threshold is chosen midway between the '1'-level and the '0'-level at the input of the sampler.

NB: The $Q(\cdot)$ function can be approximated by $Q(x) \approx \exp(-x^2/2)/(x\sqrt{2\pi})$.

13.2 A digital binary optical transmission system operates at a wavelength of 1300 nm and a bit rate of 140 Mbit s^{-1}. The mean optical power received amounts to 10 nW. The prior probabilities of sending a '1' or a '0' are independent and equal. For a '0' $E_{\min} = 0$. The thermal noise of the receiver is given to be $\sigma_{\mathrm{th}}/(eA) = 1500$. The data of the photodiode are: $\eta = 0.8$, $G = 10$, $F = 5$.

(a) What is the error probability, when the threshold is chosen so that the error probability is independent of the bit pattern?
(b) Calculate the bit error probability when the threshold is chosen midway between the '1'-level and the '0'-level at the input of the sampler.

NB: Use the same approximation for the Q function as in Problem 13.1.

13.3 Consider a receiver for digital binary optical signals. The thermal noise is negligible. The extinction ratio (the ratio of the energy content of a '0' and a '1') amounts to 0.1. The shot noise can be approximated by a Gaussian probability density function.

(a) How many photons should be contained in a '1' and a '0', respectively, in order to achieve a bit error probability of 10^{-9}?

(b) How many decibels should the mean bit energy be larger than the quantum limit?

13.4 A digital optical receiver has a sensitivity of -53 dBm at a bit rate of 140 Mbit s^{-1} and a wavelength of 850 nm.

(a) What is the difference (in dB) between this sensitivity and the quantum limit?

(b) How many photons per bit are required in order to guarantee that the receiver operates at a bit error probability of less than 10^{-9}?

(c) Answer questions (a) and (b) for a wavelength of 1300 nm.

13.5 A digital binary optical communication system operates at a bit rate of 140 Mbit s^{-1}. A PIN photodiode (area 1 mm^2) is used as the detector and this detector has the responsivity characteristic given in Problem 12.2. Background radiation with a constant spectral density of 0.2 W m^{-2} μm^{-1}, over the relevant spectral range, disturbs the receiver. In order to reduce this disturbance, the background radiation is attenuated by a neutral density filter.

How many decibels should the background radiation be attenuated in order to make its noise as small as the shot noise at the quantum limit?

14

System noise

14.1 Intensity noise of the light source
14.2 Competition noise
14.3 Partition noise
14.4 Modal noise
14.5 The signal-to-noise ratio due to system noise and receiver noise
References
Problems

The noise considerations presented in Chapters 12 and 13 only referred to noise produced in a receiver, namely the shot noise that is produced in the photo-detector and the thermal noise of the receiver amplifier. In an optical fiber communication system, however, more noise sources are present. These noise sources originate from other parts of the system or from combinations of the parts, as will become clear shortly. In certain systems, the system noise is not important, but in others it determines the system's performance. Since larger signal-to-noise ratio values are required for analog systems than for digital systems, special attention should be paid to the system noise in analog systems.

14.1 Intensity noise of the light source

Let us confine ourselves to semiconductor light sources and start by considering the LED. Due to the forward current, electrons are injected into the conduction band. When, after a random time, an injected electron recombines with a hole that has been injected into the valence band, a photon is generated (see Chapter 8). In an LED this is spontaneous recombination and the number of recombinations varies within a certain length of time. In this way the intensity of the emitted light fluctuates. Moreover, not all the photons have the same energy; in Chapter 8 we saw that the standard deviation of the photon energy is about $2k\theta$. The stochastic process that describes the LED is modelled as a Gaussian bandpass process and was given by equation (8.4). The power emitted by the LED is determined by the square of the envelope of the field and this can be deduced from the complex envelope given in equation (8.5). The random fluctuations of the emitted power of an LED is called beat noise. It will be clear that such fluctuations contribute to the noise at the receiving end of the system. The signal-to-noise ratio due to the beat noise is given by

$$\frac{S}{N} = \frac{B_0}{2B} \frac{1}{V} \tag{14.1}$$

Here B_0 is the bandwidth of the optical spectrum of the source, B is the electrical bandwidth of the receiver and V is a factor that depends on the shape of the optical spectrum and the shape of the electrical transfer function of the receiver. We shall not give the proof of this equation as it is too tedious, but in [1] the special case where the optical spectrum is an ideal rectangular bandpass characteristic and the receiver is an ideal rectangular low-pass characteristic is elaborated. Then the factor V becomes

$$V = \frac{1}{1 - (B/B_0)} \tag{14.2}$$

By means of the treatment in Chapter 8 it is easily found that $B_0 \gg B$ for an LED. Substituting this in equation (14.2) gives $V \approx 1$ and this value can be taken also for non-rectangular band limitations. From equation (14.1) it then follows that S/N becomes very large and we can conclude that with an LED as the light source, this noise source is not important.

A semiconductor laser also emits a small amount of incoherent LED light, and for this incoherent part the same conditions that were reached for the LED apply. In Section 9.7, however, we saw that the amplitude of the coherent part of the laser light also contains amplitude noise. For direct detection systems this noise does not play a significant role, but for coherent systems it can be the limiting factor. In Chapter 16 methods are explained for suppressing this amplitude noise in coherent detection systems.

14.2 Competition noise

This kind of noise occurs when a laser that oscillates in different lateral modes is used [2]. Let us again restrict ourselves to semiconductor lasers. A certain photon, which is generated by the recombination of an electron in the conduction band and a hole in the valence band, contributes to a certain mode, whereas another photon can contribute to another mode. The different laser modes are in competition for any emitted photon. At a fixed value of the laser diode current, the total emitted optical power of the laser is constant, but due to the competitive effect the power of each individual mode fluctuates. As long as all the optical power is coupled into the fiber this effect does not make any contribution to the noise. Nor is there any noise if all the modes are equally attenuated. The various lateral modes show different radiation patterns (see Chapter 8 and [3]). Spatial filtering, which can occur when the laser light is coupled into a fiber, causes selective mode attenuation and competition noise. It is therefore important to use lasers that emit only in the fundamental lateral and transverse mode: the laser then emits a Gaussian beam (see Appendix 3).

14.3 Partition noise

Partition noise originates from the fact that the power in the individual laser modes varies, notwithstanding that the total optical power of the laser is constant. Now, however, the longitudinal modes are involved. In Chapter 8 it has been shown that each longitudinal mode corresponds to a line in the optical spectrum. Partition noise is caused by the fact that the fiber has different delay times for the various waves originating from the corresponding spectral lines. This differential delay originates, in its turn, from the chromatic dispersion of the fiber material. Analogously to equation (10.48), the fiber output signal detected is written as

$$m_0(t) = \sum_{k=1}^{N} a_k \, m(t - \tau_k) \triangleq \sum_{k=1}^{N} a_k \, m_k(t) \tag{14.3}$$

where a_k is the power received from the kth spectral line (or laser mode). The coefficients a_k are stochastic variables (due to the random mode partitioning) and thus $m_0(t)$ is a stochastic process too.

For an analysis of the partition noise we consider digital transmission and use the following assumptions [4]:

1. Although the amplitudes a_k of the various modes can differ from pulse to pulse, the total optical power is constant and is assumed to be unity

$$\sum_{k=1}^{N} a_k = 1 \tag{14.4}$$

2. The optical laser spectrum has a Gaussian line shape

$$\bar{a}_k = \int a_k \, p(a_1, \ldots, a_N) \, da_1 \ldots da_N$$

$$= \frac{\exp\left[-(\lambda_k - \lambda_c)^2/2\sigma^2\right]}{\sum_{k=1}^{N} \exp\left[-(\lambda_k - \lambda_c)^2/2\sigma^2\right]} - S(\lambda) \, \delta(\lambda - \lambda_k) \tag{14.5}$$

In this equation $p(a_1, \ldots, a_N)$ is the N-dimensional joint probability density function of the set $\{a_k\}$, while λ_k is the wavelength of the kth spectral line; finally $\sum_k S(\lambda) \, \delta(\lambda - \lambda_k)$ is the laser's optical spectrum and σ is the half root mean square (r.m.s.) width of the envelope of this spectrum.

3. The pulses received consist of raised cosines in the time domain

$$s_p(t) = \cos\frac{\pi t}{T_0}, \qquad |t| < T_0 \tag{14.6}$$

with $T_0 < T$ ($1/T$ is the bit rate).

None of these assumptions means any loss of generality for the method described, but they will serve to simplify the calculations.

The differential delay between the kth laser mode with wavelength λ_k and the center mode with wavelength λ_c is denoted by $\Delta\tau_k$. Let us suppose that the pulses received are nominally sampled at the peak of the waveform described by equation (14.6), if the pulse is transmitted by the center mode. The amplitude fluctuation Δ_k at this nominal sampling moment, when the pulse is shifted in time due to transmission by laser mode k, becomes (see Figure 14.1)

$$\Delta_k = \left[1 - \cos\left(\pi\frac{\Delta\tau_k}{T_0}\right)\right] a_k \approx \frac{1}{2}\left(\pi\frac{\Delta\tau_k}{T_0}\right)^2 a_k, \qquad \text{for } \Delta\tau_k \ll \tau_0 \tag{14.7}$$

The total amplitude fluctuation with respect to the nominal pulse is given by

$$\Delta = \sum_k \Delta_k = \frac{1}{2}\left(\frac{\pi}{T_0}\right)^2 \sum_k a_k \, (\Delta\tau_k)^2 \tag{14.8}$$

From this expression the noise power due to this fluctuation is found to be

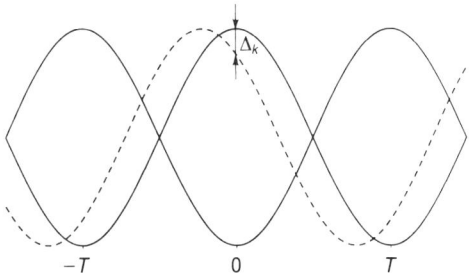

Figure 14.1 The signalling waveform $s_p(t)$ transmitted by the center wavelength (solid line) and by laser mode k (broken line).

$$\sigma^2{}_{pn} = \overline{\Delta^2} - \bar{\Delta}^2 \tag{14.9}$$

where $\sigma^2{}_{pn}$ is the average partition noise power. The values of $\overline{\sigma^2}$ and $\bar{\sigma}^2$ are calculated from equation (14.5). For the mean value of Δ it follows that

$$\bar{\Delta} = \int \Delta \, p(a_1, \ldots, a_N) \, \mathrm{d}a_1 \ldots \mathrm{d}a_N$$

$$= \frac{1}{2}\left(\frac{\pi}{T_0}\right)^2 \sum_k (\Delta\tau_k)^2 \int a_k \, p(a_1, \ldots, a_N) \, \mathrm{d}a_1 \ldots \mathrm{d}a_N$$

$$= \frac{1}{2}\left(\frac{\pi}{T_0}\right)^2 \sum_k (\Delta\tau_k)^2 \, a_k S(\lambda) \, \delta(\lambda - \lambda_k) \tag{14.10}$$

whereas for the mean squared value of Δ we get

$$\overline{\Delta^2} = \int \Delta^2 \, p(a_1, \ldots, a_N) \, \mathrm{d}a_1 \ldots \mathrm{d}a_N$$

$$= \frac{1}{4}\left(\frac{\pi}{T_0}\right)^4 \int \left[\sum_k (\Delta\tau_k)^2 a_k\right]^2 p(a_1, \ldots, a_N) \, \mathrm{d}a_1 \ldots \mathrm{d}a_N$$

$$= \frac{1}{4}\left(\frac{\pi}{T_0}\right)^4 \left[\sum_k (\Delta\tau_k)^4 \int a_k^2 \, p(a_1, \ldots, a_N) \, \mathrm{d}a_1 \ldots \mathrm{d}a_N\right.$$

$$\left. + \sum_k \sum_{l \neq k} (\Delta\tau_k)^2 (\Delta\tau_l)^2 \int a_k a_l p(a_1, \ldots, a_N) \, \mathrm{d}a_1 \ldots \mathrm{d}a_N\right] \tag{14.11}$$

In order to be able to calculate the mean value of Δ it is sufficient to know the laser spectrum (see equation 14.10); however, to calculate $\overline{\Delta^2}$ we need to know the joint probability density function $p(a_1, \ldots, a_N)$, which is generally unknown and differs from one laser diode to another. By means of substituting $a^2{}_k = a_k(1 - \Sigma_{l \neq k} a_l)$ we get

$$\overline{\Delta^2} = \frac{1}{4}\left(\frac{\pi}{T_0}\right)^4 \left[\sum_k (\Delta\tau_k)^4 \int a_k \, p(a_1, \ldots, a_N) \, \mathrm{d}a_1 \ldots \mathrm{d}a_N\right.$$

$$\left. - \sum_k \sum_{l \neq k} \{(\Delta\tau_k)^4 - (\Delta\tau_k)^2(\Delta\tau_l)^2\} \int a_k a_l \, p(a_1, \ldots, a_N) \, \mathrm{d}a_1 \ldots \mathrm{d}a_N\right]$$

$$= \frac{1}{4}\left(\frac{\pi}{T_0}\right)^4 \left[\sum_k (\Delta\tau_k)^4 \int a_k \, p(a_1, \ldots, a_N) \, \mathrm{d}a_1 \ldots \mathrm{d}a_N\right.$$

$$\left. - \sum_k \sum_{l > k} \{(\Delta\tau_k)^2 - (\Delta\tau_l)^2\}^2 \int a_k a_l \, p(a_1, \ldots, a_N) \, \mathrm{d}a_1 \ldots \mathrm{d}a_N\right]$$

$$\leq \frac{1}{4}\left(\frac{\pi}{T_0}\right)^2 \sum_k (\Delta\tau_k)^4 \int a_k \, p(a_1, \ldots, a_N) \, \mathrm{d}a_1 \ldots \mathrm{d}a_N$$

$$= \frac{1}{4}\left(\frac{\pi}{T_0}\right)^2 \sum_k (\Delta\tau_k)^4 S(\lambda) \, \delta(\lambda - \lambda_k) \tag{14.12}$$

From equation (14.12) it is seen that $\overline{\Delta^2}$, and thus σ^2_{pn}, becomes maximal if

$$\int a_k a_l \, p(a_1, \ldots, a_N) \, \mathrm{d}a_1 \ldots \mathrm{d}a_N = 0, \qquad \text{for } k \neq l \tag{14.13}$$

This condition can be physically interpreted as the situation where the laser modes within one pulse are mutually exclusive, i.e. when each data pulse consists of a single dominant longitudinal laser mode, whose wavelength may vary from pulse to pulse. For the time being, further calculations are performed using the last expression of equation (14.12). This means that it is enough to know the laser's power spectrum, while the joint probability density function $p(a_1, \ldots, a_N)$ is not needed.

The signal-to-partition-noise ratio is defined as the mean square of the signal value divided by the variance of the partition noise

$$\frac{S}{N} = \frac{(1 - \overline{\Delta})^2}{\sigma^2_{\text{pn}}} \approx \frac{1}{\sigma^2_{\text{pn}}} \tag{14.14}$$

where the approximation applies to large signal-to-noise ratios, while σ^2_{pn} follows from equations (14.9), (14.10) and (14.12). In order to simplify the calculations, the spectrum of the laser is best described by a continuous Gaussian function, rather than a discrete one

$$S(\lambda) = \frac{1}{\sigma \sqrt{2\pi}} \exp\left[\frac{-(\lambda - \lambda_c)^2}{2\sigma^2} \right] \tag{14.15}$$

As in Chapters 4 and 10 the specific phase shift of the fiber is expanded into a Taylor series about $\omega_c \triangleq 2\pi c/\lambda_c$, which leads to

$$\Delta \tau_k = \beta''(\omega_k - \omega_c)l + \tfrac{1}{2}\beta'''(\omega_k - \omega_c)^2 l \tag{14.16}$$

where l is the length of the fiber, and β'' and β''' are respectively the second and third derivatives of β with respect to ω at the point where $\omega = \omega_c$. By means of the relationship

$$\beta'' \triangleq \frac{\mathrm{d}^2\beta}{\mathrm{d}\omega^2} = \frac{-\lambda^2}{2\pi c} \frac{\mathrm{d}\tau_g}{\mathrm{d}\lambda} \tag{14.17}$$

and

$$\beta''' \triangleq \frac{\mathrm{d}^3\beta}{\mathrm{d}\omega^3} = \frac{\lambda^4}{(2\pi c)^2} \frac{\mathrm{d}^2\tau_g}{\mathrm{d}\lambda^2} + \frac{2\lambda^3}{(2\pi c)^2} \frac{\mathrm{d}\tau_g}{\mathrm{d}\lambda} \tag{14.18}$$

with τ_g the specific group delay of the fiber (see equation 4.8), the values of β'' and β''' can be determined using Figures 5.13, 5.16 and 5.17, or the results of Chapter 7.

Using equations (14.9), (14.10), (14.12) and (14.15) and replacing the summations by integrations, we arrive at the variance of the partition noise

$$\sigma^2_{\,\text{pn}} = \frac{1}{2}\left(\frac{\pi}{T_0}\right)^4\left\{A_1^4\sigma^4 + 48A_2^4\sigma^8 + \left(42 - \frac{16}{\pi}\right)A_1^2A_2^2\sigma^6\right.$$

$$\left. + 12\sqrt{\frac{2}{\pi}}\,A_1^3A_2\sigma^5 + 84\sqrt{\frac{2}{\pi}}\,A_1A_2^3\sigma^7\right\} \tag{14.19}$$

where

$$A_1 \triangleq \frac{2\pi c}{\lambda_c^2}\,\beta''l \tag{14.20}$$

$$A_2 \triangleq \frac{1}{2}\left(\frac{2\pi c}{\lambda_c^2}\right)^2\beta'''l \tag{14.21}$$

A very important conclusion can now be drawn from equations (14.14) and (14.19): the signal-to-partition-noise ratio depends on the width σ of the laser spectrum and the dispersion parameters A_1 and A_2 of the fiber, but it is independent of the signal power. This means that the signal-to-noise ratio cannot be improved by increasing the transmitting power, as in the case of the receiver noise. This property is specific to system noise. The signal-to-noise ratio decreases when the bit rate increases.

For systems operating at the wavelength of minimum dispersion, the parameter A_1 vanishes and this simplifies the expression for the signal-to-noise ratio to

$$\frac{S}{N} \approx \frac{1}{24(\pi/T_0)^4 A_2^4\sigma^8} \tag{14.22}$$

However, when the system operates at a wavelength that differs greatly from the minimum dispersion wavelength, the signal-to-noise ratio is approximated by

$$\frac{S}{N} \approx \frac{1}{\frac{1}{2}(\pi/T_0)^4 A_1^4\sigma^4} \tag{14.23}$$

since, in this case, the dispersion is dominated by the second derivative of the phase shift.

We shall now drop the assumption in equation (14.13) that shows that the laser modes within each pulse are mutually exclusive. For that purpose we consider the variance of the received signal, given by equation (14.3)

$$\sigma^2_{\,\text{pn}} = \overline{m_0^{\,2}(t)} - \overline{m_0(t)}^2 = \sum_i\sum_j m_i(t)m_j(t)\overline{a_ia_j} - \sum_i\sum_j m_i(t)m_j(t)\bar{a}_i\bar{a}_j \tag{14.24}$$

Using the relationship $a_i^2 = a_i(1 - \Sigma_{i\neq j}a_j)$ this equation can also be written as

$$\sigma^2_{\,\text{pn}} = \sum_i\sum_{j>i}[m_i(t) - m_j(t)]^2(\bar{a}_i\bar{a}_j - \overline{a_ia_j})$$

$$= k^2\sum_i\sum_{j>i}[m_i(t) - m_j(t)]^2\bar{a}_i\bar{a}_j \tag{14.25}$$

with

$$k^2 \triangleq \frac{\Sigma_i \Sigma_{j>i} [m_i(t) - m_j(t)]^2 (\bar{a}_i \bar{a}_j - \overline{a_i a_j})}{\Sigma_i \Sigma_{j>i} [m_i(t) - m_j(t)]^2 \, \bar{a}_i \bar{a}_j} \tag{14.26}$$

Simplifying this expression further, we can establish that the cross-correlation $\overline{a_i a_j}$ refers to the corresponding mean values $\bar{a}_i$ and $\bar{a}_j$ in a laser. Some of the stimulated photons in a pulse contribute to the same mode by which these photons have been stimulated. They have the property $\overline{a_i a_j} = 0$. Another part of the emission, however, consists of photons that have been stimulated by a mode that differs from that mode to which the photon contributes. This process consists of generalized Bernoulli trials. If we assume that the probability that such a photon contributes to the mode i equals $\bar{a}_i$, then for this part of the emission $\overline{a_i a_j} = \bar{a}_i \bar{a}_j$ and thus for the total emission $\overline{a_i a_j} = \alpha \bar{a}_i \bar{a}_j \ (i \neq j)$, with α being a constant that is independent of i and j [5]. If this outcome is substituted in equation (14.26) we get

$$k^2 = \frac{\Sigma_i \Sigma_{j>i} (\bar{a}_i \bar{a}_j - \overline{a_i a_j})}{\Sigma_i \Sigma_{j>i} \bar{a}_i \bar{a}_j} = \frac{\bar{a}_i \bar{a}_j - \overline{a_i a_j}}{\bar{a}_i \bar{a}_j}, \qquad i \neq j \tag{14.27}$$

By using

$$\sum_i \sum_{j>i} \bar{a}_i \bar{a}_j = \frac{1}{2} \left(\sum_i \sum_j \bar{a}_i \bar{a}_j - \sum_i \bar{a}_i^2 \right) \tag{14.28}$$

$$\sum_i \sum_{j>i} \overline{a_i a_j} = \frac{1}{2} \left(\sum_i \sum_j \overline{a_i a_j} - \sum_i \overline{a_i^2} \right) \tag{14.29}$$

and $\Sigma_i \bar{a}_i = 1$ and $\overline{(\Sigma_i a_i)^2} = 1$, which arise from equation (14.4), giving

$$k^2 = \frac{\overline{a_i^2} - \bar{a}_i^2}{\bar{a}_i - \bar{a}_i^2}, \qquad \text{for any } i \tag{14.30}$$

This result has an important practical value, namely it can be seen that the value of k can be derived from the measurement of a single longitudinal laser mode [6]. The numerator of equation (14.30) represents the variance of the ith mode, which is measured by an r.m.s. voltmeter, while the denominator is derived from the mean value, that can be measured by a DC meter. The mode partition coefficient k may vary from zero ($\overline{a_i a_j} = \bar{a}_i \bar{a}_j$) to unity ($\overline{a_i a_j} = 0$), while k^2 is the factor with which the partition noise σ^2_{pn} is reduced with respect to the worst case, given by equation (14.19), or approximated by equations (14.22) and (14.23). The case $k = 1$ was dealt with earlier in this section and $k = 0$ is interpreted as the absence of partition noise. The measured k values are in the range 0.7–0.14 [6], indicating that mode partition noise can be one of the main performance restrictions.

It becomes clear that the partition noise, like the competition noise, vanishes if either a single-mode laser or an LED is used as the light source.

14.4 Modal noise

In this section we will show that modal noise only occurs in multimode fiber systems. Monochromatic light in a fiber propagates only in a finite number of discrete ways (see below), which are called modes, or ray congruences (see Chapters 3 and 6). The maximum angle that a ray can make with the fiber axis is determined by the critical angle of the total internal reflection at the core–cladding interface. The sine of this angle multiplied by the refractive index of the medium involved is the numerical aperture NA (see Section 6.6). Figure 14.2 shows a meridional ray in a step index fiber, which has been launched with an angle θ. The wave fronts or equiphase planes of this ray are also shown, a distance of λ apart. For a consistent wave pattern the extended wavefronts of a certain ray should coincide with the wavefronts of that same ray after two successive reflections. This is only possible for discrete values of θ that correspond to the guided modes.

14.4.1 The number of modes in multimode fibers

For modal noise the number of modes in a fiber is one of the crucial parameters. In this section we shall calculate the number of guided modes that propagate in a multimode fiber at a certain wavelength. Going back to equation (6.76)

$$k_0 \int_{r_1}^{r_2} \sqrt{n^2(r) - g^2 - \left(\frac{n}{k_0 r}\right)^2} \, dr \approx m\pi \tag{14.31}$$

where m is the radial mode number and n the angular mode number. This equation holds for large values of m and can be rewritten as

$$m \approx \frac{1}{\pi} \int_{r_1}^{r_2} \sqrt{k_0^2 n^2(r) - k_0^2 g^2 - \frac{n^2}{r^2}} \, dr \tag{14.32}$$

with

$$n < r \sqrt{k_0^2 n^2(r) - k_0^2 g^2} \tag{14.33}$$

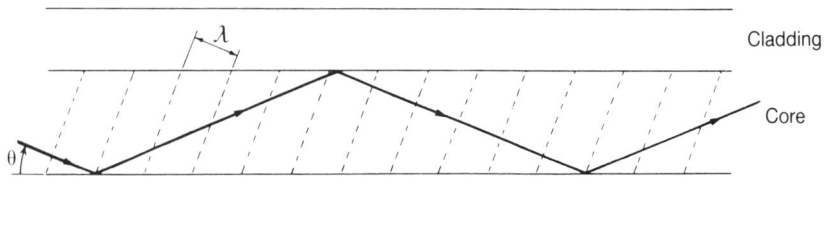

Figure 14.2 A meridional ray in a multimode step index fiber with its corresponding wavefronts (broken lines), forming a consistent wave pattern.

The maximum value of n is determined by the value of g. Each LP mode consists of four fundamental modes. If $N(g)$ denotes the number of modes that have a g value larger than a particular g, we get

$$N(g) = 4 \int_0^{n_{max}} m(n) \, dn \qquad (14.34)$$

where n has been approximated by a continuum and n_{max} is given by equation (14.33). Inserting equations (14.32) and (14.33) into equation (14.34) yields

$$N(g) = \frac{4}{\pi} \int_0^{n=r\sqrt{[k_0^2 n^2(r) - k_0^2 g^2]}} \left[\int_{r_1}^{r_2} k_0^2 n^2(r) - k_0^2 g^2 - \frac{n^2}{r^2} \, dr \right] dn$$

$$= \frac{4}{\pi} \int_0^{r_2} \left\{ \int_0^{n=rk_0\sqrt{[n^2(r) - g^2]}} \left[k_0^2 n^2(r) - k_0^2 g^2 - \frac{n^2}{r^2} \right] dn \right\} dr \qquad (14.35)$$

where r_2 is the solution of $n^2(r) = g^2$. The solution of the inner integral can be found in [7] and it follows that

$$N(g) = \int_0^{r_2} [k_0^2 n^2(r) - k_0^2 g^2] r \, dr \qquad (14.36)$$

When the well-known power-law index profile (see equation 6.58) is substituted in this equation the number of modes becomes

$$N(g) = \int_0^{r_2} \left[k_0^2 n^2(0) - 2\Delta n^2(0) k_0^2 \left(\frac{r}{a} \right)^x - k_0^2 g^2 \right] r \, dr \qquad (14.37)$$

with

$$r_2 = a \left(\frac{k_0^2 n^2(0) - k_0^2 g^2}{2\Delta n^2(0) k_0^2} \right)^{1/x} \qquad (14.38)$$

The solution of the integral in equation (14.37) is straightforward and elaborating it gives

$$N(g) = \Delta n^2(0) k_0^2 a^2 \frac{x}{x + 2} \left(\frac{k_0^2 n^2(0) - k_0^2 g^2}{2\Delta n^2(0) k_0^2} \right)^{(2/x+1)} \qquad (14.39)$$

Since $N(g)$ has been defined as the number of modes with a g value larger than the assumed g, the total number of guided modes is found by substituting the minimum value of g in equation (14.39). From Figure 6.10 it can be seen that $g_{min} = n(a)$; therefore, using equations (6.69) and (6.70) we find the total number of guided modes to be

$$N_f = \frac{x}{x + 2} a^2 k_0^2 n^2(0) \, \Delta = \frac{x}{x + 2} a^2 k_0^2 \frac{(NA)^2}{2} = \frac{x}{x + 2} \frac{v^2}{2} \qquad (14.40)$$

where v is the normalized frequency of the monochromatic light source (see equation 3.54). For a step index fiber ($x \rightarrow \infty$)

$$N_s = \frac{v^2}{2} \tag{14.41}$$

whereas a graded index fiber ($x = 2$) has half that number of modes, namely

$$N_g = \frac{v^2}{4} \tag{14.42}$$

14.4.2 Modal noise phenomena

From Figure 14.2 it follows that the different modes, which correspond to rays propagating at different angles of θ, reach different distances along the fiber and thus have different delay times. The light in a certain cross-section or end-face comprises contributions from different modes, each of which suffers a differential mode delay. Depending on the relative phases of the modes at each point, constructive or destructive interference occurs. When the light of a single-mode laser is launched into a multimode fiber, the interference phenomena cause speckles on the end-face of the fiber. To explain this effect, the wavefronts of two light rays in a step index fiber are depicted in Figure 14.3. Moreover, the relative light intensities at the cross-sections z_1 and z_2, due to the interference of the wavefronts, are presented. In these intensity curves the maxima and minima can be distinguished as a speckle pattern (see Figure 14.4). The number of degrees of freedom in this speckle pattern is equal to the number of guided modes. From equation (14.40) it follows that a fiber with a large core radius shows more speckles than a fiber with a small core radius. The same holds for a fiber with a large NA compared to a fiber with a small NA. Moreover, a step index fiber has more speckles than a graded index fiber with the same core radius and NA. Finally, the number of speckles decreases as the wavelength increases. Of course, the number of degrees of freedom of the speckle pattern is smaller than the number of modes when some of the modes are not excited.

The phase difference between two interfering modes depends on the differential mode delay and can amount to several times 2π. As a consequence, small

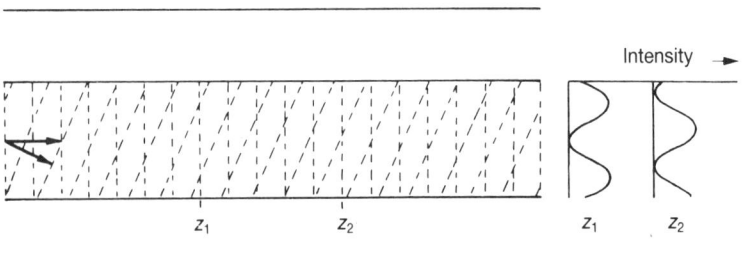

Figure 14.3 Two possible light rays in a step index fiber, with corresponding wavefronts and interference patterns at z_1 and z_2.

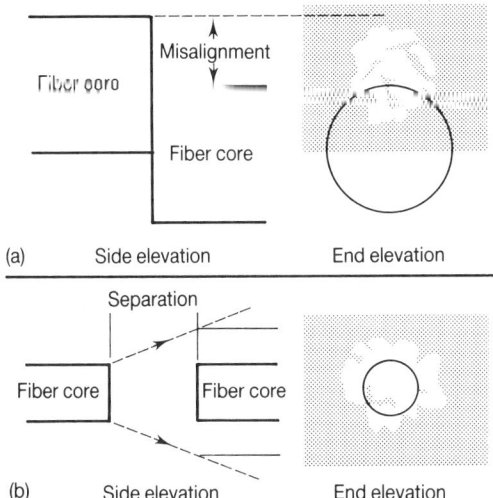

Figure 14.4 Filtering of the speckle pattern in a connector. (a) Radial misalignment of the fibers. (b) end separation of the fibers. (Source: reproduced with permission from *Laser Focus*, vol. 17, no. 9, September 1981, pp. 109–15, 'Modal noise–causes and cures', by R. E. Epworth.)

changes in the wavelength or the delay time can cause big changes in the modal phase shift. In its turn, this can cause big changes to the speckle pattern. For instance, bending the fiber by a mechanical vibration can change the delay time and thus the speckle pattern. AM to FM conversion in semiconductor lasers is a source of wavelength fluctuations in the light emitted and, therefore, also in the speckle pattern. Furthermore, fluctuations in the launching conditions can cause fluctuations in the speckle pattern. Moreover, optical feedback in the laser cavity by reflected light changes the wavelength.

In the optical fiber system a moving speckle pattern acts as a noise source in those devices – such as connectors, splices, mode mixers and coupling to a photodiode – where a part of the speckle pattern is spatially filtered due to mode-dependent losses. As an example consider Figure 14.4(a), where a part of the speckle pattern is filtered by radial displacement of the fibers in a connector. Figure 14.4(b) also shows such a spatial filtering, but in this case it is a result of end separation of the fibers. As the speckles inside the circles of Figure 14.4 vary with the changing patterns, the coupled optical power also varies. This effect alone produces modal noise.

We can estimate the modal noise after a number of assumptions have been made. We assume first that all the modes are excited, and secondly, that all the modes are excited by the same amount. The coherence time of the source is then assumed to be greater than the pulse broadening of the fiber. Finally, we assume that the speckle pattern is stationary across the cross-section of a fiber.

When the mode-dependent losses only occur after a length of fiber for which the pulse broadening is larger than the coherence time, the speckle pattern disappears there, because the modes are no longer coherent or show any interference. A splice or other mode-selective device in the optical path causes more modal noise if it is placed nearer to the source [8]. In the model we assume that the number of modes is much greater than unity. Moreover, it is assumed that the power carried by a specific mode completely passes through the connector or is completely removed at the connector. Now the problem reduces to a binomial distribution, namely the selection an average number of n modes (or speckles) from a set of N_f modes (or speckles). The probability that a speckle is inside the circle of Figure 14.4 equals

$$p = \frac{n}{N_f} = \eta \qquad (14.43)$$

which is the efficiency of the coupling.

The random variable that describes the number of speckles inside the circles is denoted by x; the expectation of this variable is then

$$E[x] = N_f p = n \qquad (14.44)$$

and its variance [5] is

$$\sigma_x^2 = N_f p (1 - p) \qquad (14.45)$$

Using these equations we can find the signal-to-modal-noise ratio to be

$$\frac{S}{N} \triangleq \frac{E^2[x]}{\sigma_x^2} = \frac{N_f^2 p^2}{N_f p (1 - p)} = \frac{n}{1 - (n/N_f)} = \frac{N_f \eta}{1 - \eta} \qquad (14.46)$$

The coupling efficiency η is determined by the mode-selective loss of the device in question. When this loss amounts to L dB, then

$$\eta = 10^{-L/10} \qquad (14.47)$$

We must stress that L contains only the mode-selective loss in this equation and not the mode-dependent loss which occurs, for example, in a neutral density filter. Equation (14.46) is used to obtain an estimate of the worst-case signal-to-noise ratio. Although many assumptions have been made, it appears that the result gives a reasonable agreement with the measured results [9]. A more thorough treatment can be found in [10] and [11], where the statistics of the speckle intensity have been taken into account.

At this point, the question arises as to whether modal noise can be prevented or removed from a system. To answer this question it should be remembered that modal noise needs the following three conditions to be fulfilled at the same time:

1. A coherent source.
2. A multimode fiber.
3. Mode-selective loss.

When one of these conditions is absent, there can be no modal noise.

Condition 1   Due to the bandwidth requirements it is not always possible to use an LED as the light source. A laser with many spectral lines reduces the modal noise [8], but cannot completely remove it. Moreover, such a laser gives rise to larger pulse broadening.

Condition 2   Using a single-mode fiber guarantees the absence of modal noise in laser-driven systems, if the fiber operates far enough beyond its cut-off wavelength.

Condition 3   When a coherent source is combined with a multimode fiber, the modal noise is minimized by minimizing the mode-selective losses in the system, especially those losses that occur near to the source, such as in the line build-out.

Just like the signal-to-partition-noise ratio, the signal-to-modal-noise ratio is independent of the power transmitted (see equation 14.46).

14.4.3 Modal noise in a link with several lossy devices

Starting with an analysis of the modal noise in a multimode fiber link with several lossy devices causing the modal noise, we consider a series connection of two mode-selective devices [12]. The randomly coupled power fractions of the two devices are denoted by x_1 and x_2, respectively. The random coupling of these two devices in series x_t is then $x_t = x_1 x_2$. The probability density function of a product of two independent random variables reads [5]

$$p(x_t) = \int_{-\infty}^{\infty} \frac{1}{|x|} \, p_{x_1}(x) \, p_{x_2}\left(\frac{x_t}{x}\right) dx \qquad (14.48)$$

With the aid of this equation it is easy to show that the average coupling efficiency η_t of the series connection is written as

$$\eta_t = E[x_t] = E[x_1 x_2] = E[x_1]E[x_2] = \eta_1 \eta_2 \qquad (14.49)$$

and the variance

$$\sigma_t^2 = \sigma_1^2 \sigma_2^2 + \eta_1^2 \sigma_2^2 + \eta_2^2 \sigma_1^2 \qquad (14.50)$$

If the mode-selective loss for each of the two devices is small, so that $\sigma_i \le \eta_i$ ($i \in \{1, 2\}$), then equation (14.50) can be approximated by

$$\sigma_t^2 \approx \eta_1^2 \sigma_2^2 + \eta_2^2 \sigma_1^2 \qquad (14.51)$$

The equations for two devices in series can be extended readily to a series connection of P mode-selective devices. With $x_t = \Pi_{i=1}^{P} x_i$ we get

$$\eta_t = E[x_t] = \prod_{i=1}^{P} \eta_i \qquad (14.52)$$

and

$$\sigma_t^2 \approx \sum_{i=1}^{P} \sigma_i^2 \prod_{\substack{j=1 \\ j \neq i}}^{P} \eta_j^2, \qquad \text{for } \sigma_i \ll \eta_i \qquad (14.53)$$

By means of equations (14.52) and (14.53) we arrive at the signal-to-noise ratio of the series connection, expressed in terms of the individual devices

$$\frac{S}{N} \triangleq \frac{E^2[x_t]}{\sigma_t^2} = \frac{\prod_{i=1}^{P} \eta_i^2}{\sum_{i=1}^{P} \sigma_i^2 \prod_{\substack{j=1 \\ j \neq i}}^{P} \eta_j^2} \qquad (14.54)$$

According to equations (14.43) and (14.45) the variance σ_i depends on the number of input modes N_i of the ith device

$$\sigma_i^2 = \frac{\eta_i(1 - \eta_i)}{N_i} \qquad (14.55)$$

The signal-to-noise ratio thus depends on the number of input modes N_i to the devices. Two cases will be dealt with, namely no mode coupling in the fibers between two successive devices and full mode coupling in those fibers. In the case of no mode coupling we assume that the number of input modes to the first device equals the maximum number of fiber modes N (see equation 14.40). At the input to the second device $N_2 = \eta_1 N_f$ and at the input to the third device $N_3 = \eta_2 \eta_1 N_f$ etc. The number of input modes to the ith devices is

$$N_i = N_f \prod_{j=1}^{i-1} \eta_j \qquad (14.56)$$

Substituting this expression into equation (14.55) and putting the result, in its turn, into equation (14.54), we get, after a lengthy but straightforward calculation

$$\frac{S}{N} = \frac{N_f \prod_{i=1}^{P} \eta_i}{1 - \prod_{i=1}^{P} \eta_i} = \frac{N_f \eta_t}{1 - \eta_t} \qquad (14.57)$$

If we compare this result with equation (14.46), in this case it can be concluded that the series connection behaves as though the total mode-selective loss η_t is concentrated into a single device that has the same loss as the complete link.

In the case of full mode coupling, all the devices have the same number of input modes, namely the same as the fiber modes N_f. Inserting this into equations (14.55) and (14.54) gives the signal-to-noise ratio for the entire link

$$\frac{S}{N} = \frac{\prod_{i=1}^{P} \eta_i^2}{\sum_i [\eta_i(1 - \eta_i)/N_f] \prod_{\substack{j=1 \\ j \neq i}}^{P} \eta_j^2} = \frac{N_f}{\sum_i [(1/\eta_i) - 1]} = \frac{N_f}{\sum_i (1/\eta_i) - P}$$

$$(14.58)$$

When considering a series connection of P devices, all having the same coupling efficiency of η_0 per device, this signal-to-noise ratio simplifies to

$$\frac{S}{N} = \frac{N_f}{P} \frac{\eta_0}{1 - \eta_0} \qquad (14.59)$$

From equation (14.46) it can be seen that the signal to noise ratio of the link is P times less than that of a single device in the link.

It can be shown easily that the signal-to-noise ratio of equation (14.57) is less than that of equation (14.59). This is further illustrated by Example 14.1.

Example 14.1
A standard graded index fiber with a core diameter of 50 μm and a numerical aperture of NA = 0.15 carries 82 modes at a wavelength of 1.3 μm (see equation 14.42). It is assumed that all modes at the input are excited with the same amount of power. Figure 14.5 shows the link signal-to-noise ratio as a function of the total mode-selective loss in the link for various values of the selective loss per device. The plots represent both full mode coupling and no mode coupling.

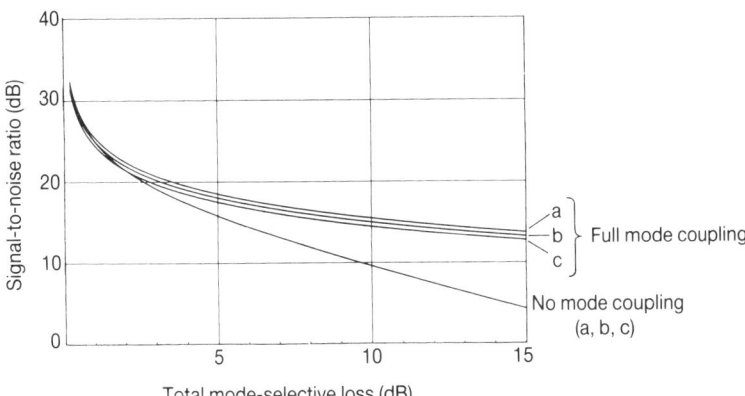

Figure 14.5 The signal-to-modal-noise ratio as a function of the total mode-selective loss in a link with several mode-selective devices (number of fiber modes $N_f = 82$). (a) Loss per device $L = 0.05$ dB ($\eta_0 = 0.9886$). (b) Loss per device $L = 1$ dB ($\eta_0 = 0.7943$). (c) Loss per device $L = 2$ dB ($\eta_0 = 0.6310$).

In practical systems the curves fall between the two extremes of no mode coupling and full mode coupling.

From equations (14.57) and (14.59) and Example 14.1 it can be concluded that the mode coupling between the mode selective devices improves the signal-to-modal-noise ratio at the end of the transmission link. This effect is greatest when the mode-selective loss is highest; for a fixed total link loss, the improvement decreases as the loss per device increases.

In single-mode fibers operating slightly above their cut-off wavelength, modal noise can be significant, due some of the power from the LP_{01} mode coupling to the LP_{11} mode and reverting to the LP_{01} mode in a second connector [13].

14.5 The signal-to-noise ratio due to system noise and receiver noise

In order to describe the combined effects of system noise and receiver noise we shall consider a configuration where the signal is transmitted by two systems. Each of the systems add some noise to the signal. The question now is: what is the signal-to-noise ratio at the end of the two systems, expressed in terms of the signal-to-noise ratios for the two individual systems? These ratios were calculated above and in Chapters 12 and 13. The individual systems have been presented in Figure 14.6(a), together with their signal-to-noise ratios $(S/N)_1$ and $(S/N)_2$, respectively. In Figure 14.6(b) the systems are shown in series, with a constant transfer α in between. The signal-to-noise ratio of the series connection becomes

$$\left(\frac{S}{N}\right)_t = \frac{\alpha^2 S_1^2}{\alpha^2 \sigma_1^2 + \sigma_2^2} = \frac{1}{(\sigma_1^2/S_1^2) + (\sigma_2^2/\alpha^2 S_1^2)} \tag{14.60}$$

provided that the noise sources are uncorrelated.

Comparing Figure 14.6(a) with Figure 14.6(b), it follows that $\alpha S_1 = S_2$, therefore

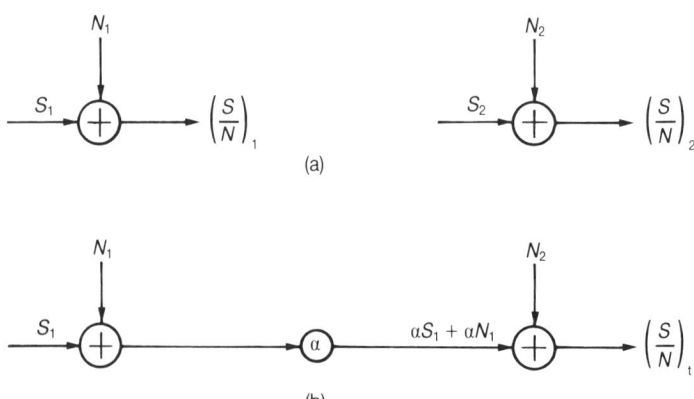

(a)

(b)

Figure 14.6 Series connection of two systems that are disturbed by noise.

$$\left(\frac{S}{N}\right)_t = \frac{1}{(N/S)_1 + (N/S)_2} = \frac{(S/N)_1(S/N)_2}{(S/N)_1 + (S/N)_2} \tag{14.61}$$

The constant transfer of α does not influence the signal-to-noise ratio at the end of the series connection, since the signal-to-noise ratio at the input of the second system does not depend on α. From equation (14.61) it can be seen that the inverse of the signal-to-noise ratio for the series connection consists of the sum of the inverted signal-to-noise ratio for the individual systems (determined in the same way as the resistance value of a parallel connection of two resistors). In these circumstances the system noise may be used for calculating $(S/N)_1$ and the receiver noise for calculating $(S/N)_2$, or the other way round. When the system noise consists of both modal noise and mode partition noise, the dominator of the center expression of equation (14.61) is the sum of three terms: $(N/S)_1$, $(N/S)_2$ and $(N/S)_3$; one accounting for the receiver noise, one for the modal noise and one for the mode partition noise. It should be borne in mind that only the noise inside the information band has to be considered. The modal noise due to AM–FM conversion in the laser generally has its spectrum (partly) inside the information band, whereas the low-frequency modal noise, due to mechanical vibrations for instance, behaves like fading.

It has previously been emphasized that the signal-to-partition-noise ratio and the signal-to-modal-noise ratio do not depend on the signal power transmitted; this contrasts with the receiver signal-to-noise ratio, which can be improved by raising the signal power. This effect leads to the typical characteristics of the bit error rate P_e versus the optical power received (see Figure 14.7). These curves show a minimum P_e (bit error rate (BER) floor), determined by the system noise. This floor cannot be lowered by increasing the power transmitted, but only by designing the system more carefully.

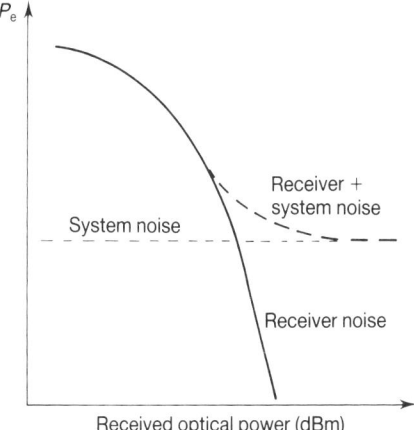

Figure 14.7 Bit error floor due to system noise.

References

[1] J. R. Pierce and E. C. Posner, *Introduction to Communication Science and Systems* (New York: Plenum Press, 1980).

[2] T. H. Zachos and J. E. Ripper, 'Resonant modes of GaAs junction lasers', *IEEE Journal of Quantum Electronics*, vol. QE-5, January 1969, pp. 29–37.

[3] H. Kogelnik and T. Li, 'Laser beams and resonators', *Applied Optics*, vol. 5, no. 10, October 1966, pp. 1550–67.

[4] K. Ogawa, 'Analysis of mode partition noise in laser transmission systems', *IEEE Journal of Quantum Electronics*, vol. QE-18, no. 5, May 1982, pp. 849–55.

[5] A. Papoulis, *Probability, Random Variables and Stochastic Processes* (New York: McGraw-Hill, 1965).

[6] K. Ogawa and R. Vodhavel, 'Measurements of mode partition noise of laser diodes', *IEEE Journal of Quantum Electronics*, vol. QE-18, no. 7, July 1982, pp. 1090–3.

[7] I. Gradshteyn and I. Ryzhik, *Tables of Integrals, Series, and Products* (Orlando: Academic Press, 1980).

[8] A. Koonen, 'Bit-error-rate degradation in a multimode fiber optic transmission link due to modal noise', *IEEE Journal on Selected Areas in Communications*, vol. SAC-4, no. 9, December 1986, pp. 1515–22.

[9] K. Hill, Y. Tremblay and B. Kawasaki, 'Modal noise in multimode fiber links: theory and experiments', *Optics Letters*, vol. 5, no. 6, June 1980, pp. 270–2.

[10] J. Goodman and E. Rawson, 'Statistics of modal noise in fibers: a case of constrained speckle', *Optics Letters*, vol. 6, no. 7, July 1981, pp. 324–6.

[11] Y. Tremblay, B. Kawasaki and K. Hill, 'Modal noise in optical fibers: open and closed speckle pattern regimes', *Applied Optics*, vol. 20, no. 9, May 1981, pp. 1652–5.

[12] A. Koonen, 'Modal noise in multimode fiber links with distributed mode-selective losses', *Journal of Optical Communications*, vol. 5, no. 4, pp. 141–3.

[13] S. Heckmann, 'Modal noise in single-mode fibers operated slightly above cutoff', *Electronics Letters*, vol. 17, no. 14, July 1981, pp. 499–500.

Problems

14.1 Deduce equations (14.17) and (14.18)

14.2 Consider a fiber communication link of length 25 km. The link uses a laser as the light source and this laser has a spectral width of 1 nm. Calculate the bit rate where the system reaches a BER floor at an signal-to-noise ratio of 20 dB, due to mode partition noise: (a) at 850 nm; (b) at the dispersionless wavelength.

14.3 Calculate the number of guided modes in a standard parabolic index fiber (core diameter 50 μm, NA = 0.17) at a wavelength of 850 nm.

14.4 A single-mode fiber has a cut-off wavelength of 1200 nm. This fiber is used for transmission at a wavelength of 633 nm. Calculate the signal-to-modal-noise ratio, when the link has a mode-selective loss of 0.5 dB, in: (a) a single device; (b) 10 equal devices with full mode coupling; (c) no mode coupling.

14.5 An optical fiber communication link has a signal-to-modal-noise ratio of 20 dB. The signal-to-partition-noise ratio amounts to 17 dB.

What value of the receiver signal-to-noise ratio can be allowed in order to arrive at an overall signal-to-noise ratio of 14 dB?

15

System components and aspects of system design

15.1 Introduction
15.2 Comparison of optical fibers and copper cables
15.3 Optical fiber cables
15.4 Splices and connectors
15.5 Optical isolators
15.6 Polarization-maintaining fiber
15.7 Wavelength multiplexing
15.8 Repeater distance and link budget
15.9 Line coding
15.10 Selection of the system components
      References

15.1 Introduction

In the preceding chapters the most important aspects of optical fiber communication systems – such as wave guiding, coupling, light sources, detectors and receivers – have been explained. When designing an optical fiber communication system, we need other components that have not yet been considered. In this chapter we shall describe these components briefly and examine some aspects that need to be taken into account.

One of the more crucial decisions to be made when planning a communication link is choosing the transmission medium. This decision will be determined by the following considerations:

1. The price. When a number of alternatives meet the technical requirements, a medium that gives the best economy should be chosen.
2. The technical feasibility. A new medium, like an optical fiber, offers more opportunities than, for example, a coaxial cable. The optical fiber should be used for applications that were out of the question a few years ago, due to limitations of the transmission media at that time.

The advantages of an optical fiber compared with a copper cable are the following:

- fewer losses;
- larger bandwidth;
- reduced bending radius for the cables;
- the transmission is unaffected by interference;
- no interference source;
- more difficult to tap illegally;
- greater potential for increased capacity.

This last advantage needs clarifying: the potentially large bandwidth of the fiber cannot be fully utilized at present due to dispersion. Advancing technology will allow fiber links to be improved simply by replacing the transmitting and receiving equipment in the system. In this way it will be possible to increase the transmission capacity of a fiber, which was initially installed to operate on a single wavelength, by adding wavelength multiplexers and demultiplexers (see Section 15.7) in order to increase the number of wavelengths available. In the case of a single-mode fiber such an increase could be achieved by packing the wavelengths more densely, in combination with coherent reception.

15.2 Comparison of optical fibers and copper cables

In the preceding chapters both analog and digital modulation in optical fiber systems were analyzed. Generally speaking, digital modulation is more suitable for use with optical fibers, whereas analog modulation lends itself better to use with copper cables. The reasons for these differing suitabilities are as follows:

1. For the same information signal, digital modulation requires a wider transmission band than analog modulation. Copper cables give more specific dispersion, whereas fibers show less specific dispersion. Or, in other words, optical fibers have a much larger bandwidth–length product.
2. The transmitters and receivers for copper cables have better linearity than those for optical fibers. Linearity is more critical for analog systems than for digital systems.

3. The potential bandwidth of optical fibers can be utilized more effectively by wavelength multiplexing combined with wavelength demultiplexing and/or coherent detection. Digital systems are more tolerant to cross-talk than analog systems, so that simpler demultiplexers are adequate in digital systems or where the channels can be more densely packed.
4. The signal-to-noise ratio of an optical fiber system is in general smaller than that of a copper cable system, as will be explained below. However, analog systems require larger signal-to-noise ratios (see Chapters 12 and 13).

Let us assume that the power coupled from an optical source to a fiber is 5 mW at a wavelength of 1.5 μm; let us also suppose that the modulation index γ is unity. Then, without attenuation, we can find that

$$\left(\frac{S}{N}\right)_f \approx \frac{2 \times 10^{16}}{BC^2} \tag{15.1}$$

for the quantum limit (see equation 12.36).

For a copper cable system the signal-to-noise ratio is

$$\left(\frac{S}{N}\right)_c = \frac{V^2}{4k\Theta BC^2 Z_0} \tag{15.2}$$

where V is the output voltage of the transmitter and Z_0 is the characteristic impendance of the cable. Let us assume that the voltage range is 6 V and the characteristic impendance is 100 ohms; it then follows that

$$\left(\frac{S}{N}\right)_c \approx \frac{2 \times 10^{19}}{BC^2} \tag{15.3}$$

Comparing equations (15.3) and (15.1) shows that the copper cable system performs 30 dB better than the fiber system. In Example 12.1 an analog FDM system for audio transmission via an optical fiber was found to have a signal-to-noise ratio of between 50 and 60 dB. Such values met the requirements only marginally, but a copper cable with an signal-to-noise ratio of 30 dB greater would be more satisfactory. From items 1–4 above, it follows that digital modulation fits optical fibers more naturally, since the signal-to-noise ratio varies inversely with bandwidth.

15.3 Optical fiber cables

Immediately after drawing a fiber, its glass cladding has to be provided with a thin polymer layer. This primary coating serves the folloing purposes:

- It protects the glass from damage. Small flaws in the cladding surface can lead to stress fractures.
- It reduces the so-called microbending losses. When a fiber is bent sharply, the losses increase greatly, but the primary coating prevents sharp bends.

- It prevents the penetration of moisture. Hydroxyl ions can react with SiO_2, to create compounds that are mechanically weaker than SiO_2 (see [1]).
- In the absence of cladding glass the primary coating acts as a cladding. Such fibers are called silicone-clad fibers (SCF) or plastic-clad fibers (PCI); in general they have a larger numerical aperture and greater losses than glass fibers.

Primary coatings can be very thin (a few micrometers) and sometimes have the ability of stripping off cladding modes.

Next, for increased mechanical strength, and as a first step towards making a fiber cable, the primary coated fiber is provided with a secondary coating. Two types of secondary coating are in use:

- A loose tube coating. The fiber lies loosely in a plastic tube which has an inner diameter considerably larger than the outer diameter of the primary coated fiber; the inner diameter of the secondary coating ranges from 0.4 to 1 mm, whereas its outer diameter is 0.6–2 mm.
- A tight coating. A plastic layer with an outer diameter of 0.4–1.5 mm is applied tightly to the primary coating.

This secondary coating is usually applied by thermoplastic extrusion around the primary coated fiber, similar to the process for making copper cables. The loose tube coating usually leads to less excess loss after cabling [2] and is the most widely used method.

The simplest process for making a fiber cable is twisting a number of secondary coated fibers in the same way as making a cable from copper wires. However, for extra strength, tensile steel or synthetic wires are included (see Figure 15.1 for an example). Some cable manufacturers have developed special fiber cables in order to facilitate splicing or connecting multi-fiber cables. Two examples are shown in Figure 15.2. For a more detailed description and pictures of fiber cables, see [1] and [2].

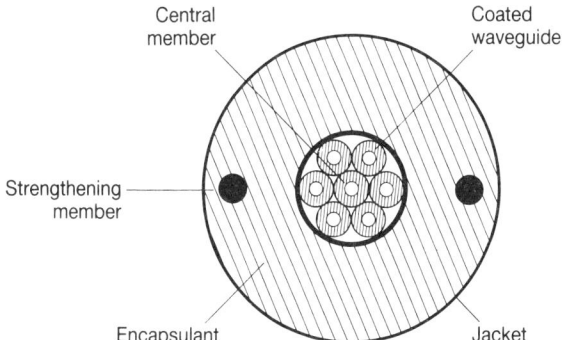

Figure 15.1 Cross-section of a fiber cable with steel or synthetic strengthening members.

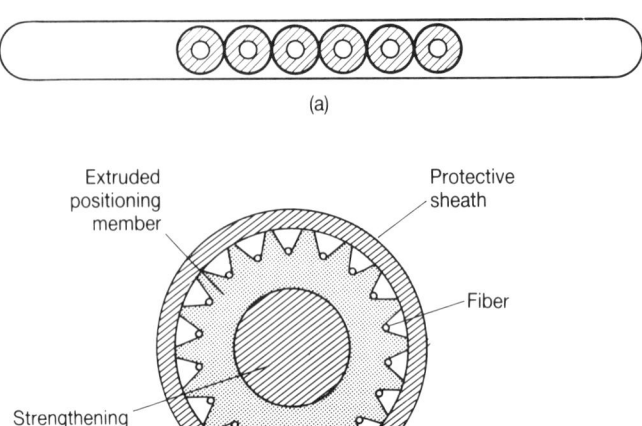

Figure 15.2 Designs to facilitate the splicing or connection of multifiber cables. (a) Flat cable. (b) Round cable with positioning member. (Source: (b) is reproduced with permission from *Proceedings of the Second European Conference on Optical Communication*, Paris, 1976, pp. 247–52, 'Structure du câble pour fibres optiques et procédés de raccordement', by G. le Noane.)

15.4 Splices and connectors

Optical fibers are made in lengths of a few kilometers. For long-distance communication, several fibers have to be joined together. A permanent joint is called a splice. Temporary joints, which have to be made and disjoined several times during the lifetime of a link, are called connectors. Each new connection must be reliable, so that the attenuation does not exceed a predetermined value.

15.4.1 Fused splices

With fused splices the two fiber ends to be joined are heated so that the glass can be fused. Before this process is started the two fibers are radially aligned, then during the fusion process they are made to approach each other axially. The optimal position can be checked with the aid of a microscope or by monitoring the coupled power. A splice is made by heating the fiber ends by an electic arc comprising two tungsten electrodes (see Figure 15.3) or with a microplasma torch. This method allows attenuations as low as 0.05 dB to be obtained, for both multimode and single-mode fiber splices.

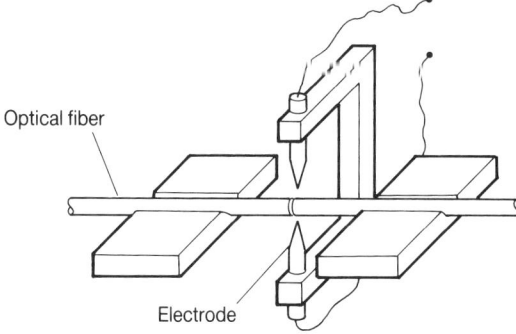

Figure 15.3 Fusion splicing by means of an electric arc between two tungsten electrodes. (Source: reproduced with permission from *Topical Meeting on Optical Fibre Transmission II*, Williamsburg, 1977, pp. WA 3-1–6, 'Optical fiber splicing', by C. Miller.)

Fusion splicing is one of the commonest splicing techniques. Modern equipment allows automatic alignment of the fibers, which are manipulated in special clamps. Light is injected into the fiber just before the joint and measured just after the joint (see Figure 15.4). A good coupling is achieved by making a kink in the fiber with a radius of about 4 mm, about 0.5 m on either side of the joint [3].

15.4.2 Other splicing techniques

In addition to the fusion splice, other slicing techniques have been developed, all of which are based on simple methods for aligning the fiber ends radially in order to bring them into contact with each other before joining them together (see [1] and [2]). A few examples are shown in Figure 15.5; all of these use glue to join the fiber ends. If the epoxy glue penetrates between the fiber end-faces, it has to have the correct optical properties regarding refractive index and

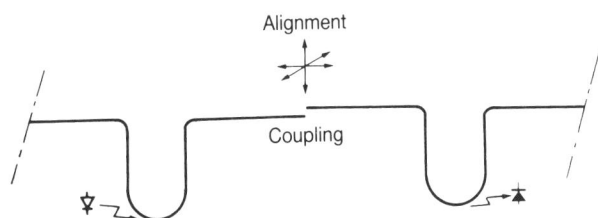

Figure 15.4 A set-up for automatic core alignment when splicing fibers.

transparency. Similar techniques have been developed to join various fibers at the same time. The splicing technique shown in Figure 15.6 is suitable for the cables shown in Figure 15.2(a).

(a)

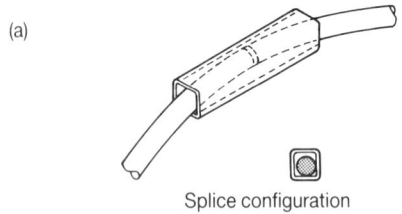

Splice configuration

(b)

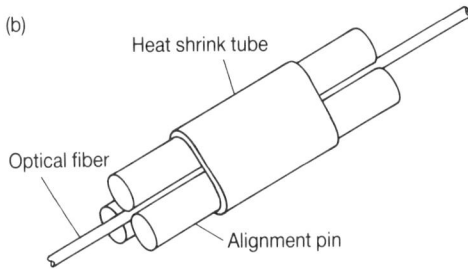

(c)

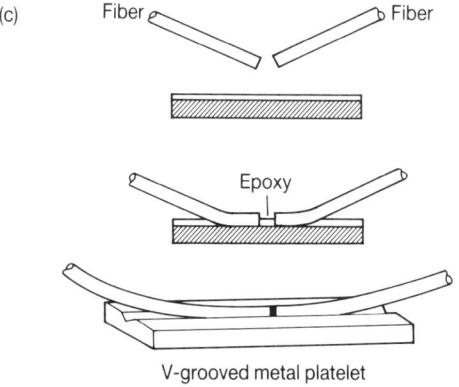

V-grooved metal platelet

Figure 15.5 Some examples of glued splices. (Source: (a) and (b) are reproduced with permission from *Topical Meeting on Optical Fibre Transmission II*, Williamsburg, 1977, pp. WA 3–6, 'Optical fiber splicing', by C. Miller.)

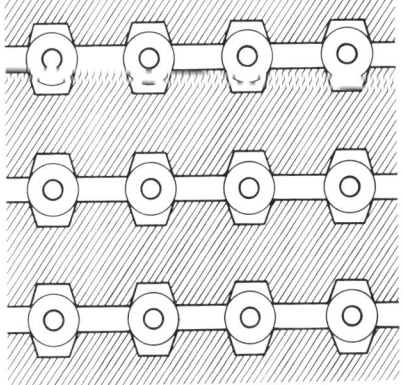

Figure 15.6 Groove splice technique for joining multifiber cables.

A somewhat different technique is shown in Figure 15.7 in which the fibers are aligned using an aluminium crimp sleeve. Although this method requires special splicing equipment, it takes less time than the other techniques.

15.4.3 Connectors

As with splicing, there are many techniques for connecting fibers (see [1] and [2]), of which we shall mention just a few. Almost all connector constructions require mechanical precision, so that each individual connector is precisely aligned in the factory. Therefore, the connectors are supplied with a pigtail and

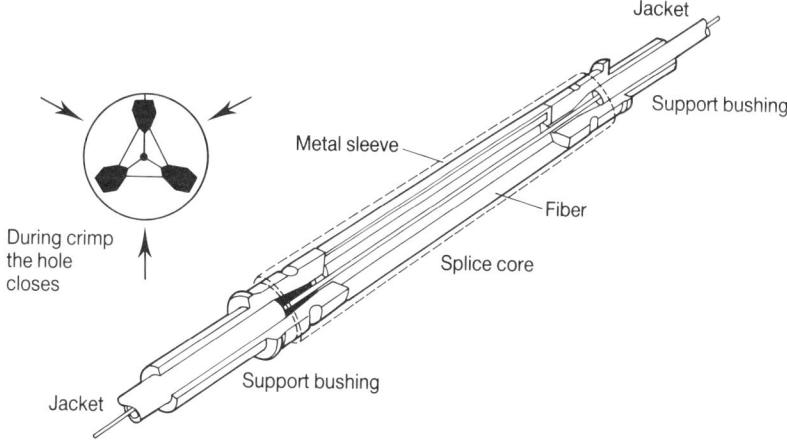

Figure 15.7 Crimped splice assembly. (Source: reproduced with permission of AMP-Holland BV, 1986.)

in this way the connector problem is reduced to a splicing problem.

The first technique uses ground cones into which the fibers are cemented. These cones are fitted to a common bushing by means of swivels (see Figure 15.8). The bushing is provided with two conical holes, so that tightening two swivels centers the fiber cores automatically. The quality of a connector obviously depends on the positioning of the cones and the precision of aligning the fibers in the cores.

The next example is the center alignment type. The fiber pigtail is cemented into a metal ferrule, then the fiber core is illuminated and the oversize ferrule is turned down to size on a precision lathe [4]. The light leaving the fiber end-face is used for centering the core in the ferrule during turning (see Figure 15.9). The two machined ferrules are joined with a close-tolerance bush.

Losses from a lens connector depend less on the precision of connecting the parts radially and axially than angularly. Figure 15.10(a) shows the principle of the connector, whereas Figure 15.10(b) shows its actual construction. On the transmitting side the fiber is aligned so that a collimated bundle of light comes out of the corresponding ball lens. In this way, the connection is less susceptible to axial misalignment. Since the diameter of the bundle is enlarged by the lens, small radial deviations have little effect on the power coupling. The lens in the receiving part focuses the light bundle on the core of the receiving fiber. In order to prevent Fresnel losses, the fiber end-faces and the ball lenses should be anti-reflection coated. The lens connector is rarely used for connecting two

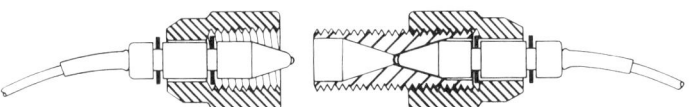

Figure 15.8 Connector with conical centering.

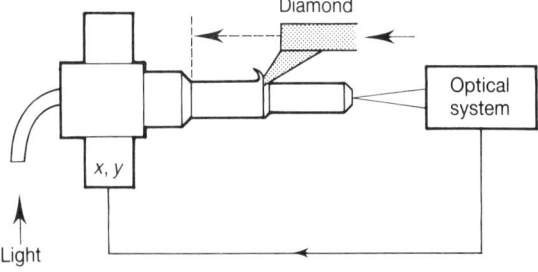

Figure 15.9 Fabrication of core-centered ferrules. (Source: reproduced with permission from *IEEE Journal on Selected Areas in Communications*, vol. SAC-4, no. 4, July 1986, pp. 457–71, 'European optical fibres and passive components: status and trends', by G. Khoe and H. Lydtin.)

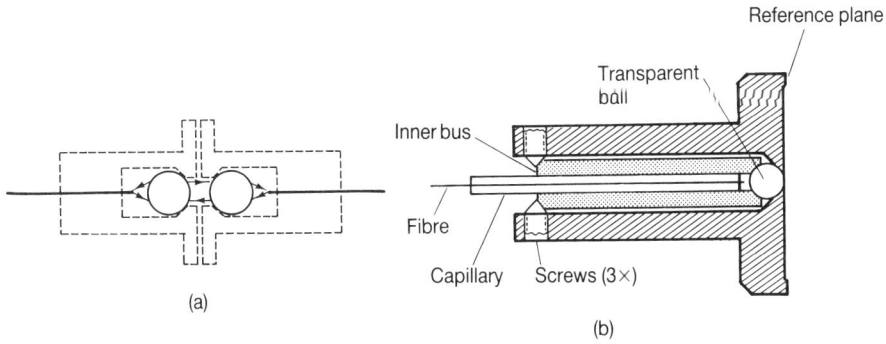

Figure 15.10 Lens connector. (a) Basic construction. (b) Implementation. (Source: reproduced with permission from *Electronics Letters*, vol. 14, no. 6, August 1978, pp. 511–12, 'Practical low-loss lens connector for optical fibres', by A. Nicia.)

fibers; it produces a collimated bundle between the two halves of the connector and provides an excellent means for placing an optical device, such as a filter or beam-splitter, in the optical path.

Losses in connectors are generally somewhat greater than for splices, namely 0.2–0.5 dB.

15.5 Optical isolators

For high bit rate systems it is necessary for the laser source to have a narrow, stable spectrum. Optical feedback from reflections in the system disturbs the stable condition; therefore, an optical isolator has to be inserted between the laser diode and the fiber. Such an isolator is explained in Figure 15.11. The light leaving the laser is collimated by the lens L_1 and impinges on polarizer 1, which is rotated until the laser light passes through it. The beam then propagates through an yttrium iron garnet (YIG) plate. This YIG material shows a strong

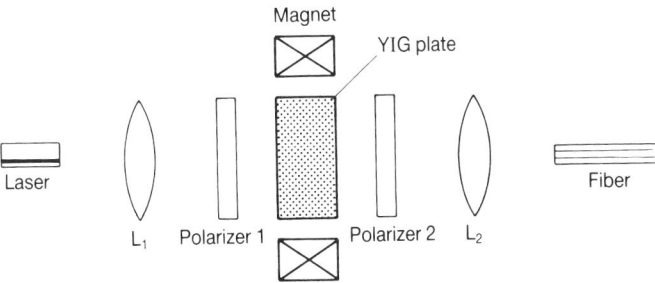

Figure 15.11 Schematic diagram of an optical isolator.

Faraday effect when influenced by a magnetic field. The thickness of the YIG plate and the magnetic field are chosen so that the linear polarization rotates $\pi/4$ when passing through the plate [4]. Polarizer 2 is then oriented until all the light passes through it. Finally, lens L_2 focuses the light bundle on the core of the fiber. Refelected light can only pass through the YIG plate when it is polarized correctly with polarizer 2. Again, this light is rotated $\pi/4$ in the same direction as the incident wave, so that the reflected light leaving the YIG plate is rotated $\pi/2$ to the initial direction. However, this polarization is now blocked by polarizer 1, which prevents the reflected light from re-entering the laser cavity. Since laser diodes only generate linearly polarized light in the plane of the junction, polarizer 1 can be omitted; these lasers are insensitive to orthogonally reflected light. Back-reflection from the isolator itself must be excluded; an anti-reflection coating or positioning the YIG plate obliquely prevents reflection from the various components. Reducing reflected light any further requires making the end-face of the fiber oblique.

The device can be simplified by shaping the YIG material into a sphere (see Figure 15.12) which can replace one or both of the lenses [6]. The polarizer consists of a calcite plate.

Optical isolators are vital components in coherent detection systems, because they can stabilize the optical spectrum of lasers that consist of DFB and DBR devices.

15.6 Polarization-maintaining fiber

Ideal single-mode fibers preserve the state of polarization (SOP) of the input light, which will be linearly polarized in case of a semiconductor laser. However,

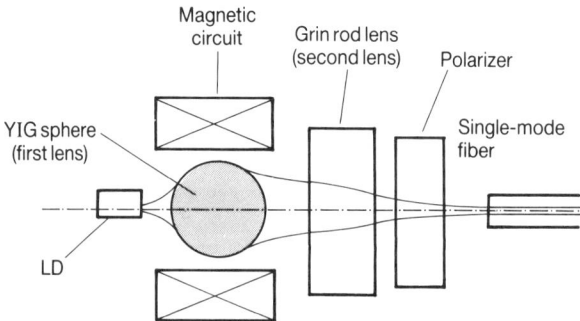

Figure 15.12 An isolator with a YIG sphere replacing one of the lenses. (Source: reproduced with permission from *IEEE/OSA Journal of Lightwave Technology*, vol. LT-1, no. 1, March 1983, pp 121–30, 'An effective nonreciprocal circuit for semiconductor laser-to-optical-fiber coupling using a YIG sphere', by T. Sugie and M. Saruwatari.

in practice, fibers do not behave ideally, but show birefringence. This birefringence is caused by the following:

1. A slightly elliptical fiber core.
2. Bending of the fiber.
3. Transverse pressure.
4. Intrinsic stress in the fiber.

Due to imperfections during the fabrication process of the fiber and to unpredictable changes in the environmental conditions, all these causes lead to randomly varying polarization changes along the fiber or in time. In direct detection systems (see Chapter 12) a randomly varying output SOP is insignificant, but in coherent systems the SOP of the local oscillator laser should match the SOP of the received signal light wave (see Chapter 16).

In Chapter 3 we saw that the fundamental HE_{11} mode can be approximated by the linearly polarized LP_{01} mode. However, when the fiber shows birefringence an x-polarized excitation results in wave propagation with a different specific phase shift β compared to a y-polarized excitation. In this way the LP_{01} mode is split into two orthogonal modes. This changes the SOP of the output light due to the randomly varying birefringence. Polarization-maintaining fiber (PMF) prevents these changes of the output SOP to a great extent. This goal is achieved by providing PMF with a deliberately produced, well-controlled large briefringence, which is much larger than the birefringence in a standard single-mode fiber. In PMFs the specific phase shift β for the x-polarized mode differs substantially from that for the y-polarized mode. The length over which the difference in phase shift between the orthogonal polarizations amounts to 2π is called the beat length and can be as small as a few millimeters in PMFs.

Among the different types of PMFs, two are most widely used, namely those with an elliptical core and those with well-controlled internal stress in the fiber. This internal stress is produced by stress-applying parts (SAPs) in the cladding. These SAPs produce an asymmetric stress distribution in the fiber core and cladding. The SAPs are positioned on opposite sides of the fiber core (see Figure 15.13), and consist of glass with a different composition compared to the cladding glass. Of course, this means that the SAPs have a different refractive index as well. PMFs are designed such that the net effect is that the specific phase shift of the x-polarized mode differs substantially from the specific phase shift of the y-polarized mode.

Figure 15.13 shows two different shapes of SAPs; the PANDA fiber has two circular SAPs (see Figure 15.13a), whereas Figure 15.13(b) shows the bow-tie fiber.

PMFs only show their polarization-maintaining ability when they are excited in one of their orthogonal modes, i.e. linearly x-polarized or linearly y-polarized as given by Figure 15.13. In practical fibers the mode coupling between orthogonal modes can be as small as $10^{-6}/m$.

PMFs generally exhibit a higher loss than standard single-mode fibers.

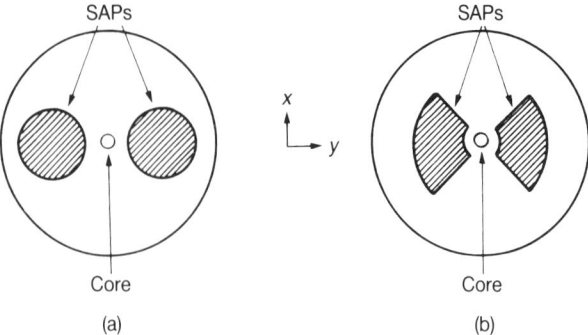

Figure 15.13 Cross-section of polarization-maintaining fibers. (a) PANDA fiber. (b) Bow-tie fiber.

15.7 Wavelength multiplexing

The potential bandwidth of an optical fiber is very great. For example, when considering wavelengths of the order of 1 μm and using the following relationship

$$|\Delta f| \approx \frac{c}{\lambda^2} |\Delta\lambda| \tag{15.4}$$

it is easy to see that a wavelength region of 100 nm corresponds to a bandwidth of approximately 3×10^{13} Hz. This is equal to a frequency-multiplexed bandwidth of three million television channels. Due to dispersion by the fiber and limitations of the processing electronics only a small part of this bandwidth can be utilized for each channel. A more efficient use of the bandwidth is possible when wavelength division multiplexing (WDM) is used. The fiber is illuminated by a number of lasers at the same time, and each laser has its own wavelength and is modulated with its own information-carrying signal. At the receiving end the receivers select their respective signal with optical filters (see Figure 15.14) or a coherent detection method (see Chapter 16). The number of channels that can be handled in this way depends on the wavelength stability of the lasers, the

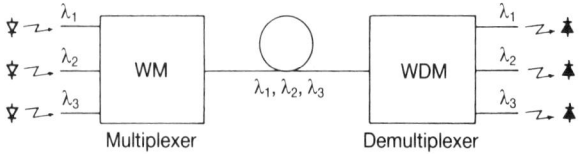

Figure 15.14 General diagram of a wavelength multiplexing and wavelength demultiplexing fiber optic system.

selectivity and stability of the multiplexers, and the cross-talk that can be permitted. Since wavelength demultiplexers are passive optical devices they can be used in the reverse direction to multiplexers. For that purpose, however, there are simpler and cheaper devices that, moreover, introduce fewer losses. The simplest wavelength multiplexer is a beam combiner in the form of a beam splitter. More than two wavelengths are multiplexed by a series connection of a number of beam splitters. These beam splitters have their fiber optic counterparts in the fused biconical taper coupler (see Figure 15.15). Two fibers are placed parallel to and in contact with each other, and are then fused by heating over a few millimeters of their length and tapered by drawing them out. In such a device part of the power of the light waves from one fiber core is coupled to the other core, and vice versa. Very few excess losses occur with this technique (0.1–0.3 dB). Fused biconical taper couplers can be made for both multimode and single-mode fibers. Moreover, the technique lends itself to producing devices with many optical inputs and outputs. They are used in fiber optic local area networks (LANs) as a reflective star; or they can serve as an multiport optical hybrid for phase diversity coherent receivers (see Chapter 16).

A wavelength demultiplexer must have an optical component that can separate the different wavelengths spatially; the following devices can serve this purpose:

- a prism;
- an interference filter;
- a grating;
- a Mach–Zehnder interferometer.

Due to their low selectivity, prisms are not used for demultiplexers.

Interference filters are based on the reflection of certain wavelengths by a stack of thin dielectric layers and the transmission of other wavelengths through a stack of layers [7, 8] (Figure 15.16 shows a demultiplexer based on this principle). The arrangement of the layers must differ from one fiber to another in order to select different wavelengths.

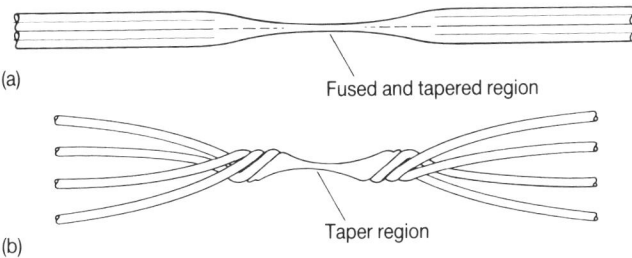

(a)

Fused and tapered region

Taper region

(b)

Figure 15.15 The fused biconical taper. (a) Two-fiber coupler. (b) Multiport coupler.

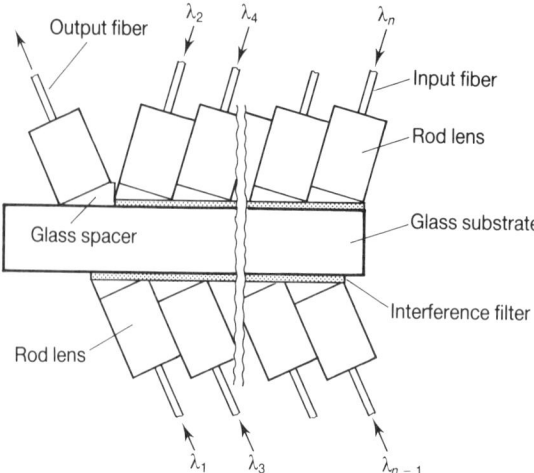

Figure 15.16 A wavelength demultiplexer with interference filters. (Source: reproduced with permission from *Proceedings of the Fifth European Conference on Optical Communication*, Amsterdam, September 1979, pp. 11.5-1–4, 'Low loss optical multi/demultiplexer using interference filters', by K. Hashimoto and K. Nosu.)

Demultiplexers with a grating often use a Littrow mounting [8]. Its single composite lens provides both collimating a decollimating of the light bundle emitted from the input fiber towards the grating, so that it is selectively reflected to the output fibers (see Figure 15.17). Figure 15.18 shows a possible loss

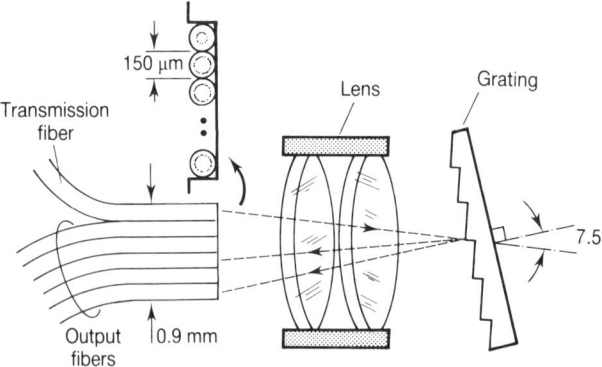

Figure 15.17 A demultiplexer with a grating in a Littrow mounting. (Source: reproduced with permission from *Applied Optics*, vol. 18, no. 16, August 1979, pp. 2834–6, 'Low-loss optical demultiplexer for WDM systems in the 0.8 µm wavelength region', by Koh-ichi Aoyama and Jun-ichiro Minowa.)

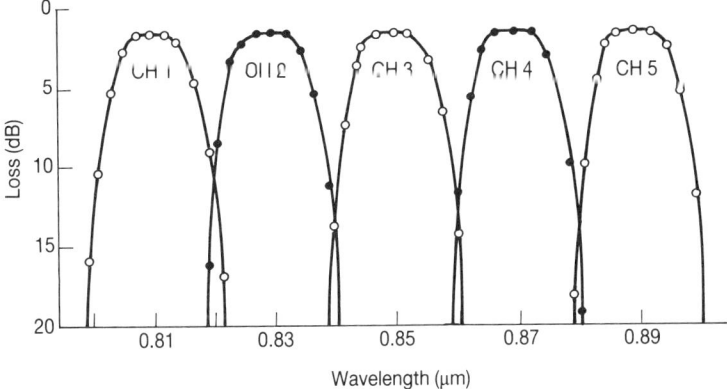

Figure 15.18 Loss characteristic of a grating wavelength demultiplexer. (Source: reproduced with permission from *Applied Optics*, vol. 18, no. 16, August 1979, pp. 2834–6, 'Low-loss optical demultiplexer for WDM systems in the 0.8 μm wavelength region', by Koh-ichi Aoyama and Jun-ichiro Minowa.)

characteristic as a function of the wavelength for that different channels. This example has channels separated by 20 nm, a separation of 2 nm has been reported. Well-designed demultiplexers with a grating can have insertion losses as low as 1.5 dB.

The next type of demultiplexer is based on the Mach–Zehnder interferometer. This interferometer consists of optical circuits in which an incoming beam is split into two beams that are recombined after a certain optical path length

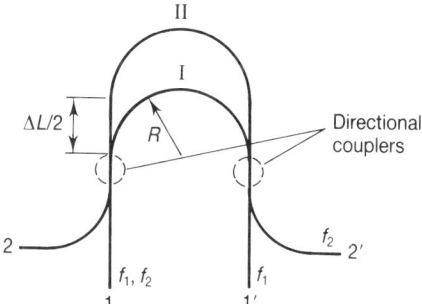

Figure 15.19 An asymmetrical Mach–Zehnder interferometer that can be used as a wavelength demultiplexer. (Source: reproduced with permission from *IEEE/OSA Journal of Lightwave Technology*, vol. LT-6, no. 6, June 1988, pp. 1003–10, 'Silica-based single-mode waveguides on silicon and their application to guided-wave optical interferometers', by N. Takato, K. Jinguji, M. Yasu, H. Toba and M. Kawachi.)

(see Figure 15.19). When the difference in path length is ΔL, we can write the transfer function from input 1 to output 1' as

$$H_{11}(\lambda_0) = \sin^2\left(\frac{\pi n}{\lambda_0}\ \Delta L\right) \tag{15.5}$$

and from input 1 to output 2'

$$H_{21}(\lambda_0) = \cos^2\left(\frac{\pi n}{\lambda_0}\ \Delta L\right) \tag{15.6}$$

where n is the refractive index of the waveguides and λ_0 is the wavelength in vacuum. The transfer functions have been plotted in Figure 15.20 for a certain configuration [9]. The remarkable characteristics of this type of demultiplexer are the closeness of the channels – as close as 10 GHz – and the steepness of the fall of the transfer function. This sharp drop can be evened out with a series of different interferometers, as shown in Figure 15.19. In this way, the demultiplexer has more outputs, and becomes a multichannel demultiplexer [9]. Evening out both the stop-band and the pass-band can be done by coupling one of the interferometer paths to a ring resonator [10].

Mach–Zehnder demultiplexers can be constructed accurately using planar waveguide techniques. Their tuning is accomplished by heating one of the paths, thereby changing the optical path length [9, 10].

Both grating and Mach–Zehnder demultiplexers show a periodic transfer function and this must be taken into account when designing a system.

Wavelength multiplexing is suitable for full-duplex transmission along a single

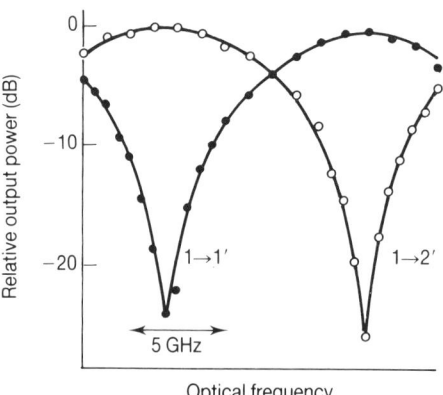

Figure 15.20 Demultiplexing characteristic of a Mach–Zehnder interferometer. (Source: reproduced with permission from *IEEE/OSA Journal of Lightwave Technology*, Vol. LT-6, no. 6, June 1988, pp. 1003–10, 'Silica-based single-mode waveguides on silicon and their application to guided-wave optical interferometers', by N. Takato, K. Jinguji, M. Yasu, H. Toba and M. Kawachi.)

fiber; each direction of transmission uses a different wavelength. However, full-duplex transmission can also be implemented with one wavelength for transmission in both directions. An optical hybrid or directional coupler is then needed, e.g. in the form of a fused biconical taper like that in Figure 15.15(a).

It will be pointed out in Chapter 16 that coherent detection is very selective for wavelength demultiplexing. Coherent systems with many densely packed channels should be provided with an optical demultiplexer, such as a grating or a Mach–Zehnder demultiplexer, in order to limit the total optical power received, otherwise the local oscillator laser power has to be unnecessarily large (see Chapter 16).

15.8 Repeater distance and link budget

When determining the repeater distance for a fiber optic digital system, two limiting factors have to be considered: dispersion and attenuation. Firstly, we consider multimode fibers operating in the short wavelength range 800–900 nm. In this case, the limits for the repeater spacing are given in Figure 15.21 as a function of the bit rate; in this figure the solid lines represent the attenuation limits and the dashed lines represent the dispersion limits (it has been assumed that the mode dispersion dominates the total dispersion and increases linearly with the fiber length). From Chapter 10 we know that this is only valid for lengths shorter than the coupling length $L_c = 1/C$ and for longer lengths the dispersion limit has a slope of $-1/2$. For the attenuation two values are given: 1 and 6 dB km^{-1}. Although 6 dB km^{-1} seems to be quite large, it should be kept in mind that splices and connectors have also to be included. If we take a standardized bit rate of 140 Mbit s^{-1} and a graded index fiber, it can seen that

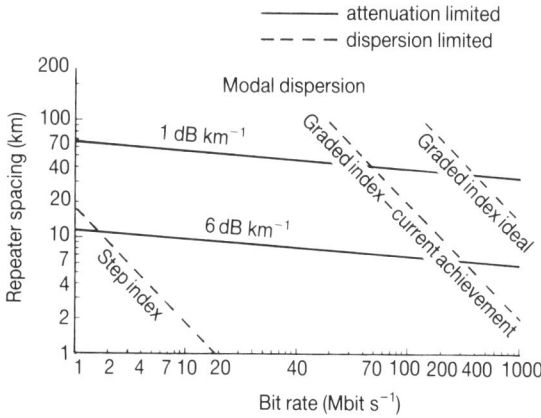

Figure 15.21 Repeater spacing as a function of the bit rate for multimode fibers operating in the wavelength region 800–900 nm.

such a system is attenuation limited. The diagram in Figure 15.21 gives only a rough description. In a more concrete situation, a link budget should be made and this topic will be dealt with below.

We next consider single-mode fibers in the long wavelength range 1.3–1.55 μm, and Figure 15.22 shows the repeater distances for this range. Characteristics are shown for a wavelength of 1.3 μm, where the fiber attenuation is 0.5 dB km^{-1}, while for a wavelength of 1.55 μm the fiber attenuation is 0.2–0.3 dB km^{-1}. Various dispersion limits are shown, namely for a Fabry–Perot laser (FP) combined with a standard single-mode fiber (SM) or a dispersion-shifted single-mode fiber (DS), and for a DFB laser combined with these two fiber types. From Figure 15.22 it follows that the FP laser systems are limited by mode partition noise (MPN); however, DFB laser designs are limited by chirp in the DFB/SM combination and by the signal bandwidth itself in the DFB/DS combination.

For a link budget we have to determine the sensitivity of the receiver and how much power can be coupled from the light source to the fiber. The difference between these two values may be lost in the fibers and the system's components such as splices, connectors, wavelength division (de)multiplexers [W(D)Ms] etc.

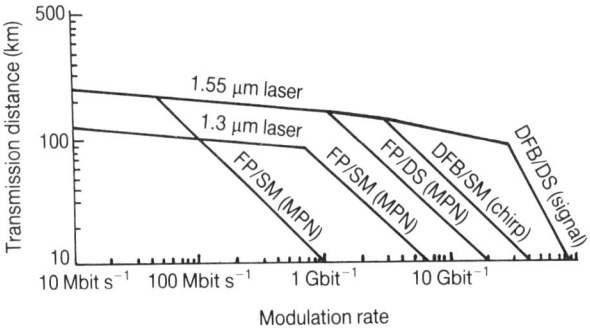

Figure 15.22 The repeater distance of single-mode fibers in the wavelength region 1.3–1.55 μm.

Example 15.1

Let us consider a fiber communication system operating in the wavelength region of 800–900 nm; it has a laser for a light source that operates at a bit rate of 140 Mbit s^{-1}. Suppose that the power coupled to the fiber amounts to 2 mW; this is equivalent to 3 dBm. For the given bit rate, the receiver sensitivity can be −47 dBm, giving a link budget of 50 dB. We then need to consider all the components of the system and to add up their losses. Let us consider a wavelength-multiplexed system, for which both the inputs and the outputs of the WDMs must be interchangeable, by means of external optical cables. For this purpose each laser pigtail has to be spliced to the pigtail of a rack-mounted connector. In the W(D)M rack again there is a connector and a splice. The pigtail of the WM output (or the respective WDM input) is spliced to a short piece of fiber, which in turn is spliced to the outsdide plant cable. Assuming a loss in a connector of 0.5 dB, in a splice of 0.2 dB and in a W(D)M of 2 dB, the total loss in the exchange amounts to 3.8 dB (see Figure 15.23).

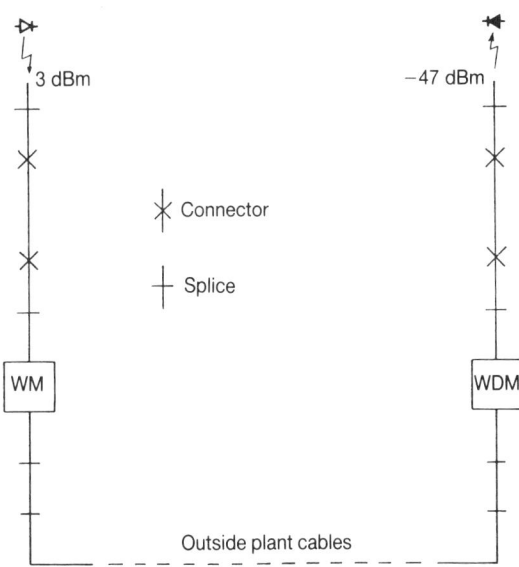

Figure 15.23 Diagram of a connection between the transmitter and receiver in two exchanges.

	dB
Budget	50
Loss in two exchanges	7.6
Available loss in outside plant cables	42.4

Assuming the loss in the fibers of the outside plant cables to be 3 dB km^{-1} and adding an excess loss of 0.2 dB km^{-1} for extra splices during the lifetime of the cables, it can be seen that a distance of 13.25 km between the exchanges can be covered. When the distance is larger than this, one or more repeaters are required.

Example 15.2

Let us now consider a single-mode fiber system operating at 1.55 μm. It is assumed that the same amount of power as before is launched into the fiber and that the same components are used as in the exchanges of Example 15.1 with the same losses. The receiver sensitivity is taken to be −35 dBm, thus the link budget becomes 38 dB.

	dB
Budget	38
Loss in two exchanges	7.6
Loss available for outside plant cables	30.4

Assume that the fiber loss is 0.25 dB km^{-1} with an excess loss of 0.2 dB km^{-1} for extra splices. This leads to an outside plant cable length of 67.6 km. It may appear strange that this is much less than would be indicated in Figure 15.22; however, in that figure no loss occurs in the exchanges and no extra splice losses were been taken into account.

15.9 Line coding

Line coding is used in digital transmission systems for various reasons. In optical fiber systems coding of the transmitted signal is needed to perform the following tasks:

1. To provide the signal with sufficient timing information.
2. To ensure that the signal is DC constrained.

At the receiving end there is a clock-regenerating circuit, e.g. a phase-locked loop, and this circuit must be able to extract a clock signal from the data signal; therefore, changes in the level of the data signal are required from time to time. In an NRZ signal consisting of long sequences of 1's or 0's, such changes are absent. The line coding has to change these sequences into sequences that show changes in level from time to time.

Although DC coupling is possible in optical fiber systems, it should be avoided for various reasons. A signal is called DC constrained if the mean value over a fixed interval length has a fixed value, independent of the bit pattern. In the case of an AC coupling, this interval should be small with respect to the time-constant of the coupling. This prevents any baseline wander, which can occur when the mean value of a signalling waveform depends on the bit transmitted (1 or 0) and a long sequence of 1 or 0 bits is generated by the information source. As an example, let us consider an NRZ rectangular pulse as the signalling waveform. In Figure 15.24(a) the data signal of a long sequence of 1's is shown; the response of the coupling is given in Figure 15.24(b). The mean value of a data signal consisting of the same number of 1's as 0's is 1/2 and this

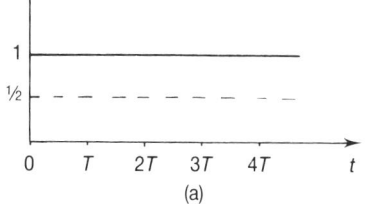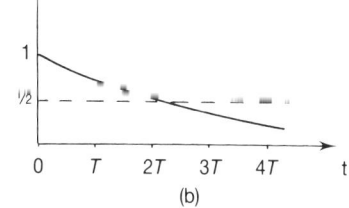

Figure 15.24 (a) NRZ data signal for a long sequence of '1's. (b) The corresponding baseline wander in an AC coupling.

forms the detection threshold. From Figure 15.24, however, it can be seen that the signal for a long sequence of 1's falls below the threshold after an interval of a few bits.

15.9.1 1B2B coding

A simple code that can meet both of the above requirements is the split-phase or Manchester code. In this code each signalling waveform consists of a double pulse (see Figure 15.25) and it can be considered as the conversion of each bit into two bits. With this code the line rate is twice the information bit rate, which can be a disadvantage. This code can also be considered as pulse position modulation or a special form of an NRZ signalling waveform. The advantages of this code are its simplicity and the fact that the signalling waveforms are themselves DC constrained over the shortest possible time. This latter reason explains why the code is well-suited for direct modulation of the laser in coherent FSK systems. Switching from one logical level to another in such a system means that the current is changed, but this implies a different value for the dissipation and the change in the chip temperature, which in its turn causes chirping of the laser frequency. The Manchester code can limit this chirping to a great extent.

Another form of 1B2B coding is the binary alternate mark inversion (AMI) coding [11]. The 1's are represented alternately by a high level and a low level, while the 0's are represented by a double pulse, as shown in Figure 15.26.

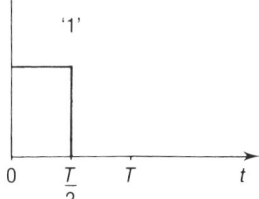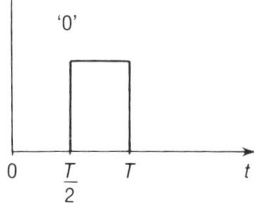

Figure 15.25 Data signalling waveforms of the split-phase or Manchester code.

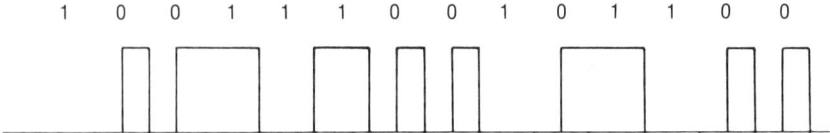

Figure 15.26 Binary AMI (alternate mark inversion) coding.

Coding and decoding is somewhat more complicated than for the split-phase code, but the bandwidth required is the same, namely twice as large as for straight binary signalling. Just like the Manchester code, the AMI code can limit chirping in direct modulation FSK coherent systems [12].

15.9.2 mBnB coding

The mBnB codes are used to encode blocks of m information bits into blocks of n line bits. Some well-known examples are the 2B3B, 3B4B and 5B6B codes [13]. These codes have the advantage that fewer bandwidths are required than for 1B2B codes; their relative bandwidth extension is n/m. One disadvantage is the need for a complicated method of encoding; in addition, decoding is very complex. Moreover, the information sequence has to be divided into blocks and this requires block synchronization in the receiver. Loss of synchronization introduces errors. The blocks of line bits are encoded so that the number of 1's within a sequence length of one or more blocks is as large as the number of 0's. In this way level changes occur regularly, which can provide the necessary clock information. These codes are not suitable for very high bit rates transmission, due to the processing time required for decoding.

15.9.3 Scrambling

A binary scrambler converts most bit sequences into pseudo-random bit sequences. The balance between 1's and 0's is not so certain as with mBnB coding, but it can be achieved in a statistical way, as the mean of a sequence

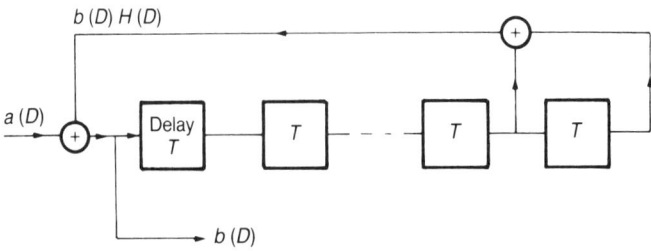

Figure 15.27 Self-synchronizing binary scrambler.

length equal to the length of the shift register of the scrambler; the same holds good for the clock information. Figure 15.27 shows an example of a self-synch-ronizing scrambler [14]; it consists of a shift register with feedback comprising a number of taps via devices that add modulo 2. The scrambled data is decoded by sending the encoded sequence to a shift register of the same length L when it is provided with the same feedback as the scrambler (see Figure 15.28). This can be described as follows. Denote the input sequence of the scrambler by $\{a_n\}$ and its corresponding D-transform by

$$a(D) \triangleq \sum_n a_n D^n \tag{15.7}$$

The transfer of the shift register with taps and modulo 2 adders, but without closing the feedback loop, is denoted by $H(D)$. If the D-transform of the output bit sequence $\{b_n\}$ of the scrambler is written as

$$b(D) \triangleq \sum_n b_n D^n \tag{15.8}$$

then it follows from Figure 15.27 that

$$b(D) = a(D) + b(D)\, H(D) \tag{15.9}$$

Special types of multiplication and addition are needed in this equation, namely those defined in the Galois field $(0, 1)$ [15]. Equation (15.9) can be rewritten as

$$b(D) = \frac{a(D)}{1 + H(D)} \tag{15.10}$$

When this sequence is supplied to the descrambler, the output $c(D)$ becomes

$$c(D) = b(D)[1 + H(D)] = a(D) \tag{15.11}$$

For the sequence $\{b_n\}$ to be pseudo-random, $1 + H(D)$ should be a primitive polynomial [15]. In that case, the free-running scrambler produces a maximum length sequence, in which all the binary words of length L occur, except the word that consists of L consecutive 0's. Consequently, it follows that it is advisable to use a shift register length L that is not too large. Although scramblers are difficult to analyze in concrete situations, it has been shown that they are satisfactory in practice. For lists of taps for different shift register lengths, see [15] or [16].

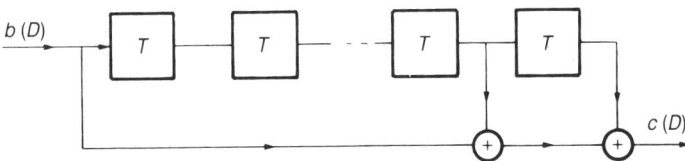

Figure 15.28 Self-synchronizing binary descrambler.

15.10 Selection of the system components

15.10.1 Selecting the operating wavelength

For long-haul transmission at high bit rates, only the long wavelength range 1.3–1.55 μm is available with modern silica fibers. This wavelength region gives low attenuation and a low dispersion for single-mode fibers.

For applications where the light-gathering capacity of a fiber from an incoherent source is an important feature and the bit rate is not very high (a few hundred Mbit s^{-1}), then multimode fibers are suitable for relatively short distances. In such cases (e.g. LANs) the short wavelength range of 800–900 nm can be used for transmission.

15.10.2 Selecting the light source

The characteristics of a laser not only change during its lifetime, but also depend on temperature. A control circuit is required to stabilize the output power and this makes laser diodes even more uneconomical. When designing a fiber optic system, it may be worthwhile to investigate whether an LED is acceptable as a light source from a technical point of view; however, when comparing its advantages and disadvantages, it should be remembered that the light gathered by a multimode fiber from an LED is 10–15 dB less than from an LD. With a single-mode fiber as the transmission medium, the difference amounts to some 20 dB. Moreover, an LD has a narrower optical spectrum, which can produce a smaller pulse broadening due to material dispersion. In a practical situation, the methods given in Chapters 4, 7 and 10 must be considered in order to decide which source should be used.

15.10.3 Selecting the detector

For short wavelengths, Si diodes are the obvious devices to use as the detector. Both PINs and APDs have excellent properties at low prices. For long wavelengths, the dark current is rather large, and therefore the performance of Ge and InGaAsP is not as good as that of Si.

When deciding whether to use an APD or a PIN diode, the criteria are much the same as for an LED versus an LD. The characteristics of APDs are not stable with changing temperatures, and stabilization circuits may be required. An APD is more expensive and requires a high voltage supply, whereas the PIN diode can be supplied with voltages in the same range as those in the rest of the receiving equipment. On the other hand, a receiver provided with an APD can be 10–15 dB more sensitive than a PIN receiver in the short-wavelength range. For long wavelengths, the dark current contributes significantly to the shot noise and an APD gives less improvement.

APDs are used for long-haul direct detection transmission systems, operating at a high bit rate. On the other hand, PIN diodes are suitable for low bit rates over shorter distances. Moreover, PIN diodes have to be used in coherent detection systems, where the received signal is boosted by the local oscillator power.

Figure 15.29 shows the receiver sensitivity as a function of the bit rate for both PIN and APD receivers. For details of the complete electronic circuits for transmitters and receivers, see [2] and [16].

15.10.4 Selecting the fiber

For long-haul transmission systems, such as the trunk network, only single-mode fibers are suitable. Plans for subscriber networks usually use this type of fiber as a starting point. In high bit rate applications such as a fiber should be combined with a laser light source. Specific services, such as signalling, telephony or low bit rate data communications, only need an LED, even for single-mode fibers, if it is used for high bit rates services at the same time. Very short distance communications, within an office for instance, can use silicone-clad or plastic-clad fibers. When only low bit rate services are involved or the distances are relatively small, as in a LAN, a multimode fiber should be selected. For short distances, mode mixing is insignificant and dispersion increases linearly with

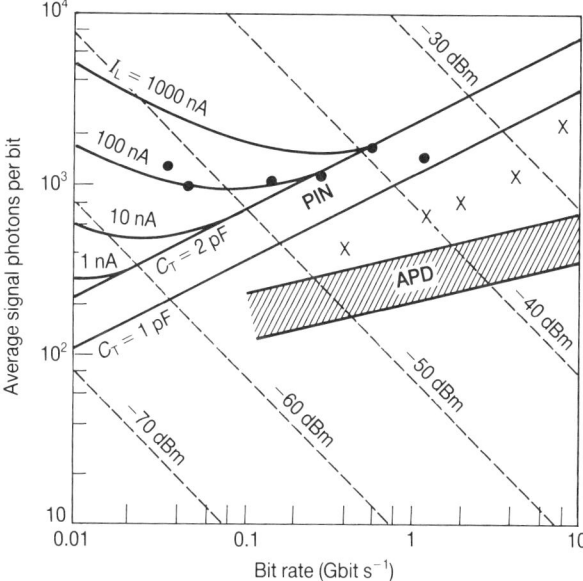

Figure 15.29 Sensitivity of optical receivers as a function of the bit rate. The dots and the crosses represent the best results for PIN and APD receivers, respectively. (Source: reproduced with permission from *IEEE Communications Magazine*, vol. 26, no. 4, April 1988, pp. 29–35, 'Multigigabit-per-second lightwave systems research for long-haul applications', by T. Li and R. Linke.)

length. If the fiber length is larger than the coupling length (see Chapter 10), then dispersion increases with the square root of the length. The coupling length can be shortened artificially by forced mode-mixing. This extra mode-mixing, however, increases the attenuation and decreases the modal noise.

References

[1] S. E. Miller and A. G. Chynoweth (eds), *Optical Fiber Telecommunications* (New York: Academic Press, 1979).

[2] Technical Staff of CSELT, *Optical Fibre Communication* (Torino: Centro Studi e Laboratori Telecomunicazioni, 1980).

[3] C. M. de Blok and P. Matthijsse, 'Core alignment procedure for single-mode-fibre jointing', *Electronics Letters*, vol. 20, no. 3, February 1984, pp. 109–10.

[4] G. D. Khoe and H. Lydtin, 'European optical fibers and passive components: status and trends', *IEEE Journal on Selected Areas in Communications,* vol. SAC-4, no. 4, July 1986, pp. 457–71.

[5] T. Sugie and M. Saruwatari, 'Distributed feedback laser diode (DFB-LD) to single-mode fiber coupling module with optical isolator for high bit rate modulation', *IEEE/OSA Journal of Lightware Technology*, vol. LT-4, no. 2, February 1986, pp. 236–45.

[6] T. Sugie and M. Saruwatari, 'An effective nonreciprocal circuit for semiconductor laser-to-optical-fiber coupling using a YIG sphere', *IEEE/OSA Journal of Lightware Technology*, vol. LT 1, no. 1, March 1983, pp. 121–30.

[7] E. Hecht and A. Zajac, *Optics* (Reading, Mass: Addison-Wesley, 1974).

[8] M. Born and E. Wolf, *Principles of Optics*, 5th edn (Oxford: Pergamon Press, 1975).

[9] N. Takato, K. Jinguji, M. Yasu, H. Toba and M. Kawachi, 'Silica-based single-mode waveguides on silicon and their application to guided-wave optical interferometers', *IEEE/OSA Journal of Lightwave Technology*, vol. 6, no. 6, June 1988, pp. 1003–10.

[10] K. Oda, N. Takato, H. Toba and K. Nosu, 'A wide-band guided-wave periodic multi/demultiplexer with a ring resonator for optical FDM transmission systems', *IEEE/OSA Journal of Lightwave Technology*, vol. 6, no. 6, June 1988, pp. 1016–23.

[11] Y. Takasaki, M. Tanaka, N. Maeda, K. Yamashita and K. Nagano, 'Optical pulse formats for fiber optic digital communications', *IEEE Transactions on Communications*, vol. COM-24, no. 4, April 1976, pp. 404–13.

[12] R. Noe, M. Maeda, S. Menocal and C. Zah, 'AMI signal format for pattern-independent FSK heterodyne transmision and two channel cross talk measurements', *Proceedings Fourteenth European Conference on Optical Communication*, Brighton, UK, September 1988.

[13] M. Rousseau, 'Block codes for optical fiber communication', *Electronics Letters*, vol. 12, no. 18, September 1976, pp. 478–9.

[14] D. Leeper 'A universal digital data scrambler', *Bell System Technical Journal*, vol. 52, no. 10, December 1973, pp. 1851–65.

[15] W. W. Peterson, *Error Correcting Codes* (New York: MIT Press and Wiley, 1961).

[16] S. D. Personick, *Optical Fiber Transmission Systems* (New York: Plenum, 1981).

16

Coherent optical fiber communication

16.1 Introduction
16.2 Basic principles of coherent optical systems
16.3 Signal-to-noise ratio of coherent optical receivers
16.4 Balanced mixing and phase diversity reception
16.5 Polarization aspects of coherent systems
16.6 Concluding remarks
References

16.1 Introduction

So far in this book only intensity-modulated (IM) optical carriers have been considered. At the receiving end of a communication system, the information-carrying signal is recovered by detecting the power of the optical wave that is received. This kind of detection is called direct detection (DD), and systems operating in this way are called IM/DD systems. Intensity modulation is a form of amplitude modulation (see also Chapter 12). Are other modulation formats, such as frequency modulation and phase modulation, suitable for optical fiber systems, and if so, do they have any advantages over IM/DD systems? These questions are answered in the affirmative, as will be shown below, but implementing these types of modulation requires more complex circuits, especially at the receiving end of the communication link. It becomes clear that FM or PM signals can never be recovered by direct detection; however, if the received optical wave is mixed with light from a local oscillator laser, an IF signal is created, which contains both the phase and frequency information of the transmitted signal. This technique, which is well known in radio reception, is called heterodyne reception when the local oscillator has a different frequency from the transmitter laser, and homodyne reception when the local oscillator is sychronized in phase with the transmitting laser. Optical homodyne reception is a very difficult technique to implement for two reasons. Firstly, it requires accurate matching of the wavefronts, and secondly, the bandwidth of the phase-locked loop and the linewidth of the lasers are critical [1]. Semiconductor lasers cannot meet the linewidth requirements without additional provisions, such as an external cavity.

Coherent detection has two big advantages:

1. Higher selectivity. Mixing the information signal to a frequency band that can be processed electronically allows a selective electrical filtering to be used, rather than optical filtering of the different multiplexed channels (see Chapter 15). It is therefore possible to place the channels close together and thereby make more efficient use of the bandwidth potential. The channel selection is easily done by tuning the local oscillator laser.
2. More sensitivity. By mixing the information signal with a much larger power from the local oscillator laser the signal received is amplified. In this way, the receiver noise (due to the dark current of the photodiode and the thermal noise of the amplifier) becomes less significant. This is of special interest in the long wavelength range, where the dark current is considerable. Coherent systems can approach the quantum limit much closer than IM/DD systems.

A disadvantage of coherent systems is that the receiver is much more complicated. In addition, laser diodes possessing extremely high spectral purity and stability are required. For a high bit rate heterodyne system, DFB or DBR lasers (see Section 8.3.1) meet the requirements.

16.2 Basic principles of coherent optical systems

Unlike direct detection, where the received optical signal is directly applied to a photodiode which generates a current proportional to the optical power received, the coherent receiver adds a locally generated optical wave to the optical received wave, before sending the combined waves to a photodiode (see Figure 16.1). In Figure 16.1 the photodiode current contains all the information of the signal, including power, phase and frequency, but converted to a lower frequency so that further signal processing can be performed easily with electronic circuitry. The combination of two optical waves is achieved with either a beam splitter or a fiber coupler (see Section 15.7).

Let us suppose that both the signal wave and local oscillator wave are

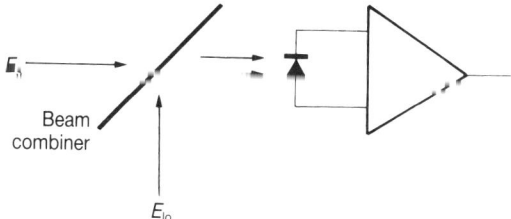

Figure 16.1 Principle of operation of the coherent optical receiver.

polarized linearly in the same direction and let us denote these two waves respectively by

$$E_s = E_o \exp\{j[\omega_o t + \phi_o(t)]\} \tag{16.1}$$

$$E_{lo} = E_1 \exp\{j[\omega_1 t + \phi_1(t)]\} \tag{16.2}$$

The addition of these two waves is done in the beam combiner and the photodiode gives a current proportional to the power of the sum of the waves

$$
\begin{aligned}
I &\propto (E_s + E_{lo})(E_s + E_{lo})^* \\
&= E_o^2 + E_1^2 + 2E_o E_1 \cos[(\omega_o - \omega_1)t + \phi_o(t) - \phi_1(t)] \\
&= P_o + P_1 + 2\sqrt{P_o P_1} \cos[(\omega_o - \omega_1)t + \phi_o(t) - \phi_1(t)]
\end{aligned}
$$

$$(16.3)$$

In this expression the first term P_o is the power of the signal received and the second term P_1 is the power of the local oscillator. The third term is the most interesting one; the first term, which provides the photocurrent for direct detection, is weaker by a factor $2\sqrt{P_1/P_o}$. Moreover, the third term also contains all the information of the received signal, namely the amplitude $\sqrt{P_o}$, the frequency ω_o and the phase $\phi_o(t)$. The magnitude of this third term can be increased by boosting the local oscillator power P_1. When the local oscillator is phase-locked to the signal wave, the third term represents a baseband signal; the receiver is then called a homodyne receiver. In the absence of this phase-lock, the third term is an IF signal and the receiver is a heterodyne type. This is further explained by means of Figure 16.2, where the solid line in (a) represents the electric field E_s of the received optical signal, whereas the broken line represents the electric field E_{lo} of the local oscillator laser. For the sake of clarity the phase noise terms $\phi_o(t)$ and $\phi_1(t)$ are taken to be zero for this diagram. The depicted time interval contains 20 cycles of the signal wave and 18 cycles of the local oscillator wave. Figure 16.2(b) shows the sum of these two waves (solid line), and the broken line represents the envelope of this sum. A simple goniometric formula shows that this envelope contains half the difference frequency of the two constituting waves, which means in this case one cycle on

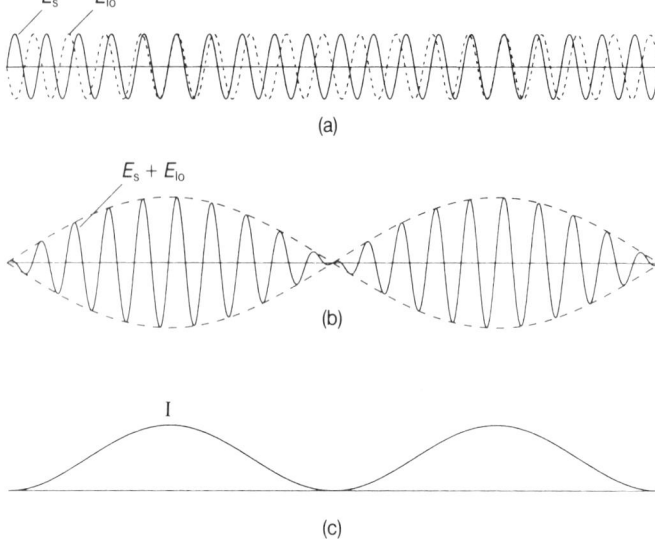

Figure 16.2 Mixing of the optical waves E_s and E_{lo}. (a) The two waves E_s and E_{lo}. (b) $E_s + E_{lo}$. (c) The resulting photodiode current.

the depicted interval. The sum of the waves impinges on the photodiode, and produces a photocurrent that is proportional to the power of the wave $E_s + E_{lo}$. This power, in its turn, is proportional to the square of the envelope of the wave, so that the photodiode current, which is depicted in Figure 16.2(c), contains a harmonic component with twice the frequency of the envelope of Figure 16.2(b). In the given example it means that this frequency is the difference of the frequency of the local oscillator and the signal frequency, and results in two cycles on the given interval.

Before practical coherent systems can be implemented, the following technical difficulties must be overcome [2, 3]:

1. Frequency-stable light sources must be available. Semiconductor lasers are most widely used as the light sources in optical fiber communication systems. Typical IF frequencies lie in the range 0.2–2 GHz, which is 10^{-6}–10^{-5} times greater than the signal frequency; this puts severe restrictions on the stability of the semiconductor laser frequency.
2. Both the amplitude and the phase of semiconductor lasers are noisy. This noise decreases the signal-to-noise ratio and reduces the performance of the system. The spectral purity of the lasers therefore has to be improved and their linewidth reduced.
3. For homodyne systems, the local oscillator laser has to be phase-locked to the optical wave received and this requires a small laser linewidth and a large

bandwidth for the phase-locked loop. For heterodyne systems, frequency-locking is sufficient [4] and easier to achieve; in the receiver, the local oscillator laser tracks the frequency of the optical wave received.

4. Since small linewidth lasers keep their spectral purity only when back reflections are prevented from entering the laser cavity, these lasers must be isolated optically. Small, cheap optical isolators (see Section 15.5), which can be built into the laser housing, must be developed.

5. Amplification of the received signal by the local oscillator is only effective when the polarization states of the two waves coincide. Random fluctuations in the polarizations can be prevented by using polarization-maintaining fibers (see Chapter 15), or handled by either a polarization control scheme or a polarization diversity scheme; they will be dealt with later on in this chapter.

6. Further research is needed to determine which modulation/demodulation scheme gives the best performance when used in an optical coherent communication system.

Modulation of the optical wave transmitted can be either direct, by means of the semiconductor laser drive current, or indirect by means of an external modulator. Although direct modulation seems to be the obvious choice for AM, it has a big disadvantage, namely that modulating the drive current not only modulates the optical power but also the laser frequency. This frequency is modulated by $0.5–3 \, \text{GHz} \, \text{mA}^{-1}$ [5], which can cause undesirable chirping of the laser frequency. Chirping causes more dispersions in the fiber and leads to serious problems in a coherent receiver, due to an undefined IF frequency. Direct FM is quite simple. Small modulations of the current can produce a relatively large frequency modulation, while modulation of the amplitude remains small. External modulation seems to be the best method for AM and PM. Several types of external modulator have been developed, either using semiconductor material [6, 7], or glass such as LiNbO_3 [8]. Modulation of the intensity may be based on one of the following physical effects, which can be varied in an electric field: loss, directional coupling of optical waveguides, Mach–Zehnder interferometry or total internal reflection. External intensity modulation does not yet guarantee the absence of chirping. The loss modulator and the directional coupler when used specifically show chirping just as in the direct modulation case. On the other hand, the Mach–Zehnder and total internal reflection modulators are chirp free, and the same holds for the directional coupler modulator when the cross-coupled signal is used as modulated output [9]. Phase modulation is accomplished by changing the refractive index of a semiconductor material or LiNbO_3 waveguide under the influence of a transverse electric field applied via electrodes on either side of the waveguide.

In any case, the temperature of a semiconductor laser has to be stabilized, since its frequency is very sensitive to temperature changes, namely in the range $10–20 \, \text{GHz} \, \text{K}^{-1}$ [5]. This type of laser requires a temperature stability of $10^{-2}–10^{-3} \text{K}$.

Due to the rigid requirements for the coherence of the optical wave that is received, only single-mode fibers can be used in coherent detection systems. The purity and stability of the laser spectrum lead to a small pulse broadening due to dispersion of the fibers. Coherent systems thus permit very large bandwidth transmission, especially when the operating wavelength is in the vicinity of the dispersionless wavelength.

16.3 Signal-to-noise ratio of coherent optical receivers

For the signal-to-noise ratio, we believe that the third term of equation (16.3) is the most interesting with regard to the information-carrying signal, although all the terms do contribute to shot noise. However, when the signal has a zero mean value, the mean value of the shot noise variance is determined by the second term, namely the power of the local oscillator. The signal term is so large that it serves no purpose to use an APD (see Chapter 12). On the basis of the signal-to-noise ratio consideration in Chapter 12, especially equation (12.20), it is concluded that the signal-to-noise ratio for the coherent detection scheme can be rewritten as

$$\frac{S}{N} = \frac{2R^2 P_o P_1}{2Be(RP_1 + e\lambda_b + e\lambda_d) + [\sigma^2_{th}/H^2(0)]} \tag{16.4}$$

When calculating this S/N, it has to be noted that the mean power of the signal term $2\sqrt{P_o P_1}\cos[(\omega_o - \omega_1)t + \phi_o(t) - \phi_1(t)]$ reads $2P_o P_1$. Equation (16.4) can be rewritten as

$$\frac{S}{N} = \frac{(RP_o)^2}{Be[RP_o + (P_o/P_1)e(\lambda_b + \lambda_d)] + [\sigma^2_{th}/H^2(0)]/(P_o/2P_1)} \tag{16.5}$$

This expression is valid for the heterodyne case. For a homodyne receiver, the cosine in the signal term always amounts to unity and so the signal power and signal-to-noise ratio double. The improved sensitivity of coherent detection becomes apparent when equation (16.5) is compared with equation (12.20), which is the signal-to-noise ratio expression for direct detection. In practical situations, the local oscillator power P_1 is much larger than the signal power P_o, when the signal has traversed a long length of fiber. The comparison shows that the photocurrent from the background radiation and, more importantly, from the dark current in the coherent system, are both reduced by a factor of P_o/P_1. Moreover, the thermal noise is reduced by a factor of $P_o/(2P_1)$. In this case, the local oscillator power acts in the same way as the avalanche gain of an APD, but it also reduces the influence of all the shot noise terms that are uncorrelated to the signal. In this way, a coherent system can approach the shot noise limit very closely. In order to calculate this shot noise limit for various modulation/demodulation methods, let us consider, firstly, a binary ASK (OOK) system with coherent (electrical) detection of the IF signal. For this purpose, let us assume that the (electrical) local oscillator has an amplitude of unity. In the noise considerations all terms in the denominator of equation (16.5) that involve the factor P_u/P_1 are neglected. Moreover, the noise is independent of the transmitting level within this approximation. Due to the large value of the shot noise variance, which is dominated by P_1, the noise can be assumed to have a Gaussian distribution. The double-sided noise spectral density of the IF signal amounts to eRP_1. After the electrical homodyning, the spectral density of the noise becomes $eRP_1/2$ [10], whereas the energy of the signal in a '1' bit interval reads $R^2 P_o P_1 T$. For a detection threshold midway between the two signal levels received and equal probabilities of the 1's and 0's, the error probability after matched filtering of the homodyned signal becomes

$$P_e = Q\left(\frac{1}{2}\sqrt{S/N}\right) = Q\left(\frac{1}{2}\sqrt{\frac{R^2 P_o P_1 T}{eRP_1/2}}\right) = Q\left(\sqrt{\frac{RP_o T}{2e}}\right) \qquad (16.6)$$

For an error probability of 10^{-9}, the argument of the Q function must equal 6. This yields

$$\frac{\eta E}{h\nu} = 72 \qquad (16.7)$$

where E is the energy received during a '1' pulse; therefore, these pulses should consist of at least 72 photons on average and, since no photons are transmitted during a '0', this gives a mean value of 36 photons per bit. Based on this calculation, it is easy to arrive at the number of photons required per bit for other modulation/demodulation methods. For dual-filter FSK the mean signal power received is twice as large as in the OOK case, but the output noise variance is also doubled, so that the signal-to-noise ratio and the number of photons required per bit are the same as for OOK. For PSK, the signal output doubles, due to the phase reversal of the received signal, whereas the noise variance remains the same. This means that, in this case, half as many photons per bit are required compared to the ASK and FSK schemes [11].

Optical homodyning instead of heterodyning results, as has already been shown, in a doubling of the signal-to-noise ratio and thus in a halving of the required number of photons per bit.

Table 16.1 summarizes the average number of photons per bit for a bit error

Table 16.1 Average number of photons per bit for a bit error probability of 10^{-9} using an ideal receiver for different modulation methods

	Heterodyne	Homodyne	Direct detection
ASK	36	18	10
FSK	36	18	–
PSK	18	9	–

probability of 10^{-9}, which can be achieved with an ideal receiver, for different modulation methods. For direct detection, a quantum limit of 10 photons per bit (see Section 13.3) is obtained. This number is based on a Poisson distribution for the shot noise, whereas a Gaussian distribution would lead to 18 photons per bit. Although the value of 10 photons per bit in Table 16.1 seems to be quite low compared to some others, it has to be emphasized that it is practically impossible to reach this limit, due to the thermal noise in the receiver and the shot noise from the dark current. For coherent receivers, both of these noise phenomena are suppressed with respect to the signal, thereby making it possible to approach the limits much more closely than in a direct detection system. The figures in Table 16.1 are based on the assumption that all the input signal power in the beam combiner (see Figure 16.1) is preserved in the output light beam. A beam splitter or optical coupler, however, has two output ports, the second of which also carries a certain amount of the signal power. The splitter or coupler can be designed so that most of the signal power is coupled to the input of the receiver; as a consequence, only a little of the power from the local oscillator is coupled to it. As long as $P_1 \gg P_0$ the performance limits of the receiver are like those given in Table 16.1. P_1 and P_0 mean, respectively, the part of the local oscillator power and the signal power at the output port that is coupled to the photodiode.

16.4 Balanced mixing and phase diversity reception

In the preceding section, we emphasized a disadvantage of using only one of the two ouputs of optical combiners such as beam splitters and optical fiber couplers. We mentioned the loss of power at the second, or unused output port. By detecting the output power from the second port, the receiver efficiency can be improved; however, using this second port may offer another advantage. Up to this point, it has been assumed that the local oscillator has no amplitude noise; nevertheless, in practice, this type of noise, also called relative intensity noise (RIN, see also Chapter 9), is often the limiting factor, because the local oscillator power is strong compared with the signal component $2 \sqrt{P_0 P_1}$. A correct combination of the power from the two output ports cancels out most of this noise term, and this technique is described below.

For optical beam splitters and fiber couplers the transmitted wave has a phase shift of $\pi/2$ with respect to the reflected and cross-coupled wave, if the components are assumed to have no losses [12, 13]. Denoting the output waves as

$$E_1 = aE_0 \exp\left\{ j\left[\omega_0 t + \phi_0(t) + \frac{\pi}{4} \right] \right\} + bE_1 \exp\left\{ j\left[\omega_1 t + \phi_1(t) - \frac{\pi}{4} \right] \right\}$$

(16.8)

$$E_2 = bE_o \exp\left\{j\left[\omega_o t + \phi_o(t) - \frac{\pi}{4}\right]\right\} + aE_1 \exp\left\{j\left[\omega_1 t + \phi_1(t) + \frac{\pi}{4}\right]\right\}$$

$$(16.9)$$

it is readily seen that detecting these output waves by means of photodiodes gives the corresponding currents

$$I_1 \propto a^2 P_o + b^2 P_1 + 2ab\sqrt{P_o P_1} \cos\left[\omega_o t - \omega_1 t + \phi_o(t) - \phi_1(t) + \frac{\pi}{2}\right]$$

$$(16.10)$$

$$I_2 \propto b^2 P_o + a^2 P_1 - 2ab\sqrt{P_o P_1} \cos\left[\omega_o t - \omega_1 t + \phi_o(t) - \phi_1(t) + \frac{\pi}{2}\right]$$

$$(16.11)$$

The RIN is contained in the terms $b^2 P_1$ and $a^2 P_1$ respectively. If we assume a symmetrical beam combiner (where $a = b$), and the currents I_1 and I_2 are subtracted, then the terms containing P_o and P_1 cancel out, and so do their amplitude fluctuations. Due to the opposite sign of the third term in equations (16.10) and (16.11), the output signal from the subtractor is doubled. Using the two outputs in this way produces a balanced mixing receiver, which is shown diagrammatically in Figure 16.3(a). Figure 16.3(a) represents a receiver with a half-reflecting and half-transmitting mirror such as a bulk beam splitter. Figure 16.3(b) shows a fiber coupler and series connection of the two photodiodes that subtract the photodiode currents automatically. Therefore, this arrangement has the additional advantage of not needing amplification of the DC terms. The shot noise terms in the two beams might have cancelled out each other, if it were not for the fact that the splitting of the optical waves can be envisaged as a random partitioning of photons, so that the sum of their shot noise terms has the same statistics as the unpartitioned beam. The receivers in Figure 16.3 do not, of course, cancel out the RIN that is included in the signal terms involving $\sqrt{P_o P_1}$.

The two-way coupler described above raises two questions: what are the phase relationships of the currents in 3×3 and 4×4 multiport couplers, respectively, and what effect may this have on coherent receivers? Multiport couplers are produced by coupling the fields mutually with three or four optical waveguides. If the waveguides consist of optical fibers, the coupling is achieved by fusing and tapering the required number of fibers and a 3×3 coupler is an appropriate configuration. Alternatively, the coupler can be made in a planar structure, either in a glass substrate or a semiconductor material. Then the 4×4 multiport coupler seems to be more suitable. The multiport couplers are also called multiport optical hybrids. When each output signal is detected by a photodiode, the signal components of the N currents ($N = 3, 4$) have a $2\pi/N$ phase difference

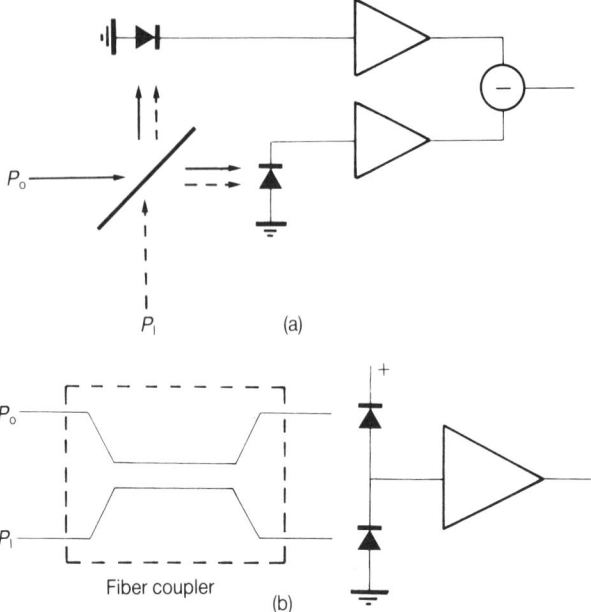

Figure 16.3 Balanced mixing receiver. (a) Using a beam splitter. (b) Diagram with fiber coupler and series connection of the photodiodes.

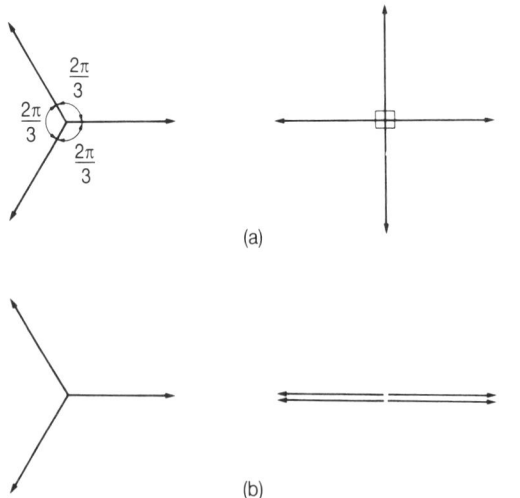

Figure 16.4 (a) The phases of the currents detected in a $N \times N$ multiport optical hybrid. (b) The phases of the double frequency terms after squaring the current terms.

between them [14], provided that each output has the same output amplitude. This has been depicted in Figure 16.4(a). When the currents are squared, a baseband term arises, as well as a double frequency term. The phases of the double frequency terms are shown in Figue 16.4(b) and it follows that adding the squared current terms cancels out the double frequency terms, whereas the baseband terms increase the signal in a single branch N times. Figure 16.5 shows a general diagram of a three-branch, multiport, optical coherent receiver. In Figure 16.6 the various signals in the system as in Figure 16.5 are given. Figure 16.6(a) shows the three photodiode currents, with a mutual phase shift of $2\pi/3$. In this figure ω_{if} is taken to be much smaller than the bit rate. For the sake of clearness we have drawn a bit pattern consisting of an alternating sequence of 1's and 0's. From this diagram it becomes clear that when in a certain branch the signal is lost, due to a zero crossing of the IF signal, the other two branches still show a significant signal contribution. This results in a constant value of the output signal when square-law detectors are used (see Figure 16.6b) and an output signal with a small ripple when envelope detectors are used (see Figure 16.6c). The four-branch receiver is very similar.

In a practical heterodyne receiver, the IF frequency $\omega_o - \omega_1 = \omega_{if}$ has to be made two or three times as great as B, where B is the baseband signal bandwidth. The bandwidths of the photodetector and the IF electronic circuits should extend to at least $\omega_{if} + B$. This condition is difficult to meet for high bit rates. Thus, ω_{if} should be kept as small as possible, so that $\omega_{if} = 0$ is the ideal case (and means homodyning). But, as can be seen from equations (16.3), $\omega_{if} = 0$ without phase-locking produces a photodiode current that is proportional to $\cos[\phi_o(t) - \phi_1(t)]$. Then, the current becomes zero when $\phi_o(t) - \phi_1(t) = \pi/2$, thus rendering the system useless. Keeping the phase difference near to zero by

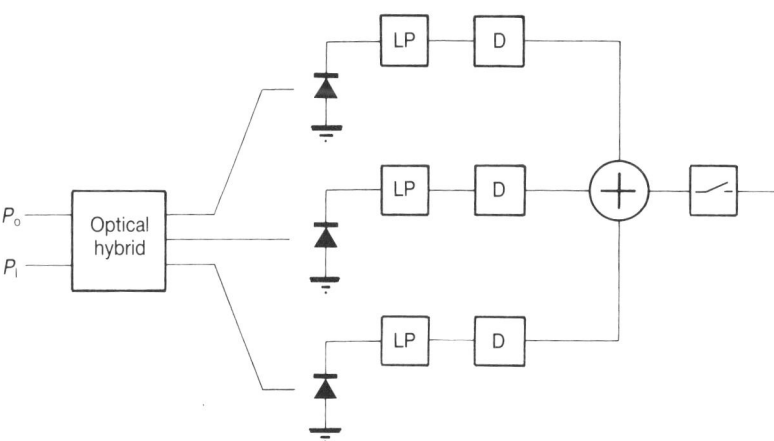

Figure 16.5 General diagram of a three-port optical coherent phase diversity receiver with low-pass filters (LIP) and detectors (D).

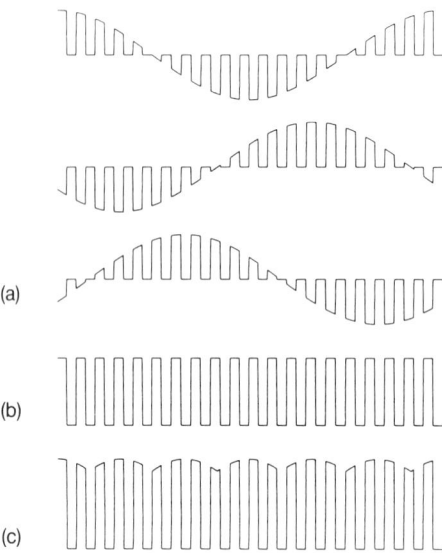

(a)

(b)

(c)

Figure 16.6 The signals in a 3 × 3 phase diversity receiver. (a) The three photodiode currents. (b) The output signal when squarers are used as the detectors. (c) The output signal when using envelope detectors.

means of phase-locking is the best option from the point of view of the receiver bandwidth and gives the most sensitive receiver. However, phase-locking requires an extremely narrow linewidth laser, and no suitable semiconductor lasers are available yet [15]; therefore a phase-locked homodyne receiver with injection lasers is impractical. But, as we have seen, the phase diversity receiver can convert the received signal to baseband without using large bandwidth IF circuits or the need for phase-locking. From the bandwidth point of view, therefore, this receiver behaves like a homodyne system. However, the optical mixing is not phase synchronized, and this makes its final performance like that of a heterodyne receiver. Many authors use the name homodyne for multiport phase diversity reception, but we prefer to use the term pseudo-homodyne receiver. The phase diversity receivers whose phasor diagrams are presented in Figure 16.4, however, cannot suppress the RIN like the balanced mixing receiver of Figure 16.3. The modified three-phase diversity scheme of Figure 16.7 combines the advantages of both receivers [16]. From the phasor diagram in Figure 16.8, it follows that the differential amplifier outputs have the same mutual phase differences as the photodiode currents. Each diode current shows the same RIN and their noise components are in phase. Subtracting the currents two at a time cancels out the RIN. A similar method can be used for a 4 × 4 optical hybrid, as shown in Figure 16.9.

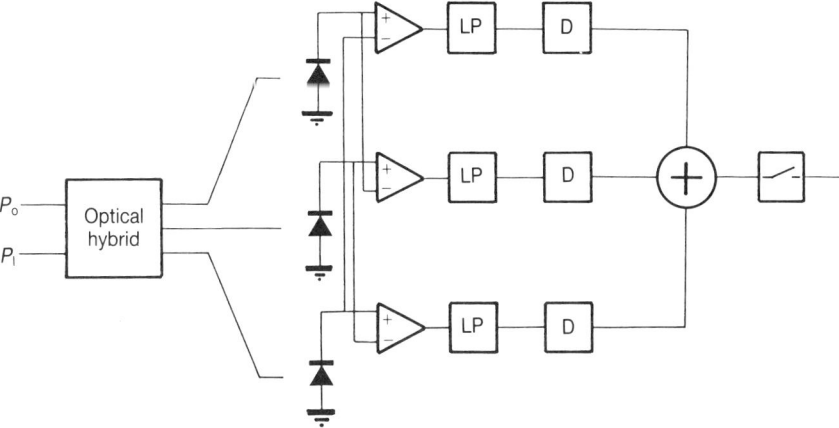

Figure 16.7 The modified 3×3 multiport phase diversity receiver which provides RIN suppression.

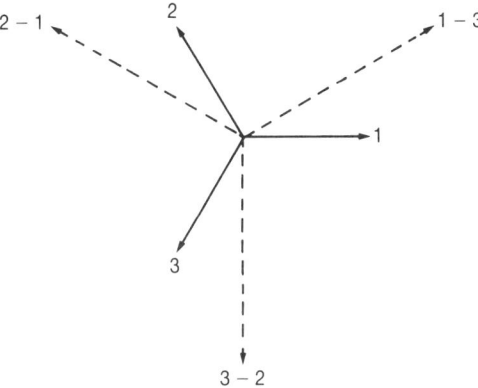

Figure 16.8 Phasor diagram for the modified 3×3 multiport receiver. The solid line arrows represent the phases of the diode currents, whereas the broken line arrows represent the outputs of the differential amplifiers.

Using these modifications leads to complicated receiver structures; however, they can be simplified considerably in the case of a 4×4 multiport. This can be explained by considering Figure 16.9. If the signal represented by the vector 3 is subtracted from vector 1 and, moreover 4 is subtracted from vector 2, then the two resulting signal vectors show a phase difference of $\pi/2$. Squaring and adding these signals results in an information baseband signal, whereas the double frequency terms cancel out, as before. Subtracting the input signal vectors can be done by two paired photodiode series connections, in the manner shown in

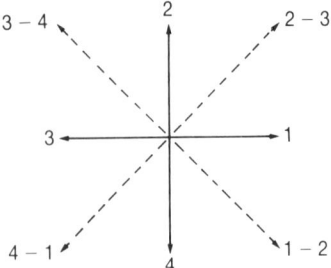

Figure 16.9 The phases in a 4×4 modified phase diversity receiver. Solid line arrows are the photodiode currents and broken line arrows are the differential amplifier outputs.

Figure 16.3(b). The resulting receiver – called a double balanced mixer receiver – is shown in Figure 16.10; and it only requires baseband electronics and suppresses the RIN.

Table 16.2 summarizes the advantages and disadvantages of balanced mixing and phase diversity.

Although 2×2 optical hybrids can also be used for phase diversity systems, they are not so attractive because equations (16.10) and (16.11) indicate that their use does not automatically produce the correct phase relationship for the diode currents. The two currents should be $\pi/2$ out of phase and this can be achieved only at the cost of additional optical processing [17], which makes the device bulky and impractical.

Phase diversity makes the receiver performance extremely tolerant to phase noise, while the linewidth requirements of the lasers are quite relaxed [18, 19].

Phase diversity can also operate with FSK and PSK modulation. Figure 16.11 shows a phase diversity receiver with a 3×3 optical hybrid and DPSK demodulation.

The phase diversity diagrams show a performance reduced by less than 1 dB compared with an ideal heterodyne receiver in the quantum limit [16].

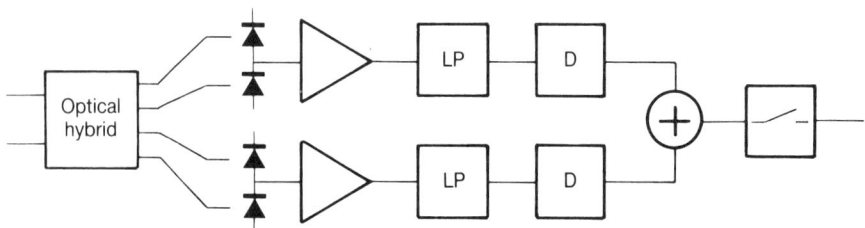

Figure 16.10 Diagram of the double balanced mixer receiver with a 4×4 optical hybrid.

Table 16.2

	Balanced mixing receiver	Phase diversity scheme
Advantages	Suppression of the RIN Efficient use of the received optical power Simple optical hybrid	Baseband circuitry is sufficient, without optical phase-locking
Disadvantages	Broadband IF circuits More complex signal processing	No RIN suppression* Complicated optical hybrid Complex signal processing

*The modified phase diversity scheme removes the disadvantages of no RIN suppression

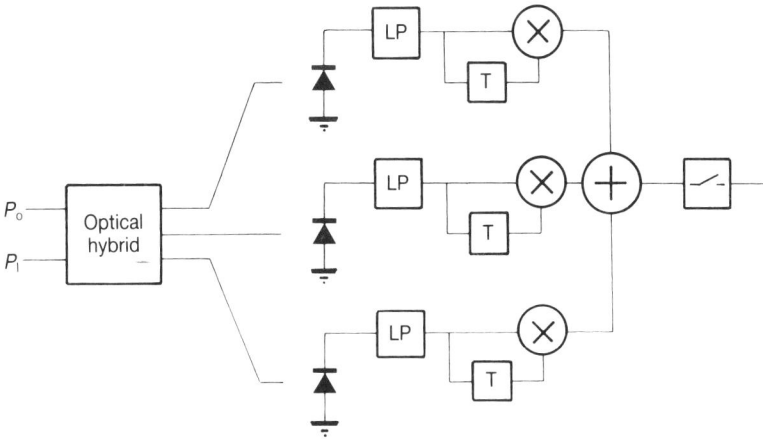

Figure 16.11 General diagram of a 3×3 DPSK phase diversity coherent receiver. LP: low-pass filter; T: delay by the bit duration T.

16.5 Polarization aspects of coherent systems

In order to make the interference of the local oscillator wave and the signal wave that is received more efficient, their polarization states must coincide. Due to random vibrations in the fiber and temperature changes, mechanical strain in the fiber introduces birefringence, which changes with time. As a consequence, the polarization state of the signal received changes randomly. The problems caused by a polarization mismatch can be overcome in the following ways:

1. By use of polarization-maintaining fibers (see Chapter 15). This solution is more expensive than the use of standard single-mode fibers. Moreover, it is

not a suitable solution for existing links that have been installed with standard single-mode fibers.

2. With polarization scrambling. In this technique the polarization state is deliberately changed at the transmitting end, so that all possible polarization states are passed along during a single bit time at the receiving end. It will be clear that this gives a sensitivity loss of at least 3 dB.

3. By using a polarization state controller. The polarization state of the received optical signal is measured in order to see if it deviates from that of the local oscillator. A control signal is generated and sent to a set of piezo-electric or magnetic fiber squeezers, which induce a mechanical strain in the fiber such that the output polarization state matches that of the local light source [20]. Although the control algorithm and equipment is rather complicated, it is nevertheless a very promising method for dealing with the polarization problem.

4. By using polarization diversity. Both the local optical wave and the signal wave received are split into two orthogonal polarization states. The power of the local oscillator, whose state of polarization can be fixed, is equally divided between two orthogonal polarization states. The signal is also decomposed into two orthogonal components, which vary in amplitude randomly. After decomposition, part of the signal is added to a polarization matched part of the local oscillator and similarly for the remaining part. The two paired orthogonal components are detected separately and added afterwards. Since the signal terms in the electrical detector are squared, then potentially no signal degradation or signal-to-noise ratio penalty occurs. Combining the phase diversity and the polarization diversity is possible [16] and Figure 16.12 shows a general diagram of such a receiver. First, both the local wave and the signal wave are decomposed into two orthogonal polarized components by a polarization beam splitter (PBS). Each of the local wave components, together with their corresponding signal components, are then applied to a phase diversity receiver (PDR). The separately processed components are added together and applied to a threshold comparator.

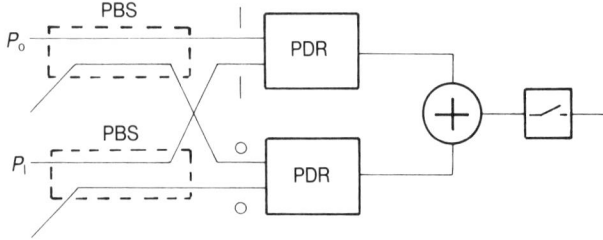

Figure 16.12 General diagram of combined phase and polarization diversity scheme. PBS: polarization beam splitter; PDR: phase diversity receiver.

In the quantum limit, the combined phase and polarization receivers have a sensitivity that is about 1 dB (maximum 1.2 dB) worse than the ideal heterodyne receiver [16].

16.6 Concluding remarks

Coherent optical detection is still at the experimental stage and much research remains to be done on such items as laser stability and spectral purity, optical phase-locked loops, polarization tracking or polarization diversity, and optimum modulation/demodulation. In this chapter we have shown that the sensitivity of coherent receivers can approach the quantum limit quite closely, especially if the thermal noise and shot noise of the dark current can be suppressed in order to improve sensitivity by some 10–20 dB. Coherent systems are most suitable for high bit rate systems which operate in the long-wavelength region (1300–1500 nm), where single-mode fibers have a low attenuation and reduced dispersion. The larger dark current of photodiodes, due to their smaller bandgap, can be suppressed when the local oscillator power amplifies the signal received.

Unfortunately, such elegant receiver diagrams as those for phase and polarization diversity can lead to very complicated signal processing, both optically and electronically. For the time being, this prevents any practical application being developed on a large scale. However, when opto-electronic integrated circuits (OEIC) for combining optical and electronical functions on a single chip become more common, it will be possible to make these receivers cheaply, and they will become standard equipment for optical fiber communication systems.

References

[1] L. Kazovsky, 'Balanced phase-locked loops for optical homodyne receivers: performance analysis, design considerations, and laser linewidth requirements', *OSA/IEEE Journal of Lightwave Technology*, vol. LT-4, no. 2, February 1987, pp. 182–95.

[2] T. Okoshi, 'Heterodyne and coherent optical fiber communications: recent progress', *IEEE Transactions on Microwave Theory and Techniques*, vol. MTT-30, no. 8, August 1982, pp. 1138–49.

[3] T. Okoshi and K. Kikuchi, 'Heterodyne-type optical fiber communications', *Journal of Optical Communications*, vol. 2, no. 3, 1981, pp. 82–8.

[4] R. Steele and S. Wright, 'Electronic control of the difference frequency between two semiconductor lasers', *Annales des Télécommunications*, vol. 38, no. 1–2, 1983, pp. 53–7.

[5] T. Kimura, 'Coherent optical fiber transmission', *IEEE/OSA Journal of Lightwave Technology*, vol. LT-5, no. 4, April 1987, pp. 414–28.

[6] T. Wood, 'Multiple quantum well (MQW) waveguide modulators', *IEEE/OSA Journal of Lightwave Technology*, vol. LT-6, no. 6, June 1988, pp. 743–57.

[7] S. Wang and S. Lin, 'High speed III–V electro-optic waveguide modulators at $\lambda = 1.3 \, \mu m$', *IEEE/OSA Journal of Lightwave Technology*, Vol. LT-6, no. 6, June 1988, pp.758–71.

[8] L. Thylén, 'Integrated optics in LiNbO$_3$: recent developments in devices for telecommunications', *IEEE/OSA Journal of Lightwave Technology*, Vol. LT-6, no. 6, June 1988, pp. 847–61.

[9] F. Koyama and K. Iga, 'Frequency chirping in external modulators', *IEEE/OSA Journal of Lightwave Technology*, vol. LT-6, no. 1, January 1988, pp. 87–93.

[10] P. Z. Peebles, Jr, *Probability, Random Variables, and Random Signal Principles*, 2nd edn (Auckland: McGraw-Hill, 1987).

[11] A. B. Carlson, *Communication Systems*, 3rd edn (New York: McGraw-Hill, 1986).

[12] T. Waite, 'A balanced mixer for optical heterodyning: the magic T optical mixer', *Proceedings of the IEEE*, vol. 54, February 1966, pp. 334–5.

[13] H. Unger, *Planar Optical Waveguides and Fibres* (Oxford: Clarendon Press, 1977).

[14] J. Siuzdak, *Optical Couplers for Coherent Optical Phase Diversity Systems*, Eindhoven University of Technology, EUT Report 88-E-190, March 1988, Eindhoven, The Netherlands.

[15] L. Kazovsky, 'Recent progress in phase and polarization diversity coherent optical techniques', *Proceedings of the 13th European Conference on Optical Communication*, Helsinki, 13–17 September 1987, vol. II, pp. 83–90.

[16] J. Siuzdak and W. van Etten, 'BER evaluation for phase and polarization diversity optical homodyne receivers using non-coherent ASK and DPSK demodulation', *IEEE/OSA Journal of Lightwave Technology*, vol. LT-7, no. 4, April 1989, pp. 584–99.

[17] L. Kazovsky, L. Curtis, W. Young and N. Cheung, 'All fiber 90° optical hybrid for coherent communications', *Applied Optics*, vol. 26, no. 3, February 1987, pp. 437–9.

[18] A. Davis, M. Pettitt, J. King and S. Wright, 'Phase diversity techniques for coherent optical receivers', *IEEE/OSA Journal of Lightwave Technology*, vol. LT-5, no. 4, April 1987, pp. 561–72.

[19] L. Kazovsky, P. Meissner and E. Patzak, 'ASK multiport optical homodyne receiver', *IEEE/OSA Journal of Lightwave Technology*, vol. LT-5, no. 6, June 1987, pp. 770–91.

[20] R. Noé, 'Endless polarization control in coherent optical communications', *Electronics Letters*, vol. 22, no. 15, July 1988, pp. 772–3.

Appendix 1

Bessel functions

Bessel functions, denoted as $Z_\nu(z)$, are solutions of the linear differential equation (Bessel's differential equation)

$$z^2 \frac{d^2 Z_\nu}{dz^2} + z \frac{dZ_\nu}{dz} + (z^2 - \nu^2)Z_\nu = 0 \tag{A1.1}$$

In general z, ν and Z_ν are complex. ν is called the order of the Bessel function; if ν is an integer we put $\nu = n$.

Special Bessel functions are the following:

Bessel functions:	$J_\nu(z)$	are functions of the first kind
Neumann functions:	$N_\nu(z)$	are functions of the second kind
Hankel functions:	$H_\nu^{(1)}(z)$ $H_\nu^{(2)}(z)$	are functions of the third kind

$$H_\nu^{(1)}(z) = J_\nu(z) + jN_\nu(z)$$

$$H_\nu^{(2)}(z) = J\nu(z) - jN\nu(z)$$

$N_\nu(z)$ is sometimes denoted as $Y_\nu(z)$

If ν is not an integer, then $J_\nu(z)$ and $J_{-\nu}(z)$ are linearly independent solutions of the second-order differential equation (A1.1); the general solution is in that case

$$Z_\nu(z) = C_1 J_\nu(z) + C_2 J_{-\nu}(z) \tag{A1.2}$$

If $\nu = n$ is an integer, then $J_n(z)$ and $J_{-n}(z)$ are dependent and the following relation holds:

$$J_{-n}(z) = (-1)^n J_n(z) \tag{A1.3}$$

The Neumann function $N_\nu(z)$ is defined as

$$N_\nu(z) = \frac{J_\nu(z)\cos \nu\pi - J_{-\nu}(z)}{\sin \nu\pi} \tag{A1.4}$$

The Neumann function is independent of $J_\nu(z)$ for integer values of ν. If $\nu = n$, then

$$N_n(z) = \lim_{\nu \to n} N_\nu(z) = \frac{1}{\pi} \left[\frac{\partial}{\partial \nu} J_\nu(z) - (-1)^n \frac{\partial}{\partial \nu} J_{-\nu}(z) \right] \qquad (A1.5)$$

The general solution of equation (A1.1) for $\nu = n$ is given by

$$Z_n(z) = C_1 J_n(z) + C_2 N_n(z) \qquad (A1.6)$$

or

$$Z_n(z) = C_3 H_n^{(1)}(z) + C_4 H_n^{(2)}(z) \qquad (A1.7)$$

Series definitions of Bessel functions
The series definitions of $J_n(z)$ and $N_n(z)$ are given by

$$J_n(z) + \sum_{s=0}^{\infty} \frac{(-1)^s}{s!(n+s)!} \left(\frac{z}{2}\right)^{n+2s} \qquad (A1.8)$$

and

$$N_\nu(z) = \frac{2}{\pi} \left(C + \ln \frac{z}{2} \right) J_n(z) - \frac{1}{\pi} \left(\frac{z}{2}\right)^n$$

$$\times \sum_{k=0}^{\infty} \frac{(-1)^k}{k!(n+k)!} \left(\frac{z}{2}\right)^{2k} \left(\sum_{l=1}^{k} \frac{1}{l} + \sum_{l=1}^{k+n} \frac{1}{l} \right)$$

$$- \frac{1}{\pi} \left(\frac{z}{2}\right)^{-n} \sum_{k=0}^{n-1} \frac{(n-k-1)!}{k!} \left(\frac{z}{2}\right)^{2k} \qquad (A1.9)$$

where C is Euler's constant

$$C \triangleq \lim_{m \to \infty} \left(1 + \frac{1}{2} + \frac{1}{3} + \ldots + \frac{1}{m} - \ln m \right) = 0.577215665 \ldots \qquad (A1.10)$$

Modified Bessel functions
Modified Bessel functions $M_\nu(z)$ are solutions of the linear differential equation

$$z^2 \frac{d^2 M_\nu}{dz^2} + z \frac{dM_\nu}{dz} - (z^2 + \nu^2) M_\nu = 0 \qquad (A1.11)$$

$I_n(z)$ and $K_n(z)$ are modified Bessel functions of order n and the first and second kinds, respectively. The general solution of equation (A1.11) for integer values of ν is given by

$$M_\nu(z) = C_5 I_n(z) + C_6 K_n(z) \qquad (A1.12)$$

Recurrence relations for Bessel functions

$$Z_{n-1}(z) + Z_{n+1}(z) = \frac{2n}{z} Z_n(z) \qquad (A1.13)$$

$$Z_{n-1}(z) - Z_{n+1}(z) = 2Z'_n(z) \tag{A1.14}$$

where

$$Z'_n(z) = \frac{dZ_n(z)}{dz} \tag{A1.15}$$

From equations (A1.13) and (A1.14) it can be deduced that

$$Z'_n(z) = -\frac{n}{z} Z_n(z) + Z_{n-1}(z) = \frac{n}{z} Z_n(z) - Z_{n+1}(z) \tag{A1.16}$$

$$Z'_0(z) = -Z_1(z) \tag{A1.17}$$

$$Z'_1(z) = Z_0(z) - \frac{1}{z} Z_1(z) \tag{A1.18}$$

Recurrence relations for modified Bessel functions
For the modified Bessel function $I_n(z)$:

$$I_{n-1}(z) - I_{n+1}(z) = \frac{2n}{z} I_n(z) \tag{A1.19}$$

$$I_{n-1}(z) + I_{n+1}(z) = 2I'_n(z) \tag{A1.20}$$

where

$$I'_n(z) = \frac{dI_n(z)}{dz} \tag{A1.21}$$

From equations (A1.19) and (A1.20) it is deduced that

$$I'_n(z) = -\frac{n}{z} I_n(z) + I_{n-1}(z) = \frac{n}{z} I_n(z) + I_{n+1}(z) \tag{A1.22}$$

$$I'_0(z) = I_1(z) \tag{A1.23}$$

$$I'_1(z) = I_0(z) - \frac{1}{z} I_1(z) \tag{A1.24}$$

For the modified Bessel function $K_n(z)$:

$$K_{n+1}(z) - K_{n-1}(z) = \frac{2n}{z} K_n(z) \tag{A1.25}$$

$$K_{n+1}(z) + Z_{n-1}(z) = -2n K'_n(z) \tag{A1.26}$$

where

$$K'_n(z) = \frac{dK_n(z)}{dz} \tag{A1.27}$$

From equations (A1.25) and (A1.26) it is deduced that

$$K'_n(z) = -\frac{n}{z} K_n(z) - K_{n-1}(z) = \frac{n}{z} K_n(z) - K_{n+1}(z) \qquad \text{(A1.28)}$$

$$K'_0(z) = -K_1(z) \qquad \text{(A1.29)}$$

$$K'_1(z) = -K_0(z) - \frac{1}{z} K_1(z) \qquad \text{(A1.30)}$$

Limiting forms for small arguments

For small values of the arguments we can make use of limiting forms $(z \to 0)$. If $z \ll 1$, then

$$J_n(z) \approx \frac{1}{n!} \left(\frac{z}{2} \right)^n \qquad \text{(A1.31)}$$

$$J_0(z) \approx 1 - \frac{z^2}{4} \qquad \text{(A1.32)}$$

$$J_1(z) \approx \frac{z}{2} - \frac{z^3}{16} \qquad \text{(A1.33)}$$

$$N_0(z) \approx \frac{2}{\pi} \left(C - \ln \frac{2}{z} \right) \qquad \text{(A1.34)}$$

where C is Euler's constant.

$$N_n(z) \approx -\frac{(n-1)!}{\pi} \left(\frac{2}{z} \right)^n, \qquad n > 0 \qquad \text{(A1.35)}$$

$$I_n(z) \approx \frac{1}{n!} \left(\frac{z}{2} \right)^n \qquad \text{(A1.36)}$$

$$I_0(z) \approx 1 + \frac{z^2}{4} \qquad \text{(A1.37)}$$

$$I_1(z) \approx \frac{z}{2} + \frac{z^3}{16} \qquad \text{(A1.38)}$$

$$K_n(z) \approx \frac{(n-1)!}{2} \left(\frac{2}{z} \right)^n \qquad \text{(A1.39)}$$

$$K_0(z) \approx -\ln z + (\ln 2 - C) \qquad \text{(A1.40)}$$

where C is Euler's constant.

Asymptotic expansions for large arguments

For large values of the arguments, the following asymptotic expansions hold $(z \to \infty)$:

$$J_n(z) \approx \sqrt{\frac{2}{\pi z}} \cos \left(z - n \frac{\pi}{2} - \frac{\pi}{4} \right) \qquad \text{(A1.41)}$$

$$N_n(z) \approx \sqrt{\frac{2}{\pi z}} \sin\left(z - n\frac{\pi}{2} - \frac{\pi}{4}\right) \tag{A1.42}$$

$$I_n(z) \approx \sqrt{\frac{1}{2\pi z}} \left(1 - \frac{4n^2 - 1}{8z}\right) \exp(z) \tag{A1.43}$$

$$K_n(z) \approx \sqrt{\frac{\pi}{2z}} \left(1 + \frac{4n^2 - 1}{8z}\right) \exp(-z) \tag{A1.44}$$

Some indefinite integrals

$$\int z J_0^2(az)\,\mathrm{d}z = \frac{z^2}{2}\left[J_0^2(az) + J_1^2(az)\right] \tag{A1.45}$$

$$\int z K_0^2(az)\,\mathrm{d}z = \frac{z^2}{2}\left[K_0^2(az) - K_1^2(az)\right] \tag{A1.46}$$

Appendix 2

Transmission of modulated signals via bandpass systems

When the current of a semiconductor laser is modulated, the optical signal generated by the laser can be described by an amplitude-modulated signal. The transfer function of the optical fiber behaves like a bandpass filter. In order to be able to deal with the transmission of optical signals through optical fibers, we must first analyze the general problem of the transmission of modulated signals via bandpass filters.

We consider signals that consist of a high-frequency carrier modulated in amplitude or phase by a time function which varies much more slowly than the period of the carrier. For instance AM signals are written as

$$s(t) = A[1 + m(t)] \cos \omega_c t \tag{A2.1}$$

where A is the amplitude of the unmodulated carrier, $m(t)$ is the low-frequency modulating signal and ω_c is the carrier frequency [1]. Assuming that $[1 + m(t)]$ is never negative, then $s(t)$ looks like a harmonic signal whose amplitude varies with the modulating signal.

A frequency-modulated signal reads

$$s(t) = A \cos \left\{ \omega_c t + \int_0^t \psi(s) \, ds \right\} \tag{A2.2}$$

The instantaneous frequency of this signal is $\omega_c + \psi(t)$, while the slowly varying function $\psi(t)$ contains the information to be transmitted. Such a signal has a constant amplitude, but the zero-crossings change with the modulating signal.

The most general form of a modulated signal is given by

$$s(t) = a(t) \cos [\omega_c t + \phi(t)] \tag{A2.3}$$

In this equation $a(t)$ is the amplitude modulation and $\phi(t)$ is the phase modulation, while the derivative $d\phi/dt$ is the frequency modulation of the signal. Expanding the cosine of equation (A2.3) gives

$$s(t) = a(t)[\cos \phi(t) \cos \omega_c t - \sin \phi(t) \sin \omega_c t]$$
$$= x(t) \cos \omega_c t - y(t) \sin \omega_c t \tag{A2.4}$$

with

$$x(t) \triangleq a(t) \cos \phi(t)$$

$$y(t) \triangleq a(t) \sin \phi(t) \tag{A2.5}$$

The functions $x(t)$ and $y(t)$ are called the quadrature components of the signal, and they vary little during one period of the carrier. Combining the quadrature components to produce a complex function enables a modulated signal to be represented [2] in terms of the complex envelope

$$f(t) \triangleq x(t) + \mathrm{j}y(t) = a(t) \exp[\mathrm{j}\phi(t)] \tag{A2.6}$$

namely

$$s(t) = \mathrm{Re}\,[f(t) \exp(\mathrm{j}\omega_c t)] \tag{A2.7}$$

The complex function $f(t)$ can be regarded as a phasor in the x–y-plane. The end of the phasor moves around in the complex plane, whereas the plane itself rotates with an angular frequency of ω_c and the signal $s(t)$ is the projection of the rotating phasor on a fixed line. If the movement of the phasor $f(t)$ with respect to the rotating plane is much slower than the speed of rotation of the plane, then the signal is quasi-harmonic. When it is assumed that the Fourier transform of the complex envelope of the signal exists, namely

$$F(\omega) = \int_{-\infty}^{\infty} f(t) \exp(-\mathrm{j}\omega t)\,\mathrm{d}t \tag{A2.8}$$

then the spectrum $S(\omega)$ of the signal is written as

$$S(\omega) = \frac{1}{2} \int_{-\infty}^{\infty} [f(t) \exp(\mathrm{j}\omega_c t) + f^*(t) \exp(-\mathrm{j}\omega_c t)] \exp(-\mathrm{j}\omega t)\,\mathrm{d}t$$

$$= \frac{1}{2}[F(\omega - \omega_c) + F^*(-\omega - \omega_c)] \tag{A2.9}$$

where * denotes the complex conjugate. Since the quadrature components of $f(t)$ vary much more slowly than the carrier $\cos \omega_c t$, the spectral width of $F(\omega)$ is much smaller than ω_c. The modulus of the spectrum of the signal, $|S(\omega)|$, shows two narrow peaks, one at the frequency ω_c and the other at $-\omega_c$. Consequently, $s(t)$ is called a narrowband signal. The spectrum of equation (A2.9) is Hermitian, i.e. $S(-\omega) = S^*(\omega)$, a condition imposed by the fact that $s(t)$ is real. The Fourier transform $F(\omega)$ of the complex envelope satisfies the same condition, if $f(t)$ is real and the signal is only modulated in amplitude. In that case, the modulus $|F(\omega)|$ is an even function and the peaks of $|S(\omega)|$ exhibit symmetry about the carrier frequency ω_c. This carrier frequency is chosen arbitrarily as a characteristic frequency in the passband of the bandpass filter. A shift by just ω_0 only introduces a factor of $\exp(-\mathrm{j}\omega_0 t)$ in the complex envelope, without changing the signal $s(t)$, so that

$$f(t) \exp(\mathrm{j}\omega_c t) = [f(t) \exp(-\mathrm{j}\omega_0 t)] \exp[\mathrm{j}(\omega_c + \omega_0)t] \tag{A2.10}$$

The calculations can sometimes be simplified by choosing the characteristic frequency correctly.

As far as detection is concerned, the function $x(t)$ is restored by multiplying $s(t)$ by $\cos \omega_c t$ and removing the double frequency components with a lowpass filter

$$s(t) \cos \omega_c t = \tfrac{1}{2} x(t)(1 + \cos 2\omega_c t) - \tfrac{1}{2} y(t) \sin 2\omega_c t \qquad (A2.11)$$

After filtering, the signal produced is $x(t)/2$. The second quadrature component $y(t)$ is restored in a similar way by multiplying $s(t)$ by $\sin \omega_c t$ and using a low-pass filter to remove the double frequency components afterwards.

A circuit that delivers an output signal that is a function of the amplitude modulation is called a rectifier and such a circuit always involves a nonlinear operation. The quadratic rectifier is a typical rectifier; it has an output signal in proportion to the square of the envelope. This output is achieved by squaring the signal and reads

$$s^2(t) = \tfrac{1}{2}[x^2(t) + y^2(t)] + \tfrac{1}{2}[x^2(t) - y^2(t)] \cos 2\omega_c t - x(t)\, y(t) \sin 2\omega_c t$$

$$(A2.12)$$

By means of a low-pass filter the frequency terms in the vicinity of $2\omega_c$ are removed, so that the output is proportional to $|f(t)|^2 = a^2(t) = x^2(t) + y^2(t)$. A linear rectifier, which may consist of a diode and a low-pass filter, yields $a(t)$.

A circuit giving an output signal that is proportional to the instantaneous frequency deviation $\phi'(t)$, is known as a discriminator, the output of which is proportional to $\mathrm{d}[\mathrm{Im}\{\ln f(t)\}]/\mathrm{d}t$

Quasi-harmonic signals are often filtered by bandpass filters, i.e. filters that pass frequency components in the vicinity of the carrier frequency and attenuate other frequency components. The transfer function of such a filter may be written as

$$H(\omega) = H_1(\omega - \omega_c) + H_1^*(-\omega - \omega_c) \qquad (A2.13)$$

where the function $H_1(\omega)$ is called the equivalent baseband transfer function. Equation (A2.13) is Hermitian, because the impulse response $h(t)$ is a real function; however, in general the equivalent baseband function $H_1(\omega)$ is not Hermitian. If the spectra of the input and output signal are denoted by $S_i(\omega)$ and $S_o(\omega)$, respectively, then

$$S_o(\omega) = H(\omega)\, S_i(\omega) \qquad (A2.14)$$

Using the notation introduced in equations (A2.9) and (A2.13) it follows that

$$\tfrac{1}{2}[F_o(\omega - \omega_c) + F_o^*(-\omega - \omega_c)]$$

$$= [H_1(\omega - \omega_c) + H_1^*(-\omega - \omega_c)][\tfrac{1}{2}F_i(\omega - \omega_c) + \tfrac{1}{2}F_i^*(-\omega - \omega_c)]$$

$$= \tfrac{1}{2}H_1(\omega - \omega_c)\, F_i(\omega - \omega_c) + \tfrac{1}{2}H_1^*(-\omega - \omega_c)F_i^*(-\omega - \omega_c)$$

$$+ \tfrac{1}{2}H_1^*(-\omega - \omega_c)\, F_i(\omega - \omega_c) + \tfrac{1}{2}H_1(\omega - \omega_c)\, F_i^*(-\omega - \omega_c) \quad (A2.15)$$

In the case of a bandpass input signal and a bandpass filter the last two terms in Equation (A2.15) vanish and it follows that

$$F_o(\omega - \omega_c) = H_1(\omega - \omega_c) F_i(\omega - \omega_c)$$ (A2.16)

or

$$F_o(\omega) = H_1(\omega) F_i(\omega)$$ (A2.17)

By means of the convolution theorem, equation (A2.17) has the following equivalent in the time domain:

$$f_o(t) = \int_{-\infty}^{\infty} h_1(\tau) f_i(t - \tau) \, d\tau$$ (A2.18)

where $f_o(t)$ is the complex envelope of the output, $f_i(t)$ is the complex envelope of the input and $h_1(t)$ is the complex impulse response of the equivalent baseband system.

References

[1] A. B. Carlson, *Communication Systems*, 3rd edn (New York: McGraw-Hill, 1986) chap. 6.
[2] C. W. Helstrom, *Statistical Theory of Signal Detection*, 2nd edn (Oxford: Pergamon Press, 1968) chap. 1.

Appendix 3

The propagation of Gaussian beams in free space and optical systems

Gaussian beams crop up many times in optical fiber communications. The light from a laser emitting in the transverse fundamental mode is approximated by a Gaussian function if a cross-section is taken perpendicular to the direction of propagation [1]. Moreover, the field in a single-mode fiber can be approximated in many practical situations by a Gaussian beam (see Chapter 11). Now the question arises as to how Gaussian beams propagate in an homogeneous medium, such as free space, and how such a beam behaves when transmitted through an optical system, e.g. a lens.

A3.1 Propagation of Gaussian beams in homogeneous media

Coherent Gaussian beams spread in homogeneous media as a result of diffraction. In this section we shall show how the beam keeps its Gaussian intensity distribution when this spread occurs. In other words, we shall show that an expanding Gaussian beam is an approximate solution to Maxwell's equations in an homogeneous medium. In such a medium these equations reduce to the so-called Helmholtz equation

$$\nabla^2 \mathbf{E} + k^2 \mathbf{E} = 0 \tag{A3.1}$$

Cylindrical coordinates are chosen to describe the fields and we only consider rotationally symmetrical field solutions. For a transverse field component the scalar equation obtained is

$$\frac{\partial^2 E}{\partial r^2} + \frac{1}{r} \frac{\partial E}{\partial r} + \frac{\partial^2 E}{\partial z^2} + k^2 E = 0 \tag{A3.2}$$

Let us assume that this equation has the following solution:

$$E(r, z) = \psi(r, z) \exp(-jkz) \tag{A3.3}$$

Furthermore, we assume that ψ varies only slightly when the wave has travelled over a distance equal to its wavelength, so that the term containing $\partial^2 \psi / \partial z^2$ may be ignored with respect to the term containing $\partial \psi / \partial z$; equation (A3.2) then changes to

$$\frac{\partial^2 \psi}{\partial r^2} + \frac{1}{r} \frac{\partial \psi}{\partial r} - 2jk \frac{\partial \psi}{\partial z} = 0 \tag{A3.4}$$

Further, ψ is written as

$$\psi(r, z) = A \exp\left\{-j\left[P(z) + \frac{kr^2}{2q(z)}\right]\right\} \tag{A3.5}$$

When this function is inserted into equation (A3.4) we get

$$-\left(\frac{k^2}{q^2}\right)r^2 - 2j\left(\frac{k}{q}\right) + \frac{k^2r^2q'}{q^2} - 2kP' = 0 \tag{A3.6}$$

As the solution has to be valid for all values of r, the coefficients of the individual powers must be zero, so that

$$\frac{k^2}{q^2} = \frac{k^2}{q^2}q' \tag{A3.7}$$

and

$$P' = \frac{-j}{q} \tag{A3.8}$$

The solution of equation (A3.7) reads

$$q(z) = q_0 + z \tag{A3.9}$$

Inserting equation (A3.9) into equation (A3.8) yields

$$P(z) = -j\ln\left(1 + \frac{z}{q_0}\right) \tag{A3.10}$$

The beam parameter $q(z)$ is in general complex. When

$$\frac{1}{q(z)} = \frac{1}{R(z)} - \frac{2j}{kw^2(z)} \tag{A3.11}$$

then the deduced parameters $R(z)$ and $w(z)$ can be easily interpreted: equation (A3.11) is inserted into equation (A3.5) in order to give $w(z)$ as a measure of the beam width at position z. The parameter $R(z)$ is the radius of curvature of the nearly spherical wave fronts. This can be seen from Figure A3.1 and the following expression:

$$kR = k\sqrt{z^2 + r^2}$$
$$= kz\sqrt{1 + \frac{r^2}{z^2}} \approx kz\left(1 + \frac{r^2}{2z^2}\right) \approx k\left(z + \frac{r^2}{2R}\right), \qquad r \ll z \tag{A3.12}$$

The point $z = 0$ is defined where the radius of curvature of the wave front becomes infinity. Inserting this into equation (A3.11) gives

$$q_0 = \frac{jkw^2(0)}{2} \tag{A3.13}$$

Inserting equations (A3.9) and (A3.10) into equations (A3.5) and (A3.3) leads, after some manipulation, to

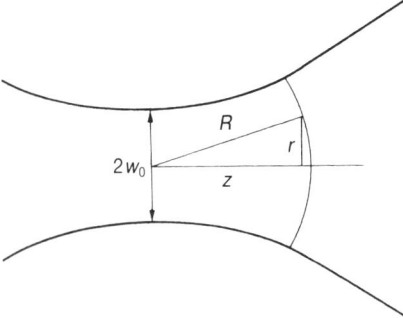

Figure A3.1 The radius of curvature of the wave fronts for a Gaussian beam.

$$E(r, z) = A \frac{w(0)}{w(z)} \exp\left[-r^2/w^2(z)\right] \exp\left\{-j[kz + kr^2/2R(z) - \phi(z)]\right\}$$

$$(A3.14)$$

with

$$w(z) \stackrel{\triangle}{=} w_0 \sqrt{1 + \frac{z^2}{z_0^2}} \qquad (A3.15)$$

$$R(z) \stackrel{\triangle}{=} z + \frac{z_0^2}{z} \qquad (A3.16)$$

$$\phi(z) \stackrel{\triangle}{=} \arctan\left(\frac{z}{z_0}\right) \qquad (A3.17)$$

$$z_0 \stackrel{\triangle}{=} \frac{kw_0^2}{2} \qquad (A3.18)$$

For a fixed value of z the field is a Gaussian function of r. The beam width parameter $w(z)$ is defined as the distance, in the transverse direction, at which the field amplitude decays to $\exp(-1)$ of its maximum value. The half-width $w(z)$ of this Gaussian function depends on z linearly for large values of z and has a minimum at $z = 0$. The cross-section of the beam at its minimum width is called the beam waist. For negative values of z the beam is described by equation (A3.14).

From the phase of this equation it follows that the wave front is almost spherical for $r \ll z$ and this can be seen when comparing the first two terms of the phase of equation (A3.14) with equation (A3.12). This means that, under the given condition, the light rays of a Gaussian beam emerge virtually from a single point. Except for the first two terms, the phase consists of the term $\phi(z)$, which has a maximum value of $\pi/2$ and changes substantially only in the vicinity of the waist (i.e. for $z < z_0$). For $z \gg z_0$ this phase shift is nearly constant, although negligible with respect to the other phase terms.

As mentioned above, the half beam width $w(z)$ varies nearly linearly with z for a large z; this means that the divergence angle becomes constant (see Figure A3.2)

$$\theta_0 \triangleq \frac{w(z)}{z} \approx \frac{w_0}{z_0} = \frac{2w_0}{kw_0^2} = \frac{2}{kw_0} = \frac{\lambda}{\pi w_0} \tag{A3.19}$$

The characteristic value z_0 can also be written as

$$z_0 = \frac{\pi w_0^2}{\lambda} \tag{A3.20}$$

From equation (A3.19) it follows that the product of the half beam width w_0 in the waist and the asymptotic divergence angle is constant for a given wavelength. Consequently, a beam that has a large half beam width w_0 in the waist has a small divergence angle and vice versa.

By taking the square of equation (A3.14) and dividing it by twice the wave impedance, the beam power per unit area is obtained. This is called the irradiance, and has dimensions of $\mathrm{W\,m^{-2}}$.

For some applications it is more convenient to use the radiant intensity, which is defined as the power per unit of solid angle in a certain direction (dimension $\mathrm{W\,sr^{-1}}$). The radiant intensity can be deduced from equation (A3.14). Consider Figure A3.3. The power that passes through the annular region is given by

$$\mathrm{d}P \approx c\,\frac{z_0^2}{z^2}\exp\left(-2\,\frac{r^2}{z^2}\,z_0^2\right)2\pi r\,\frac{R\,\mathrm{d}\theta}{\cos\theta} \approx c\,\frac{z_0^2}{z^2}\exp\left(-2\,\frac{\theta^2}{\theta_0^2}\right)\frac{2\pi r R^2\,\mathrm{d}\theta}{z} \tag{A3.21}$$

where c is a constant. The approximations are based on the fact that only paraxial rays are considered; this assumption plays a major role in the treatment of Gaussian beams described here.

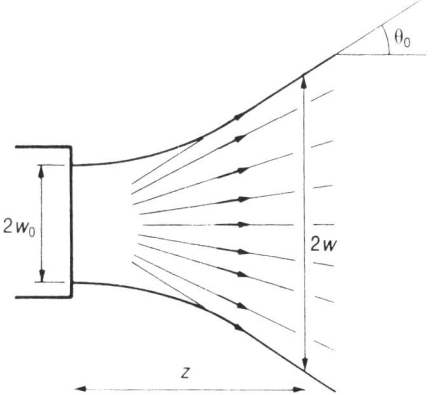

Figure A3.2 The divergence angle θ_0 of a Gaussian beam.

Figure A3.3 An annular region $\{r, r + dr\}$ at distance z.

A solid angle is associated with the annular region and this solid angle reads

$$d\Omega = \frac{2\pi r R\, d\theta}{R^2} \tag{A3.22}$$

leading to

$$I = \frac{dP}{d\Omega} \approx cz_0^2 \exp\left(-2\,\frac{\theta^2}{\theta_0^2}\right) = I_0 \exp\left(-2\,\frac{\theta^2}{\theta_0^2}\right) \tag{A3.23}$$

For the sake of completeness it must be realized that the Gaussian beam is not the only possible solution to equation (A3.1). Gausian beams multiplied by Hermitian or Laguerre polynomials also satisfy this equation (see [2]).

A3.2 Gaussian beams in optical systems

Once more, paraxial rays are considered, i.e. those rays that make small angles with the axis of symmetry. One way of describing light rays is by means of their place coordinates and angles with the optical axis. Instead of angles it is often more convenient to take an appropriate function of the angles, such as the sine or tangent. This kind of coordinate system is called the phase space and, for the sake of simplicity, it will be restricted to one place coordinate x and one angle coordinate $u = \tan\theta$ (see Figure A3.4).

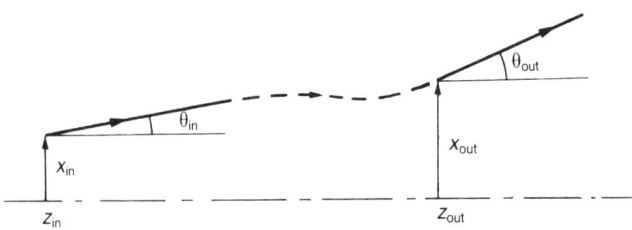

Figure A3.4 Parameters that describe light rays in the phase space.

For a number of optical components and systems the coordinates x and u of the output can be related to the coordinates of the input by means of a matrix description for characterizing the system

$$\begin{pmatrix} x_{out} \\ u_{out} \end{pmatrix} = \begin{pmatrix} A & B \\ C & D \end{pmatrix} \begin{pmatrix} x_{in} \\ u_{in} \end{pmatrix} \tag{A3.24}$$

This transfer matrix is called the *ABCD* matrix. It is easy to see that the following *ABCD* matrices can be found for the given systems.

Homogeneous medium

$$\begin{pmatrix} A & B \\ C & D \end{pmatrix} = \begin{pmatrix} 1 & d \\ 0 & 1 \end{pmatrix} \tag{A3.25}$$

where d is the distance between input and output.

Thin lens

$$\begin{pmatrix} A & B \\ C & D \end{pmatrix} = \begin{pmatrix} 1 & 0 \\ -\dfrac{1}{f} & 1 \end{pmatrix} \tag{A3.26}$$

where f is the focal distance of the lens.

Discontinuity between two plane dielectrics

$$\begin{pmatrix} A & B \\ C & D \end{pmatrix} = \begin{pmatrix} 1 & 0 \\ 0 & \dfrac{n_1}{n_2} \end{pmatrix} \tag{A3.27}$$

where n_1 is the refractive index of the input medium and n_2 is the refractive index of the output medium.

The matrix notation offers a big advantage for describing the series connection of various subsystems, since the overall transfer matrix consists of a multiplication of transfer matrices that describe the individual subsystems.

The matrix description for optical components can also be used for the transformation of Gaussian beams that pass through these components. The complex Gaussian beam parameter $q(z)$ can be considered as a generalization of the radius of curvature of the wave fronts associated with the light rays in Gaussian optics [3, 4]. It is easily shown that the relationship for ray optics is

$$R_{out} = \frac{AR_{in} + B}{CR_{in} + D} \tag{A3.28}$$

We shall show below that for the optical components dealt with so far, the actual $ABCD$ matrices produced can be applied to a Gaussian beam transformation, if the radius of curvature R is replaced in equation (A3.28) by the complex beam parameter q.

Homogeneous medium
From the ABCD matrix (equation A3.25) and equation (A3.28) while using $d = z$ it follows that

$$q_{out} = q_{in} + z \tag{A3.29}$$

This agrees with equation (A3.9). If z is measured from the waist, $q_{in} = q_0 = jkw_0^2/2$ is obtained and

$$\frac{1}{q_{out}} = \frac{1}{z + jkw_0^2/2} = \frac{1}{R(z)} - \frac{1}{kw^2(z)} \tag{A3.30}$$

where $R(z)$ and $w^2(z)$ are given by equations (A3.16) and (A3.17), respectively.

Thin lens
A thin lens does not change the beam width $2w$, but it does change the inverse of the radius of curvature of the spherical wave fronts by an amount $-1/f$. The following transformation is then expected

$$\frac{1}{q_{out}} = \frac{1}{q_{in}} - \frac{1}{f} \tag{A3.31}$$

This is exactly the same as the transformation that follows from equations (A3.26) and (A3.28).

Discontinuity between two plane dielectrics
In the output medium, the radius of curvature of a spherical wave front becomes

$$R_2 = \frac{n_2}{n_1} R_1 \tag{A3.32}$$

and since the beam width remains unchanged, it follows that

$$q_{out} = \frac{n_2}{n_1} q_{in} \tag{A3.33}$$

From equation (A3.27) the same relationship results. It is easy to verify that the transformation also holds for the imaginary part of $1/q$.

References

[1] I. H. Zaclus and I. L. Ripper, 'Resonant modes of GaAs junction lasers', *IEEE Journal of Quantum Electronics,* vol. QE-5, January 1969, pp. 29–31.

[2] S. Ramo, J. Whinnery and T. van Duzer, *Fields and Waves in Communication Electronics,* 2nd edn (New York: Wiley, 1984).

[3] H. Kogelnik, 'On the propagation of Gaussian beams of light through lenslike media including those with a loss or gain variation', *Applied Optics,* vol. 4, no. 12, December 1965, pp. 1562–9.

[4] H. Kogelnik and T. Li, 'Laser beams and resonators', *Applied Optics,* vol. 5, no. 10, October 1966, pp. 1550–67.

Appendix 4

Poisson processes

A4.1 Introduction

In some of the chapters we have spoken about two ways of describing optical signals, namely by means of electromagnetic waves and by means of photons. Some phenomena can be described accurately by means of the electromagnetic wave model; however, others are better described by the photon model. In particular, detection, i.e. the interaction between optical power and the carriers in a semiconductor, can be described satisfactorily by the photon model. In this model the optical power consists of packets of quantized energy which amount to $h\nu$, where h is Planck's constant and ν is the optical frequency. These packets are supposed to arrive at random points in time and in that way lead to a Poisson distribution for the number of arrivals in a fixed time interval.

In this appendix we shall deal with the most important parameters of homogeneous Poisson processes (i.e. Poisson processes with a constant expectation) and inhomogeneous Poisson processes (i.e. processes with an expectation that is a function of time); moreover, we shall develop some theory about stochastic processes, as far as it relates to Poisson processes. Finally, we shall consider marked Poisson processes, which are processes that arise in APDs and photomultipliers, where each primarily created electron–hole pair is multiplied by a factor that itself behaves as a stochastic variable.

The realization of a Poisson process consists of randomly distributed points in time $\{t_j\}$. If we consider a limited part $(0, T)$ of the time axis, then $|t_j| \leq T$, $j = 1, \ldots, M$, with M being a stochastic variable. The description of Poisson processes given in this appendix is not based on the number of events in a given interval, but on the probability of an arbitrary realization of the random points $\{t_j\}$. By means of this we can derive the M-dimensional joint probability density function of the random points [1]. Poisson processes are also called random point processes.

A4.1.1 The characteristic function

The characteristic function of a random variable X is defined as

$$\Phi(\omega) \triangleq E[\exp(j\omega X)] = \int_{-\infty}^{\infty} \exp(j\omega X)\, p(X)\, dX \qquad (A4.1)$$

where $p(X)$ is the probability density function of X. However, when X is a discrete variable with the possible realizations $\{X_k\}$, then the definition becomes

$$\Phi(\omega) \triangleq \sum_k \exp(j\omega X_k) \Pr\{X = X_k\} \tag{A4.2}$$

Sometimes it is useful to consider the logarithm of the characteristic function

$$\Psi(\omega) \triangleq \ln \Phi(\omega) \tag{A4.3}$$

This function is called the second characteristic function of the random variable X. From equation (A4.1) it follows that

$$\Phi(0) = \int_{-\infty}^{\infty} p(X)\, dX = 1 \tag{A4.4}$$

so that

$$\Psi(0) = 0 \tag{A4.5}$$

For the random variable $Y = aX + b$, with a and b as constants, it follows that

$$\Phi_Y(\omega) = E[\exp(j\omega Y)] = \exp(j\omega b)\, \Phi_X(a\omega) \tag{A4.6}$$

In this appendix we are primarily interested in the Poisson distribution given by

$$\Pr\{X = k\} = \exp(-\lambda)\frac{\lambda^k}{k!}, \qquad k = 0, 1, 2, \ldots \tag{A4.7}$$

For this distribution it can be seen that

$$\Phi(\omega) = \exp(-\lambda) \sum_{k=0}^{\infty} \exp(j\omega k)\frac{\lambda^k}{k!} = \exp\{\lambda[\exp(j\omega) - 1]\} \tag{A4.8}$$

and

$$\Psi(\omega) = \lambda[\exp(j\omega) - 1] \tag{A4.9}$$

On the other hand, the probability density function can be restored using the characteristic function in the integral

$$p(X) = \frac{1}{2\pi} \int_{-\infty}^{\infty} \Phi(\omega) \exp(-j\omega X)\, d\omega \tag{A4.10}$$

The moments of the random variable are defined by

$$E[X^n] \triangleq m_n \tag{A4.11}$$

It follows, using the Fourier transform theory, that a relationship between the derivatives of the characteristic function and these moments can be established. This relationship is found by expanding the exponential of the integrand of equation (A4.1) as follows

$$\Phi(\omega) = \int_{-\infty}^{\infty} p(X)\left(1 + j\omega X + \cdots + \frac{(j\omega X)^n}{n!} + \cdots\right) dX \qquad (A4.12)$$

Assuming that integrating term by term is allowed, it follows that

$$\Phi(\omega) = 1 + j\omega m_1 + \cdots + \frac{(j\omega)^n}{n!} m_n + \cdots \qquad (A4.13)$$

It follows from this equation that

$$\frac{d^n \Phi(0)}{d\omega^n} = j^n m_m \qquad (A4.14)$$

The operations leading to equation (A4.14) are allowed if all the moments m_n exist and the series expansion of equation (A4.13) converges absolutely at $\omega = 0$. In this case $p(X)$ is uniquely determined by the moments $\{m_n\}$.

Sometimes it is more interesting to consider the central moments, e.g. the second central moment or variance. Then the preceding operations are applied to the random variable $X - E[X]$

A4.1.2 Cumulants

Consider a probability distribution of which all the moments of arbitrary order exist. In the characteristic function $j\omega$ is replaced by p and the function that results is called the moment-generating function. The logarithm of this function becomes

$$\Psi(p) = \ln \Phi(p)$$

$$= \ln \int p(X)\left[1 + pX + \frac{p^2 X^2}{2!} + \cdots + \frac{p^n X^n}{n!} + \cdots\right] dX$$

$$= \ln\left(1 + pm_1 + \frac{p^2 m_2}{2!} + \cdots + \frac{p^n m_n}{n!} + \cdots\right) \qquad (A4.15)$$

Now, develop $\Psi(p)$ into a Taylor series about $p = 0$, remembering that

$$\ln(1 + x) = x - \frac{x^2}{2} + \frac{x^3}{3} - \cdots \qquad (A4.16)$$

it is then found

$$\Psi(p) = m_1 p + \frac{p^2 m_2}{2!} + \cdots - \frac{p^2 m_1^2}{2} + \cdots$$

$$= m_1 p + \frac{p^2}{2}(m_2 - m_1^2) + \cdots$$

$$= m_1 p + \frac{\sigma^2}{2} p^2 + \sum_{k=3}^{\infty} \frac{\gamma_k p^k}{k!} = \sum_{k=1}^{\infty} \frac{\gamma_k p^k}{k!} \qquad (A4.17)$$

where σ^2 is the variance of X and γ_k is the kth cumulant or semi-invariant [2, 3].

By means of equations (A4.9) and (A4.7) it can be shown easily that both the expectation and the variance of an homogeneous Poisson process equal λ.

A4.2 Inhomogeneous Poisson processes

A4.2.1 The probability density function of the random points

Consider a Poisson process with a time-varying expectation of $\lambda(t) \geq 0$, then the probability that the number of points in the time interval (a, b) equals the integer value of k is given by [4]

$$\Pr\{k, (a, b)\} = \left\{\exp\left[-\int_a^b \lambda(t)\,dt\right]\right\}\left\{\int_a^b \lambda(t)\,dt\right\}^k /k! \qquad (A4.18)$$

A complete description of the process at the interval $(0, T)$ depends upon including all the possible subdivisions $0 = a_0 < a_1 < \ldots < a_k = T$ and observing the number of points in each interval (a_i, a_{i+1}), $i = 0, 1, \ldots, k-1$. When considering all the possible events, k becomes infinitely large. Another way of describing the process fits better in the nature of the process and uses the probability density function of the points t_j.

The following assumptions have been made in order to arrive at equation (A4.18)

1. The probability $\Pr\{1, \Delta t\}$ of a single point in an infinitesimal interval Δt is given by

$$\Pr\{1, \Delta t\} = \lambda(t)\Delta t, \qquad \Delta t \to 0 \qquad (A4.19)$$

2. The probability that more than one random point in Δt occurs is zero for $\Delta t \to 0$.
3. The number of random points in each arbitrary interval is independent of the number of random points in other intervals that do not overlap.

The special case $k = 0$ simplifies equation (A4.18) to

$$\Pr\{0, (a, b)\} = \exp\left[-\int_a^b \lambda(t)\,dt\right] \qquad (A4.20)$$

Consider a possible realization of the Poisson process shown in Figure A4.1. The set of M random points $0 \leq t_1 < t_2 < \ldots < t_M \leq T$ is denoted by $\{t_M\}$. The M-dimensional probability density function $p\{t_M\}$ of such a realization is obtained from considering infinitesimal intervals of a width $\Delta t_j/2$ on both sides of t_j. The probability that exactly one point occurs in each of these intervals and nowhere else is: $p\{t_M\} \Delta t_1 \Delta t_2 \ldots \Delta t_M$, and it is found from equations (A4.19) and (A4.20), so that

Figure A4.1 Possible realization of the Poisson process.

$$p\{t_M\}\,\Delta t_1\,\Delta t_2\,\ldots\,\Delta t_M = \Pr\{0,(0,\,t_1 - \tfrac{1}{2}\Delta t_1)\}$$
$$\times\,\lambda(t_1)\,\Delta t_1\,\Pr\{0,(t_1 + \tfrac{1}{2}\Delta t_1,\,t_2 - \tfrac{1}{2}\Delta t_2)\}\,\ldots$$
$$\times\,\lambda(t_M)\,\Delta t_M\,\Pr\{0,(t_M + \tfrac{1}{2}\Delta t_M,\,T)\}\quad(\text{A4.21})$$

With the limit of $\Delta t_j \to 0$ for all j, and for $M \ge 1$, it follows that

$$p\{t_M\} = \exp(-Q)\prod_{j=1}^{M}\lambda(t_j),\quad 0 \le t_1 < t_2 \ldots < t_M \le T\quad(\text{A4.22})$$

where

$$Q \triangleq \int_0^T \lambda(t)\,\mathrm{d}t\quad(\text{A4.23})$$

For $M = 0$ the set $\{t_M\}$ is empty, i.e. there are no random points in the interval $0 \le t \le T$, and equation (A4.20) gives

$$\Pr\{0,(0,\,T)\} = \exp(-Q)\quad(\text{A4.24})$$

For $M = 0$ the probability $p\{t_M\}$ can be interpreted as the probability of there being no random points in $(0, T)$; then, $\Pi_{j=1}^{M}\lambda(t_j)$ is defined as unity. Using these interpretations, equation (A4.22) gives a complete statistical description of the Poisson process in the interval $(0, T)$.

It must be emphasized that $p\{t_M\}$ does not represent a conditional density, because M is a stochastic variable too. The event space is shown in Figure A4.2,

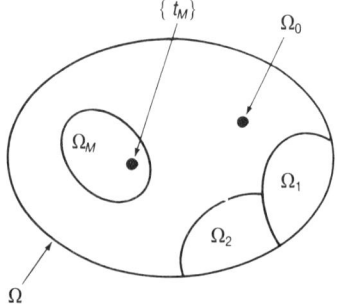

Figure A4.2 The event space of the Poisson process with random point density function $p\{t_M\}$.

where Ω_M is the set of all possible sets $\{t_M\}$, while Ω_0 is the event that no points are in the interval $(0, T)$ and Ω is the certain event.

A4.2.2 The expectation of product functions

In this section the expectation of product functions $f_\pi(\{t_M\})$ with probability density function $p\{t_M\}$ of the random points will be encountered when they are defined by

$$f_\pi(\{t_M\}) \triangleq \begin{cases} \displaystyle\prod_{j=1}^{M} f(t_j), & j = 1, 2, \ldots, M, \quad M \ge 1 \\ \\ 1, & M = 0 \end{cases} \tag{A4.25}$$

The expectation of this function reads

$$E[f_\pi] = \int_\Omega p\{t_M\} \prod_{j=1}^{M} f(t_j) = \sum_{M=0}^{\infty} \int_{\Omega_M} p\{t_M\} \prod_{j=1}^{M} f(t_j)$$

$$\triangleq \sum_{M=0}^{\infty} E_M[f_\pi] \tag{A4.26}$$

Here, the expectation $E[f_\pi]$ is written as the sum of the expectations $E_M[f_\pi]$, the expectation of $f_\pi(\{t_M\})$, where a number of M random points is observed in the interval $(0, T)$. These M points, however, can occur at arbitrary times, so that

$$E_M[f_\pi] = \int_0^T \int_{t_1}^T \ldots \int_{t_{M-1}}^T dt_1\, dt_2 \ldots dt_M\, p\{t_M\} \prod_{j=1}^{M} f(t_j) \tag{A4.27}$$

The M-fold integral of equation (A4.27) can be found when it is realized that t_1 can be situated anywhere between 0 and T. However, once t_1 has been fixed, t_2 can only be found at the interval (t_1, T), and so on. Using equation (A4.22) the integral of equation (A4.27) is written as

$$E_M[f_\pi] = \exp(-Q) \int_0^T \int_{t_1}^T \ldots \int_{t_{M-1}}^T dt_1\, dt_2 \ldots dt_M\, p\{t_M\} \prod_{j=1}^{M} g(t_j) \tag{A4.28}$$

with

$$g(t) \triangleq f(t)\, \lambda(t) \tag{A4.29}$$

It will be shown that

$$\int_0^T \int_{t_1}^T \ldots \int_{t_{M-1}}^T dt_1\, dt_2 \ldots dt_M \prod_{j=1}^{M} g(t_j) = \frac{G^M(T)}{M!} \tag{A4.30}$$

with

$$G(T) \triangleq \int_0^T g(t)\, dt \tag{A4.31}$$

Proof
Consider the integral

$$I = \int_x^T g(y) \left[\int_y^T g(z)\,\mathrm{d}z \right]^N \mathrm{d}y \tag{A4.32}$$

Let

$$\int_y^T g(z)\,\mathrm{d}z \triangleq f(y) \tag{A4.33}$$

then

$$I = -\int_{f(x)}^{f(T)} f(y)^N\,\mathrm{d}f(y) \tag{A4.34}$$

because

$$\frac{\mathrm{d}}{\mathrm{d}y} \int_y^T g(z)\,\mathrm{d}z = -g(y) \tag{A4.35}$$

see [5, p. 320], in order to show that

$$I = -\frac{1}{N+1}\, [f(y)^{N+1}] \Big|_{f(x)}^{f(T)} = -\frac{1}{N+1}\,[f(T)^{N+1} - f(x)^{N+1}]$$

$$= +\frac{1}{N+1}\, f(x)^{N+1} = \frac{1}{N+1} \left[\int_x^T g(z)\,\mathrm{d}z \right]^{N+1} \tag{A4.36}$$

Repeated use of equations (A4.32)–(A4.36) can produce equation (A4.30) after starting with $x = t_{M-2}$, $y = t_{M-1}$, $z = t_M$ and $N = 1$.

Including equation (A4.30) in equation (A4.28) yields

$$E_M[f_\pi] = \frac{\exp(-Q)\,G^M(T)}{M!} \tag{A4.37}$$

and including this expression, in its turn, in equation (A4.26) we get the interesting but quite simple result

$$E[f_\pi] = \exp(-Q) \sum_{M=0}^\infty \frac{G^M(T)}{M!} = \exp[-Q + G(T)] \tag{A4.38}$$

In the special case of $f(t) = 1$ the expectation $E_M[1]$ becomes the probability of Ω_M, the event that M points are realized in the interval $(0, T)$. It then follows from equations (A4.29), (A4.31) and (A4.23) that $G(T) = Q$ and equation (A4.37) must be equivalent to equation (A4.18); in this way we have found an alternative derivation of equation (A4.18). For this case, equation (A4.38) gives the probability of a certain event being equal to unity.

A4.2.3 Conditional probabilities

The probability density function $p\{t_M\}$ that was derived in the preceding section is a multidimensional joint probability density function

$$p\{t_M\} = p[\{t_M\}, M] \tag{A4.39}$$

It is obvious that for a given set $\{t_M\}$ the conditional probability that a realization contains M points, reads

$$\Pr[M|\{t_M\}] = 1 \tag{A4.40}$$

The conditional probability density function $p[\{t_M\}|M]$ equals the quotient of equations (A4.22) and (A4.37), while $G(T) = Q$, so that

$$p[\{t_M\}|M] = Q^{-M}M! \prod_{j=1}^{M} \lambda(t_j) \tag{A4.41}$$

This follows from the mixed form of Bayes' rule [6].

A4.2.4 Filtered Poisson processes

Consider a linear, time-invariant filter with impulse response $h(t)$ and suppose that the input of the filter consists of a Poisson series of impulses (see Figure A4.3)

$$x(t) = \sum_{n} \delta(t - t_n) \tag{A4.42}$$

which has the variable expectation $\lambda(t)$. In this instance, we are looking for particular parameters of the output process $s(t)$ and the search will be confined to the expectation and the variance, or second central moment, of this process.

The expectation and the variance of a Poisson process are known to equal $\lambda(t)$, i.e. both parameters are time-variant. This means that the input process is non-stationary; consequently, the output process is also non-stationary. The output process is described by

$$s(t) = \sum_{n} h(t - t_n) \tag{A4.43}$$

In the same way as for a homogeneous Poisson process, the expectation and the variance of $s(t)$ are found via the second characteristic function. This function becomes

$$\Phi_s(\omega) \triangleq \mathrm{E}[\exp\{j\omega s(t)\}] = \mathrm{E}\left[\exp\left\{j\omega \sum_{n} h(t - t_n)\right\}\right]$$
$$= \mathrm{E}\left[\prod_{n=1}^{M} \exp\{j\omega h(t - t_n)\}\right] \tag{A4.44}$$

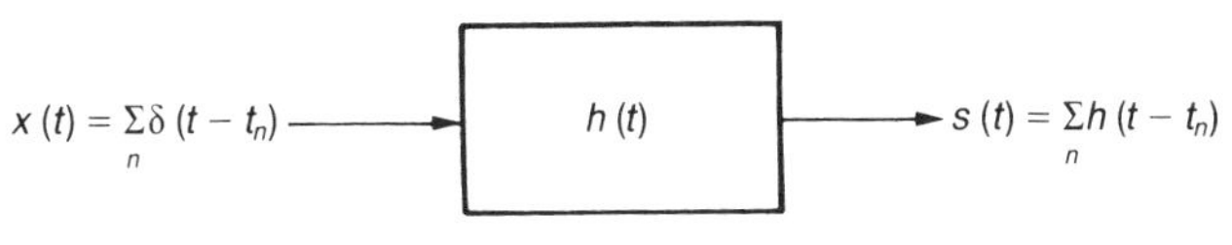

Figure A4.3 A linear, time-invariant system with a Poisson process as input signal.

With the aid of equation (A4.38) and making the substitution $f(\tau) = \exp[j\omega h(t - \tau)]$, this equation is written as

$$\Phi_s(\omega) = \exp\left\{-Q + \int \lambda(\tau) \exp[j\omega h(t - \tau)]\,d\tau\right\} \tag{A4.45}$$

The second characteristic function becomes

$$\Psi_s(\omega) = -Q + \int \lambda(\tau) \exp[j\omega h(t - \tau)]\,d\tau$$

$$= \int \lambda(\tau) \{\exp[j\omega h(t - \tau)] - 1\}\,d\tau \tag{A4.46}$$

Expanding this function into a McLaurin series with $p = j\omega$ yields

$$\Psi_s(p) = \Psi_s(0) + \Psi_s'(0)\,p + \Psi_s''(0)\,\frac{p^2}{2!} + \cdots$$

$$= \sum_{n=1}^{\infty} \int \lambda(\tau) h^n(t - \tau)\,\frac{p^n}{n!}\,d\tau = \sum_{n=1}^{\infty} \gamma_n\,\frac{p^n}{n!} \tag{A4.47}$$

where γ_n is the nth cumulant, so that

$$\gamma_n = \int \lambda(\tau)\,h^n(t - \tau)\,d\tau \tag{A4.48}$$

The first cumulant equals the expectation and the second cumulant gives the variance (see also equation A4.17)

$$\gamma_1 = E[s(t)] = \int \lambda(\tau)\,h(t - \tau)\,d\tau \tag{A4.49a}$$

$$\gamma_2 = \sigma_s^2 = \int \lambda(\tau)\,h^2(t - \tau)\,d\tau \tag{A4.49b}$$

For homogeneous processes, in which $\lambda(t) = \lambda = $ constant, equations (A4.49a) and (A4.49b) are together known as Campbell's theorem [4]. In fact, we have derived a generalization of Campbell's theorem for inhomogeneous processes.

A4.3 Marked Poisson processes

A4.3.1 The cumulants of a marked and filtered Poisson process

In an avalanche photodiode the initial carrier pairs multiply randomly. When the primary carriers show a Poisson statistic, the statistic of the secondary carriers is called a marked Poisson process [7]. In this section we shall establish the cumulants of the process that is found at the output of a linear, time-invariant filter when a marked Poisson process is fed to the input of this filter. This situation is encountered in an optical receiver, where the opto-electrical converter consists of an APD. A diagram of such a receiver is shown in Figure A4.4. The process that represents the primary carriers is seen at point 1 and the signal at this point is given by $\sum_n \delta(t - t_n)$, a series of impulses, randomly distributed over the time axis. The times t_n show a Poisson distribution. At point 2 a marked Poisson process is found and it is written as $\sum_n g_n \delta(t - t_n)$, where g_n is the random multiplication of the impulse at t_n. Finally, the output signal $s(t)$ is a marked and filtered Poisson process and is given by

$$s(t) = \sum_n g_n h(t - t_n) \tag{A4.50}$$

The characteristic function of the process $s(t)$ reads

$$\Phi_s(\omega) = \mathrm{E}[\exp\{j\omega s(t)\}] = \mathrm{E}\left[\exp\left\{j\omega \sum_n g_n h(t - t_n)\right\}\right]$$

$$= \mathrm{E}\left[\prod_n \exp\{j\omega g_n h(t - t_n)\}\right] \tag{A4.51}$$

We assume that the variables g_n are mutually independent and taken from the same distribution. Furthermore, it is assumed that M primary carriers are generated in the observation interval and the number M is independent of g_n. A process that satisfies these assumptions is called a compound Poisson process [7] and with the aid of the assumptions we find the characteristic function of such a process

$$\Phi_s(\omega) = \mathrm{E}\left[\prod_{n=1}^{M} \exp\{j\omega g_n h(t - t_n)\}\right]$$

$$= \mathrm{E}\left[\prod_{n=1}^{M} \mathrm{E}_g[\exp\{j\omega g_n h(t - t_n)\}]\right]$$

$$= \mathrm{E}\left[\prod_{n=1}^{M} \int_0^\infty p(g) \exp\{j\omega g h(t - t_n)\}\,dg\right] \tag{A4.52}$$

where E_g is the expected value with respect to g and $p(g)$ is the probability density function of g. By means of equation (A4.38) and making $f(\tau) = \int_0^\infty p(g) \exp[j\omega g h(t - \tau)]\,dg$, we find

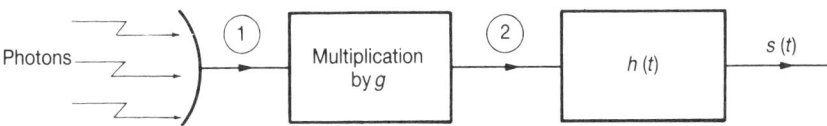

Figure A4.4 Model of an optical receiver, consisting of an APD followed by a linear, time-invariant filter.

$$\Phi_s(\omega) = \exp\left\{-Q + \int_0^\infty p(g)\left[\int \lambda(\tau)\exp\{j\omega gh(t-\tau)\}\,d\tau\right]dg\right\}$$

$$= \exp\left\{-Q + \int \lambda(\tau)\left[\int_0^\infty p(g)\exp\{j\omega gh(t-\tau)\}\,dg\right]d\tau\right\}$$

$$= \exp\left\{-Q + \int \lambda(\tau)[\Phi_g\{\omega h(t-\tau)\}]\,d\tau\right\} \qquad (A4.53)$$

where Φ_g is the characteristic function of g. The second characteristic function of $s(t)$ becomes

$$\Psi_s(\omega) = -Q + \int \lambda(\tau)[\Phi_g\{\omega h(t-\tau)\}]\,d\tau$$

$$= \int \lambda(\tau)[\Phi_g\{\omega h(t-\tau)\} - 1]\,d\tau \qquad (A4.54)$$

Now, expand $\Psi_s(\omega)$ into a McLaurin series with $p = j\omega$

$$\Psi_s(p) = \Psi_s(0) + \Psi_s'(0)\,p + \Psi_s''(0)\frac{p^2}{2!} + \cdots$$

$$= \sum_{n=1}^\infty \int_0^\infty p(g)g^n\left[\int \lambda(\tau)\,h^n(t-\tau)\,d\tau\right]dg\,\frac{p^n}{n!} = \sum_{n=1}^\infty \gamma_n\frac{p^n}{n!} \qquad (A4.55)$$

where γ_n is the nth cumulant, so that

$$\gamma_n = E[g^n]\int \lambda(\tau)\,h^n(t-\tau)\,d\tau \qquad (A4.56)$$

and

$$\gamma_1 = E[s(t)] = E[g]\int \lambda(\tau)\,h(t-\tau)\,d\tau \qquad (A4.57a)$$

$$\gamma_2 = E[\{s(t) - E[s(t)]\}^2] = \sigma_s^2 = E[g^2]\int \lambda(\tau)\,h^2(t-\tau)\,d\tau \qquad (A4.57b)$$

The equations (A4.57a) and (A4.57b) represent a further generalization of Campbell's theorem.

A4.3.2 The autocorrelation function and spectrum of marked and filtered Poisson processes

A marked Poisson process, in which the expected value varies with time, is a non-stationary process; therefore, the autocorrelation function cannot be written as a function of one single variable. Of course, the same holds for a marked and filtered Poisson process. A general relationship exists between the autocorrelation function $R_s(t_1, t_2)$ and the autocovariance function $C_s(t_1, t_2)$, which is [4]

$$C_s(t_1, t_2) = R_s(t_1, t_2) - \mathrm{E}[s(t_1)]\,\mathrm{E}[s(t_2)] \tag{A4.58}$$

so that

$$R_s(t_1, t_2) = C_s(t_1, t_2) + \mathrm{E}[s(t_1)]\,\mathrm{E}[s(t_2)] \tag{A4.59}$$

The auto-covariance function is obtained from the second characteristic function, which is a function of two variables in this case [4]

$$C_s(t_1, t_2) = -\frac{\partial^2 \Psi_s(0, 0)}{\partial \omega_1 \partial \omega_2} \tag{A4.60}$$

Here $\Psi_s(\omega_1, \omega_2)$ is defined as

$$\Psi_s(\omega_1, \omega_2) \triangleq \ln \Phi_s(\omega_1, \omega_2) \triangleq \ln \mathrm{E}[\exp\{j\omega_1 s(t_1) + j\omega_2 s(t_2)\}] \tag{A4.61}$$

This equation is used in order to calculate the characteristic function of a marked and filtered Poisson process. As for equation (A4.52), this function becomes

$$\Phi_s(\omega_1, \omega_2) = \mathrm{E}\left[\prod_{m=1}^{M} \exp\{j\omega_1 g_m h(t_1 - t_m) + j\omega_2 g_m h(t_2 - t_m)\}\right]$$

$$= \mathrm{E}\left[\prod_{m=1}^{M} \int p(g)\exp\{j\omega_1 gh(t_1 - t_m) + j\omega_2 gh(t_2 - t_m)\}\,dg\right]$$

$$= \exp\left[-Q + \int p(g)\left[\int \lambda(\tau)\exp\{j\omega_1 gh(t_1 - \tau)\right.\right.$$

$$\left.\left. + j\omega_2 gh(t_2 - \tau)]\,d\tau\right]dg\right] \tag{A4.62}$$

This equation can then be used to find the second characteristic function so that

$$\Psi_s(\omega_1, \omega_2) =$$

$$-Q + \int p(g)\left[\int \lambda(\tau)\exp\{j\omega_1 gh(t_1 - \tau) + j\omega_2 gh(t_2 - \tau)\}\,d\tau\right]dg \tag{A4.63}$$

Finally, from equations (A4.59), (A4.60), (A4.63) and (A4.49a) it follows

$$R_s(t_1, t_2) = \mathrm{E}[g^2]\int \lambda(\alpha)\,h(t_1 - \alpha)\,h(t_2 - \alpha)\,d\alpha$$

$$+ \mathrm{E}^2[g]\int \lambda(\alpha)\,h(t_1 - \alpha)\,d\alpha\int \lambda(\alpha)\,h(t_2 - \alpha)\,d\alpha \tag{A4.64}$$

Putting $t_1 = t$ and $t_2 = t - \tau$ into this equation gives

$$R_s(t, t - \tau) = E[g^2] \int \lambda(\alpha)\, h(t - \alpha)\, h(t - \tau - \alpha)\, d\alpha$$

$$+ E^2[g] \int \lambda(\alpha)\, h(t - \alpha)\, d\alpha \int \lambda(\alpha)\, h(t - \tau - \alpha)\, d\alpha$$

$$= E[g^2] \int \lambda(\alpha)\, h(t - \alpha) h(t - \tau - \alpha)\, d\alpha$$

$$+ E[s(t)]\, E[s(t - \tau)] \tag{A4.65}$$

Referring to an optical communication system, $\lambda(t)$ contains the information to be transmitted and, usually, the most convenient way of describing this information signal is as a stochastic process. An inhomogeneous Poisson process whose function $\lambda(t)$ is itself also a stochastic process, is called a doubly stochastic Poisson process [7]. The autocorrelation function of such a process can be determined from the expected value of equation (A4.65) with respect to the stochastic process $\lambda(t)$, which yields

$$R_s(t, t - \tau) = E[g^2]E[\lambda] \int h(t - \alpha)\, h(t - \tau - \alpha)\, d\alpha$$

$$+ E^2[g] \int \int R_\lambda(\alpha_1, \alpha_2)\, h(t - \alpha_1)\, h(t - \tau - \alpha_2)\, d\alpha_1\, d\alpha_2$$

$$\tag{A4.66}$$

In this equation $R_\lambda(\cdot)$ is the autocorrelation function of $\lambda(t)$. If this process is at least stationary in the wide sense, then the autocorrelation function of $s(t)$ is written as

$$R_s(\tau) = E[g^2]\, E[\lambda] \int \int h(\xi)\, h(\xi - \tau)\, d\xi$$

$$+ E^2[g] \int \int R_\lambda(\alpha_1 - \alpha_2)\, h(\xi_1)\, h(\xi_2)\, d\xi_1\, d\xi_2$$

$$= E[g^2]\, E[\lambda]\, h(\tau) * h(-\tau)$$

$$+ E^2[g] \int \int R_\lambda(\tau + \xi_1 - \xi_2)\, h(\xi_1)\, h(\xi_2)\, d\xi_1\, d\xi_2$$

$$= E[g^2]E[\lambda]\, h(\tau) * h(-\tau) + E^2[g]\, R_\lambda(\tau) * h(\tau) * h(-\tau) \tag{A4.67}$$

From both this equation and equation (A4.57) it follows that the output process $s(t)$ also becomes stationary in the wide sense. The spectrum of $s(t)$ is achieved by transforming equation (A4.67) to the frequency domain

$$S_s(\omega) = |H(\omega)|^2 \{ E[g^2]\, E[\lambda] + E^2[g]\, S_\lambda(\omega) \} \tag{A4.68}$$

In this equation $S_\lambda(\omega)$ is the power spectrum of the process $\lambda(t)$. The first term of this expression behaves like filtered white noise, whereas the second term is a filtered version of the power spectrum of $\lambda(t)$, which generally includes the spectrum of the information signal.

If $\lambda(t) = \lambda_0$ in equation (A4.68) is made constant, while $H(\omega)$ and $E[g^2] = E^2[g]$ are made unity, then the spectrum of an homogeneous Poisson process appears (see [4])

$$S(\omega) = 2\pi\lambda_0^2\delta(\omega) + \lambda_0 \qquad (A4.69)$$

This consists of an impulse at $\omega = 0$ plus a white noise term and is the expected result.

References

[1] I. Bar-David, 'Communication under the Poisson regime', *IEEE Transactions on Information Theory*, vol. IT-15, no. 1, January 1969, pp. 31–7.

[2] C. Helstrom, *Statistical Theory of Signal Detection*, 2nd edn (Oxford: Pergamon Press, 1968).

[3] A. Whalen, *Detection of Signals in Noise* (New York: Academic Press, 1971).

[4] A. Papoulis, *Probability, Random Variables and Stochastic Processes*, 2nd edn (New York: McGraw-Hill, 1984).

[5] L. Pipes, *Applied Mathematics for Engineers and Physicists*, (New York: McGraw-Hill, 1958).

[6] J. Wozencraft and I. Jacobs, *Principles of Communication Engineering* (New York: Wiley, 1965), p. 75.

[7] D. L. Snyder, *Random Point Processes* (New York: Wiley, 1975).

Appendix 5

Some physical constants

$$c_0 \qquad = \text{speed of light } \textit{in vacue} = 2.99792458 \times 10^8 \text{ m s}^{-1}$$

$$e \qquad = \text{electron charge} = 1.60210 \times 10^{-19} \text{ C}$$

$$h \qquad = \text{Planck's constant} = 6.62559 \times 10^{-34} \text{ J s}$$

$$\frac{1}{h\nu} = \frac{1}{hc}\lambda = 5.03448 \times 10^{24} \lambda_0 \text{ J}^{-1}$$

$$\frac{e}{h\nu} = \frac{e}{hc}\lambda = 8.066 \times 10^5 \lambda_0 \text{ A W}^{-1}$$

$$k \qquad = \text{Boltzmann's constant} = 1.38 \times 10^{-23} \text{ J K}^{-1}$$

$$k\theta \qquad \approx 4 \times 10^{-21} \text{ W s for } \theta = 290 \text{ K}$$

$$\mu_0 \qquad = 4\pi \times 10^{-7} \text{ Wb A}^{-1}\text{m}^{-1}$$

$$\varepsilon_0 \qquad = \frac{1}{\mu_0 c_0^2} \approx \frac{10^{-9}}{36\pi} \text{ C}^2\text{ N}^{-1}\text{m}^{-2}$$

Data of semiconductor materials

	Bandgap (eV) at 300 K	Refractive index, n
Si	1.14 (indirect)	3.5
Ge	0.67 (indirect)	4
GaAs	1.43 (direct)	3.6
InAs	0.35 (direct)	3.5
InP	1.35 (direct)	3.45
GaSb	0.73 (direct)	3.8
$Al_{0.55}Ga_{0.45}As$	2 (direct)	3.3

Index

ABCD matrix, 383
absorption, light, 11–12
active layer, LED, 145
AlGaAs lasers, 145
alternate mark inversion coding (AMI), 345
AM, 374
amplifier
  front-end, 273
  high-impedance, 258
  transimpedance, 258, 278
amplitude
  characteristic, 67, 68
  modulation, 63–8, 374
analog receivers, 269–73
angular misalignment, 239–43, 253
antisymmetric field, 24
APD, 164, 260, 267, 288, 348, 349, 395
ASK, 357
attenuation, 61, 138
  constant, 17, 28, 36, 61
  due to hydroxyl groups, 11–12
  of light in glassfibers, 9–13
  measurement of, 10–11
  due to metals, 11–12
  versus wavelength, 9–10
autocorrelation function of Poisson
    processes, 396
avalanche
  gain, 298
  photodiode *see* APD

background radiation, 265
balanced mixing, 359
bandgap energy, 158
bandpass filter, 66, 374
bandwidth, 62
  equivalent noise, 267
  modulation bandwidth of LD, 181–2
  modulation bandwidth of LED, 178
  of photodiodes, 157, 163
  receiver, 290
  system, 193, 196

beam splitter, 337, 352, 358
BER floor, 321
Bessel
  differential equation, 36
  function, 36, 37
    modified, 38, 39
BH laser, 149
binary communication, 286
bipolar input, 276
birefringence, 335
bit rate, 290
boundary condition, 35, 40, 41, 46, 47
bow-tie fiber, 335
buried-heterostructure laser, 149
butt joint, 216

Campbell's theorem, 394, 396
carrier
  confinement, 145, 149
  density, 170
caustic
  line, 105
  surface, 105, 110
characteristic equation
  for hybrid modes, 45, 47
  for LP modes, 52, 53
  solution of, 26, 42, 48–51
  for TE waves, 25, 26, 41
characteristic function, 386, 396, 397
  second, 387, 393–4, 397
chirp, 355
cladding, 4, 5, 34, 36
coating, 4, 110
  loose tube, 327
  primary, 326
  secondary, 327
  tight, 327
coding
  AMI, 345
  Manchester, 345
  mBnB, 346
  scrambling, 346
  split-phase, 345

coherence length, 152
coherent
  detection, 352
  scattering, 11
competition noise, 306
complex
  envelope, 375
  ray, 110
conduction band, 140
conductivity, 21, 22
confinement
  carrier, 145, 149
  current, 149
connector mismatches, 222
connectors, 331–3
consistent wave pattern, 312
constant
  attenuation, 17, 28, 36, 61
  phase, 17, 36, 61, 68–70
  propagation, 5, 17, 28, 35, 61, 68
continuity condition, 25
core, 4, 5, 34, 36
coupled power equations, 201
coupling
  coefficient, 201
  efficiency, 87
  laser to multimode fiber, 218
  laser to single-mode fiber, 247
  LED to multimode fiber, 216
  LED to single-mode fiber, 248
  length, 201
coupling loss
  extrinsic, 221, 223
  intrinsic, 221, 222
crest factor, 270
critical angle, 30, 31
cumulant, 389, 394, 396
current confinement, 149
cut-off frequency, 27, 28, 30, 44, 69, 70
cylindrical polar coordinates, 35

dark current, 157, 260, 294, 358
DBR laser, 150
DC-PBH laser, 149
demultiplexer, 337
depletion layer, 157
detection
  coherent, 352
  direct, 258, 351
  threshold of, 287
DFB laser, 150
differential delay, 307
digital receivers, 283–301

direct
  bandgap semiconductor, 140
  detection, 258, 351
  modulation, 355
dispersion, 33, 138
  material, 67, 69, 72–4, 78, 90–4
  multimode, 75, 76
  pulse, 90
  in step index fiber, 60–75
  waveguide, 69–74, 78, 88–94
dispersion-flattened fiber, 78, 92–4
dispersion-shifted fiber, 78, 92–4
distributed Bragg reflection laser, 150
distributed feedback laser, 150
divergence angle, Gaussian beam, 381
diversity
  phase, 361
  polarization, 366
dopants for III–V semiconductors, 145
double heterostructure (DH), 142
doubly stochastic Poisson process, 398
DPSK, 364

edge emitter, 142
efficiency of lasers, 175
eigen equation, 41
eigenvalue, 42
eikonal equation, 21, 96–8
emission wavelength, AlGaAs, 145
end separation, 227–33, 252
envelope
  delay, 65, 66
  Gaussian, 65
  velocity, 65
equalization, 283
equalizer, 274, 291, 293
equiphase surface *see* wavefront
equivalent
  baseband, 376
  noise bandwidth, 267
error probability, 286
evanescent
  field, 36, 110
  wave, 27
excess noise, 164
  factor, 267, 295
external modulation, 355
extinction ratio, 303
extrinsic loss, 221, 223
eye opening, 296–8

Fabry–Perot resonator, 147
far-field pattern, 155

Faraday effect, 334
FET input, 276
fiber
  cables, 326
  coupler, 352, 358
  graded index, 6, 21
  monomode, 5, 78–94
  multimode, 5
  multimode graded index, 96–119
  self-focusing, 109
  single-mode, 5
    see also monomode fiber
  step index, 5
  weakly guiding, 33, 52
fiber manufacture, 6–9
  CVD method, 6
  double crucible, 6
  PCVD method, 8
  rod in tube, 6
  VAD method, 8–9
field
  antisymmetric, 24
  characteristic distribution, 5, 17
  evanescent, 36, 110
  oscillating, 36
  symmetric, 24
filtered Poisson process, 265, 393
FM, 271, 355, 374
focal line, 104
Fourier transform, 65
frequency
  cut-off, 27, 28, 30, 44, 69, 70
  modulation, 271, 355, 374
  normalized, 26, 28, 42
Fresnel reflection, 160
front-end
  amplifiers, 273
  circuit, 259
front-illuminated photodiode, 162–3, 263
FSK, 357, 364
fused biconical taper, 337

GaAs, 145
gain-guided laser, 149
GaInPAs, 145
Gaussian beam, 86, 87, 218, 245, 378
  divergence angle, 381
  in homogeneous media, 378
  waist, 219, 380
  wavefronts, 380
Gaussian pulse, 187, 298
Ge photodiode, 157, 161–6
geometric optics, 21, 97

glassfiber
  cladding, 4, 6
  coating, 4
  configuration, 4–6
  core, 4, 34
  cross-sections, 5
graded index
  fiber, 6
  optical waveguide, 21
group
  delay, 63, 71
  index, 72–4, 152
  velocity, 62, 63, 65
guided ray, 110, 111

Helmholtz equation, 378
Hermitian spectrum, 375
heterodyne receiver, 361
heterodyning, 357
homodyne receiver, 362
homodyning, 357
homojunction, 142
hybrid
  mode, 40, 45–51
  optical, 359
  wave, 33, 40
hydroxyl ions, 138, 327

improper mode, 27
impulse response, 125
index-guided laser, 149
indirect bandgap semiconductor, 140
infra-red absorption, 138
InGaAsP
  lasers, 145
  photodiodes, 157, 161–6
inhomogeneous Poisson process, 386
integrate and dump receiver, 285
integrated optics, 31
intensity
  modulation, 137, 351
  noise, 305
interference, 105
  in connector, 233–9
  constructive, 106
  filter, 338
intersymbol interference, 283, 295
intrinsic
  efficiency, 254
  loss, 221–2
ionization
  coefficient, 267
  rate, 164, 267

irradiance, 218

joint losses, 88

Lambertian source, 143, 214
laser
    AlGaAs, 145
    BH, 149
    conditions, 152, 173
    coupling to multimode fiber, 218
    coupling to single-mode fiber, 247
    DBR, 150
    DC-PBH, 149
    DFB, 150
    diodes, 147–56, 348
    direct modulation, 355
    external modulation, 355
    gain-guided, 149
    index-guided, 149
    InGaAsP, 145
    spectrum, 151–3
    strip-geometry, 149
    threshold, 174
laser modes
    lateral, 154
    longitudinal, 152, 306
    transverse, 154
lateral laser modes, 154
leaky
    mode, 33
    ray, 110–13
LED, 139–47, 348
    active layer of, 145
    coupling to multimode fiber, 216
    coupling to single-mode fiber, 248
    spectra, 146
light-emitting diode, 139–47, 348
    see also LED
light ray, 30, 97, 98
line coding, 344
    see also coding
linearly polarized mode, 33, 52–8
link budget, 341
Littrow mounting, 338
local numerical aperture, 96, 114–15
longitudinal laser modes, 152, 306
loose tube coating, 327
Lorentz profile, 152
LP mode, 33, 52–8

Mach–Zehnder interferometer, 339
magnetic permeability, 22, 34
Manchester code, 345

marked and filtered Poisson process, 260, 394
marked Poisson process, 260
material dispersion, 67, 69, 72–4, 78, 90–4, 189
Maxwell equations, 22
mBnB coding, 346
medium
    homogeneous, 1, 20, 21, 34
    isotropic, 1, 21, 34
    lineair, 21, 34
    time-invariant, 21
meridional
    plane, 103
    ray, 101–9
microbending loss, 88, 326
misalignment in fiber connections
    angular, 239–43, 253
    end separation, 227–33, 252
    radial offset, 223–4, 251
modal noise, 312–20
mode, 5, 17, 26
    degenerate, 17
    in fiber, 5
    hybrid, 40, 45–51
    improper, 27
    leaky, 33
    linearly polarized (LP), 33, 52–8
    proper, 27
    transmission, 42
mode conversion, 76, 121, 201
mode coupling, 200, 320
mode field, 33
    diameter, 78, 86–8
mode hopping, 154
mode mixing, 350
mode-selective loss, 316, 317
modulated signals, 374
modulation
    ASK, 357
    bandwidth, 62, 68, 93, 94
    bandwidth of LED, 178
    characteristic of lasers, 182
    FM, 355, 374
    FSK, 357, 364
    index, 269, 271
    PM, 374
    PSK, 357, 364
    subcarrier, 270
moment-generating function, 388
moments, 388
monochromatic
    carrier, 63
    source, 122

monomode fiber, 78–94
multimode dispersion, 75, 76
multimode fiber–fiber coupling, 221
multiplexer, 337
multiplication
　gain, 267, 295
　region, 164
multiport optical hybrid, 359

narrowband signal, 375
near-field pattern, 155
noise
　competition, 306
　modal, 312–20
　partition, 306
　Poisson, 283
　relative intensity, 183, 358, 362
　shot, 266, 283, 286, 293, 298
　system, 304
normalized
　frequency, 26, 28, 42
　intensity, 85
NRZ signalling, 345
number of modes in multimode fiber, 312
numerical aperture, 96, 114–15, 217

optical
　hybrid, 359
　isolator, 333, 355
　length, 99
optics
　geometric, 21
　ray, 21
　wave, 21
oscillating field, 36

PANDA fiber, 335
paraxial rays, 382
partial coherence, 190
partition noise, 306
pattern
　far-field, LD, 155
　near-field, LD, 155
　speckle, 314–15
permeability, 21, 22
　free space, 34
permittivity, 21, 22
　free space, 26
　relative, 26
phase
　change on reflection, 31
　characteristic, 28, 60, 62, 63, 66

constant, 17, 28, 36, 61, 62, 68–70
delay, 63
diversity, 361
equal, 30
modulation, 67, 374
propagation coefficient, 62
shift, 61, 66
space, 214–16
velocity, 24, 37, 38, 62, 63, 65, 70
photoconductive mode, 157, 258
photodetector, 156
　see also photodiodes
photodiodes, 156–67
　APD, 164, 260, 267, 288, 348, 349, 395
　front-illuminated, 162–3, 263
　Ge, 157, 161–6
　InGaAs, 157, 161–6
　PIN, 162, 348, 349
　Si, 157, 161–6
　side-illuminated, 163, 261
photon
　density, 170
　energy, 158
　lifetime, 171
photovoltaic mode, 157, 258
P–I characteristic
　LD, 156
　LED, 156
PIN
　diode, 162, 348, 349
　photodiode impulse response, 260–4
plastic clad fiber (PCF), 327
PM, 374
p–n diode, 162
Poisson
　distribution, 387
　noise, 283
　process, 260, 386
　　autocorrelation function, 396
　　doubly stochastic, 398
　　homogeneous, 386
　　inhomogeneous, 386, 389, 394
　　marked, 260
　　marked and filtered, 260, 394
polarization
　beam splitter, 366
　diversity, 366
　scrambling, 366
　state, 365
　state controller, 366
polarization-maintaining fiber, 334, 355, 365
　bow-tie fiber, 335–6
　PANDA fiber, 335–6

population inversion, 147
power
  density, 87
  distribution, 86
  flow, 78, 84, 85
Poynting vector, 30
primary
  coating, 326
  signal current, 267
propagation constant, 5, 17, 28, 35, 61, 68
proper mode, 27
pseudo-random bit sequence, 346
PSK, 357, 364
pulse
  dispersion, 90
  Gaussian, 187, 298
  width, 209

quadrature components, 375
quantum efficiency
  lasers, 151
  photodiodes, 160
quantum limit, 272, 289, 352

radial offset, 223–4, 251
radiant intensity, 219, 381
radiated ray, 103, 110, 113, 114
raised cosine pulse, 296, 307
random
  multiplication, 395
  point processes, 386
rate equations, 170
ray
  congruence, 99
  optics, 21
  tracing, 98, 243
Rayleigh scattering, 11, 138
real ray, 110
receiver
  analog, 269–73
  bandwidth, 290
  digital, 283–301
  sensitivity, 273, 349
recombination
  radiation, 140
  spontaneous, 139–40
  stimulated, 170
reduced wave equation, 34
refractive index, AlGaAs, 145
refractive index profile, 5
  parabolic, 101, 122
  power law, 6
relative intensity noise, 183, 358, 362

repeater distance, 341
responsivity, 160
RIN, 183, 358, 362

scattering
  coherent, 11
  elastic, 11
  Rayleigh, 11
SCF, 327
scrambling, 346
second characteristic function, 387, 393–4, 397
self-focusing fiber, 109
semi-invariant, 389
separation of variables, 35
shot noise, 266, 283, 286, 293, 298
  limit, 288
Si photodiode, 157, 161–6
side-illuminated photodiode, 163, 261
signal-to-noise ratio, 264, 266, 271, 283, 293, 305, 310, 320, 326, 356
silicone clad fiber, 327
single-mode fiber *see* monomode fiber
single-mode fiber–fiber coupling, 251
slab waveguide, 21–32
  structure, 21–2
specific
  attenuation, 61
  group delay, 63, 71, 72, 309
  phase delay, 63, 72
  phase shift, 61
speckle pattern, 314, 315
spectral
  lines, 153
  width of LED, 146
spectrum
  of electromagnetic waves, 1
  of LD, 151–3
  of LED, 146
  of marked and filtered Poisson process, 398
splices, 328–31
split phase code, 345
spontaneous emission, 140
spot size *see* mode field diameter
state of polarization, 334
step index fiber, 5, 21
  analysis of, 33
stimulated
  photons, 171
  recombination, 170
stripe-geometry LD, 149
subcarrier modulation, 270

superluminescence, 175
surface emitter, 142
symmetric field, 24
system
  bandwidth, 193, 196
  noise, 304
  transfer function, 193

TE mode, 26
TE wave *see* transverse electric wave
thermal noise, 265, 273–80, 286, 300, 358
  limit, 273, 290
threshold current, 147, 149, 174
tight coating, 327
TM mode, 28
TM wave *see* transverse magnetic wave
total internal reflection, 30, 31
transfer function
  of fiber, 187–200
  of an LD, 181
  of an LED, 178
transient phenomena in lasers, 178
transimpedance amplifier, 258, 278
transmission
  of information, 3
  mode, 42
transverse
  electric wave, 22–8, 40–4
  laser modes, 154
  magnetic wave, 28, 40, 45
  mode, 40
  wave, 40
turning point, 119
turn-on delay, 177–8

valence band, 140
velocity
  of envelope, 65
  group, 62, 63, 63
  phase, 24, 37, 38, 65, 70

waist of Gaussian beam, 219, 380
wave
  evanescent, 27
  harmonic, 2, 20
  hybrid, 33
  linearly polarized, 1
  plane, 1, 30
  transverse electric, 22–8
  transverse magnetic, 28
wave equation, 22, 34
  reduced, 34
wave optics, 21
wavefront, 30, 97, 98
  Gaussian beam, 380
waveguide
  slab, 21–32
  dispersion, 69–74, 78, 88–94
wavelength
  dependence of photodiodes, 161
  of lasers, 145, 150
  of LED, 140, 145
  multiplexing, 336
weakly guiding fiber, 33, 52
WKB method, 117

yttrium iron garnet, 333